Cable Structures

Max Irvine

Professor of Civil Engineering
The University of South Wales
Australia

Dover Publications, Inc., New York

Published in Canada by General Publishing Company, Ltd., 30 Lesmill Road, Don Mills, Toronto, Ontario.
Published in the United Kingdom by Constable and Company, Ltd., 3 The Lanchesters, 162–164 Fulham Palace Road, London W6 9ER.

This Dover edition, first published in 1992, is an unabridged republication of the work originally published by The MIT Press, Cambridge, Massachusetts, 1981, as the first volume in "The MIT Press Series in Structural Mechanics."

Manufactured in the United States of America
Dover Publications, Inc., 31 East 2nd Street, Mineola, N.Y. 11501

Library of Congress Cataloging-in-Publication Data

Irvine, H. Max.
Cable structures / Max Irvine.
p. cm.
Originally published: Cambridge, Mass. : MIT Press, 1981, in series: MIT Press series in structural mechanics.
Includes bibliographical references and index.
ISBN 0-486-67127-5 (pbk.)
1. Cable structures. 2. Bridges, Suspension. I. Title.
TA660.C6I78 1992
624′.5—dc20 91-48224
CIP

The schooner Wyvern, built 1845, passing under the Menai Straits suspension bridge, built 1826, which spans 580 ft (courtesy Hart Nautical Museum, MIT)

Contents

Preface

Few books on structural mechanics deal even in part with cable structures, and fewer confine their attention to this one area. One exception is Frei Otto's *Tensile Structures* (The MIT Press, 1967), which is devoted mainly to a discussion of architectural form and to the mathematics of the structural response of various types of suspension roofs and membranes. Necessarily, there is some overlap with that comprehensive work in this book, but it is slight.

My aim has been to give a unified account of cable structures from the point of view of a structural engineer who has strong interests in both the theoretical and practical aspects of the subject. Therefore, the outcome is a work that stresses the practical applications but does not shy away from a certain amount of rigor, as defined in a loose mathematical sense, when that is necessary to the treatment. In short, the book draws on some of my research work over the past decade and some of my experience as a consulting engineer.

I want the practicing engineer who is confronted with, say, a mooring line problem, or who wishes to undertake a dynamic analysis of a cable-suspended roof, to use this book; and at the same time I want the teacher, research worker, or graduate student to find pointers here that may help with some aspect of their work. Being all things to all men is a somewhat difficult task, however, and I will be delighted with even limited success in achieving this goal.

Some readers may be disquieted to find very little recourse to matrix methods of solution, or to computational techniques in general. There are two reasons for this. First, others are far better qualified to attempt this task (see, for example, *The Analysis of Hanging Roofs by Means of the Displacement Method* by H. Møllmann, Polyteknisk Forlag, Lyngby, 1974). Second, and more important, I feel strongly that the vast majority of problems in cable structures are best attempted by hand, at least for preliminary purposes. Not only is one able to see directly the effect of various changes made but one is able to explore the range of applicability and the character of solutions in a way that can be time consuming, expensive, and possibly unproductive by computer. Needless to say, I have not followed my own advice exclusively on this matter, but the reader will be exposed to a philosophy wherein simple approximate analytical techniques are exploited for all they are worth (with an occasional exact solution making an appearance). By and large, cable structures are simple—the layman as well as the engineer can see how they work—so simple hand techniques, wisely used, are hardly out of place.

In each of the five chapters that comprise the work a balance has had to be struck between depth and breadth of treatment and the interests of brevity. If I have erred on the side of the latter, then the references cited and the exercises set at the end of the chapters may help redress that balance.

Although the list of topics covered is by no means exhaustive, the list does include a chapter on classical, or exact, theories for the static response of single cables to various applied loads, together with a historical introduction and applications of the theories. A little arch theory is also included since the temptation proved too strong to resist. This paves the way for a chapter on the static response of single cables, the profiles of which are sufficiently flat for engineering approximations to be applied to the theory. The third chapter is devoted to an account of the dynamic response of cables. A lot of this work has direct structural applications. The fourth chapter deals with aspects of guyed masts, cable trusses, and suspension bridges. Three-dimensional surfaces, in the form of tension membranes of revolution, are then treated using the equilibrium theory with extensions to cover small elastic deformations. The remainder of this final chapter is devoted to an investigation, using appropriate engineering theories of the static and dynamic response of cable networks that cover regular plan forms. Numerous worked examples are presented throughout the chapters, and several of the exercises sets cover topics that could not be addressed in the main body of the book.

Acknowledgments

Henry J. Hopkins, who for many years headed the Department of Civil Engineering at the University of Canterbury, first excited my interest in cable structures with lectures on the theory of suspension bridges. Most people, I guess, find the suspension bridge attractive, and, for me, the added technical interest stems from those undergraduate days. I am particularly grateful to Michael F. Parsons, Partner, Freeman, Fox & Partners, London, for a short but stimulating period of practical experience on the Bosporus and Humber bridges.

I became seriously interested in the theory of cable structures while a graduate student at the California Institute of Technology. I owe a debt of gratitude to my thesis advisers, Paul C. Jennings and Thomas K. Caughey. I also had the benefit of frequent discussions with colleagues Cecil M. Segedin, Jerry H. Griffin, Michael J. O'Sullivan, and Glenn B. Sinclair at the University of Auckland.

Material from the book has been presented in courses at both the University of Auckland and the Massachusetts Institute of Technology. The manuscript was read by John M. Biggs, and two of my graduate students, Hooshang Banon and Pierre Montauban, assisted me with some of the work. To them, to Jessica Malinofsky, who did the typing, and to Paul Fallon, who traced the figures, I express my thanks.

And to Sue, my wife, goes my heartfelt appreciation for her support while this hang-up of mine worked itself out.

1

Historical and Classical Matters

1.1 Introduction

It is a somewhat surprising fact that various cable problems have been the vehicle by which some well-known techniques, equations, and mathematical functions were introduced. For example, Stevin [15] in 1586 established the triangle of forces by experimenting with loaded strings, although Leonardo da Vinci's [24] fifteenth-century sketches anticipated this result and several others, including the catenary and the concept of the collapse mechanism in a voussoir arch. It seems likely, according to Truesdell [30], that Beeckman had by 1615 solved the suspension bridge problem, namely, that in responding to load uniformly distributed in plan, a cable hangs in a parabolic arc. But nearly two centuries were to pass before this solution became well known; it was rediscovered in 1794 by the Russian engineer Fuss, Euler's son-in-law, who was charged with the responsibility of attempting to span the Neva River, at St. Petersburg [29].

Galileo, in *Discourses on Two New Sciences*, published in 1638, muses on the shape of a hanging chain and concludes that it is parabolic—primarily by an analogy to the flight of a projectile. That this view was incorrect was certainly known by Huygens in the mid-seventeenth century but not proved until the years 1690 and 1691 when the Bernoullis (James and his brother, John), Leibnitz, and Huygens more or less jointly discovered the catenary [30], the word coming from the Latin for chain and meaning universally the form of a chain hanging between two points. The word funicular, based on the Latin for cord, was also used; so today the phrase funicular polygon refers generally to the profile of a string supporting concentrated loads or to an ideal arch profile under concentrated loads. Perhaps its widest use relates to the funicular construction for the bending moment diagram for a beam under loads.

In the discovery of the catenary different approaches were employed, with Huygens relying on geometrical principles, and Leibnitz and the Bernoullis using the calculus, then a comparatively recent invention. It is difficult to decide who did it first, for a certain amount of acrimony—although no more than was par for the course at that time—existed between the participants in this great contest to find the catenary [30]. Between them the Bernoulli brothers also formulated the general differential equations of equilibrium of a chain element under various loadings and allowed for the effects of stretch by incorporating Hooke's law in these equations. Hooke enunciated that law as an anagram in the minutes of a meeting of the Royal Society in 1675; appearing in the same entry was his famous dictum that the shape of an arch may be found by inverting the analogously loaded cable [15]. In addition James Bernoulli helped lay the groundwork for the calculus of variations by showing that in keeping the center of gravity of the chain as low as possible, the same equations arose. Early investigations into the principle of virtual work, using an arch as a working example, were also his doing.

The vibration of taut strings was studied extensively in the early part of the eighteenth century. As to the importance of this topic we can do no better than quote Lord Rayleigh [27], who wrote, "To the mathematician they must always possess a peculiar interest as the battlefield on which were fought out the controversies of D'Alembert, Euler, [Daniel] Bernoulli, and Lagrange, relating to the nature of the solution of partial differential equations."

In 1738 Daniel Bernoulli, the son of John, published a solution for the natural frequencies of a chain that hangs from one end. The solution was in the form of an infinite series which is now written as a Bessel function of zero order [32]. By 1764 Euler had obtained the equation of motion for a vibrating taut membrane in terms of polar coordinates, and he then proceeded to separate the variables and arrive at a solution, part of which was an infinite series. This series represents the first appearance of a Bessel function of the first kind and of general integral order [32]. (In fact Euler did not proceed then to solve the circular membrane; a partial solution was given by Poisson in 1829, and Clebsch, in 1862, finished it off by including the antisymmetric modes.) Euler's derivation in 1766 of the equations of motion in terms of Cartesian coordinates led, in 1829, to Poisson's solution of the rectangular membrane, this being one of the earlier applications of the Fourier series to problems other than heat flow [19]. Lagrange used a discrete, string of beads model of the taut string as an illustration of the application of his equations of motion. This work, among the first on the solution of vibrational problems by difference equations, was done in 1760 but was included in his celebrated *Mécanique analytique* of 1788 [33]. Mathieu functions first arose in connection with Mathieu's solution of the elliptical membrane in 1868.

One could go on, but the list is sufficiently complete to illustrate the point. Later in this work we will address other historical aspects of the theory and practice of cable structures. In any event, the discussion to date indicates that in Europe cable theory had at least a firm grounding by the nineteenth century (and indeed was understood somewhat earlier than beam theory). Outside Europe the practical aspects had received considerable attention at a much earlier date. Unlikely as it is that much, if any, of the theory was understood, the early civilizations of the Far East and Central and South America mastered suspension bridge construction. Needham [20] records the existence of sophisticated suspension bridges in China before the start of the Christian era. The native Americans did, and still do, make intricate rope pathways, and perfected these with some ingenious ways of improving the tensile properties of vine ropes. Iron chain suspension bridges date from A.D. 65, when a bridge was built in Yunnan, China; but records are not reliable as to who built these early bridges. By contrast, much more is known about the Tibetan monk, Thang-stong rGyal-po, who in the fifteenth century was renowned as a builder of iron chain suspension bridges (see figures 1.1 and 1.2). The evolution of the suspension bridge, surely the oldest and most recognizable species of cable structure, has been given an admirable treatment by Hauri and Peters [10], and there is no point in repeating it here.

With this brief historical introduction we may move to develop and explore the nature and applications of some exact solutions for the static response of a single cable—some of this material may be considered classical. The more formal stance that will be adopted will be relaxed somewhat when we come to the approximate theories that comprise most of the remainder of the work.

1.2 The Catenary and the Arch

The Catenary

We consider here the profile adopted by, and other properties of, a uniform inextensible cable, or chain, that hangs between two fixed points that are at the same level. The elements of the cable are assumed to be perfectly flexible—they are devoid of flexural rigidity—and it is further assumed that the cable can sustain only tensile forces. An elegant way of describing this pure cable behavior is that attributed to James Bernoulli [30], "The action of any part of the line upon its neighbor is purely tangential."

Consider now the sketches in figure 1.3. Vertical equilibrium of the isolated element of the cable located at (x, z) requires that

1.1 The Tibetan monk Thang-stong rGyal-po (1385–1464), builder of iron chain suspension bridges (from Hauri and Peters [10], courtesy ASCE)

$$\frac{d}{ds}\left(T\frac{dz}{ds}\right) = -mg, \tag{1.1}$$

where T is the tension in the cable, dz/ds is the sine of the angle subtended to the horizontal by the tangent to the profile, and mg is the self-weight of the cable per unit length. Horizontal equilibrium of the element yields

$$\frac{d}{ds}\left(T\frac{dx}{ds}\right) = 0, \tag{1.2}$$

where dx/ds is the cosine of the angle of inclination. Equation (1.2) may be integrated directly to

1.2 Thang-stong rGyal-po's Paro River bridge near Tamchogang Gompa, Bhutan (from Hauri and Peters [10], courtesy ASCE)

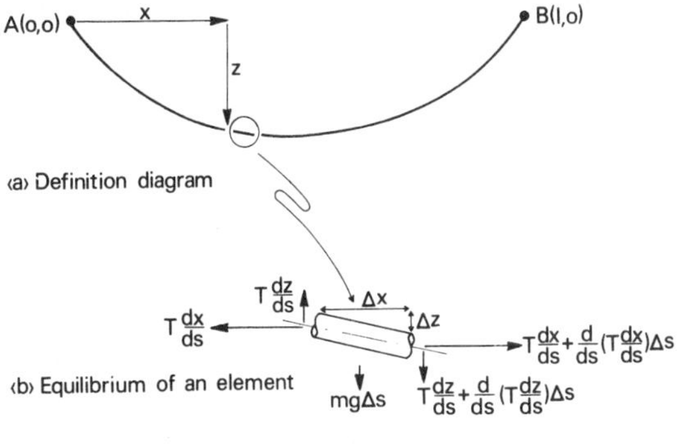

1.3

$$T\frac{dx}{ds} = H, \tag{1.3}$$

where H is the horizontal component of cable tension which is constant everywhere since no longitudinal loads are acting. Consequently (1.1) may be reduced to

$$H\frac{d^2z}{dx^2} = -mg\frac{ds}{dx}, \tag{1.4}$$

and it may be noted that, when $mg\frac{ds}{dx}$, the intensity of load per unit span is constant the resulting profile is parabolic.

Because the following geometric constraint must be satisfied, namely,

$$\left(\frac{dx}{ds}\right)^2 + \left(\frac{dz}{ds}\right)^2 = 1, \tag{1.5}$$

the governing differential equation of the catenary takes the form

$$H\frac{d^2z}{dx^2} = -mg\left\{1 + \left(\frac{dz}{dx}\right)^2\right\}^{1/2}. \tag{1.6}$$

Armed with the following identities for the hyperbolic functions

$$\cosh^2 t - \sinh^2 t = 1,$$

$$\frac{d}{dt}(\cosh t) = \sinh t,$$

$$\frac{d}{dt}(\sinh t) = \cosh t,$$

it can be seen that a solution that satisfies (1.6) and the boundary conditions is

$$z = \frac{H}{mg}\left\{\cosh\left(\frac{mgl}{2H}\right) - \cosh\frac{mg}{H}\left(\frac{l}{2} - x\right)\right\}. \tag{1.7}$$

An expression for the length of a portion of the cable is

$$s = \int_0^x \left\{1 + \left(\frac{dz}{dx}\right)^2\right\}^{1/2} dx = \frac{H}{mg}\left\{\sinh\left(\frac{mgl}{2H}\right) - \sinh\frac{mg}{H}\left(\frac{l}{2} - x\right)\right\}, \tag{1.8}$$

so that, if a cable of length L_0 is used to span between the supports, the horizontal component of cable tension may be found by solving the equation

$$\sinh\left(\frac{mgl}{2H}\right) = \frac{mgL_0}{2H}, \tag{1.9}$$

for H, it being tacitly assumed that mg and l are known beforehand. Unless the condition of inextensibility is relaxed, a solution cannot exist if L_0 is not greater than l. The tension at any point is

$$T = H \cosh \frac{mg}{H}\left(\frac{l}{2} - x\right). \tag{1.10}$$

We may note that mgl/H is small when the cable length is only fractionally longer than the span and that then the substitution of power series approximations for the hyperbolic functions yields the properties associated with a parabola—the limiting form of the catenary as the profile flattens.

Several other catenary problems could be pursued; most, however, are of only academic interest. One that is now in that category, but which at the time of its discovery was of some practical interest, is the catenary of uniform strength, whereby the cross-sectional area is varied to allow the stress to remain constant along the profile. The solution was obtained by Davies Gilbert in 1826 [29] in connection with Telford's design of the Menai Straits suspension bridge. Even in those days when the cables often consisted of arrays of massive wrought iron eye bars, in which the cross-sectional area could be varied reasonably easily, and for which the dead weight of the whole suspended structure was not much greater than the cables themselves, there was little advantage in adopting the theory for design. In fact it has never been used, and it remains as a historical curiosity.

Example 1.1 Mooring Calculations for a Floating Dock

Floating docks are moored by a series of heavy relatively slack chains which are attached at regular intervals around the perimeter of the structure and dip down to the seabed some distance away. Under quiescent conditions a significant portion of the chain will lie along the seabed before being fixed to an anchor block or pile cap. Under design conditions, which consider wind plus current due to a running tide, the windward chains will progressively lift off and tighten while the leeward chains will slacken. A typical design constraint is that where under the full drag the windward chains are just about to lift off the seabed at the pile caps, and the information sought includes the chain length, the location of the pile cap, and the likely drift of the vessel in responding to the design load. Data supplied will include the water depth (taken as constant but usually variable in practice on account of seabed topography), the submerged self-weight of the chains and the tension in them when they are fixed to the dock. Even under the design load the stresses in the chains are low, and they may be safely assumed inextensible; therefore the catenary equations may be used. Figure 1.4 illustrates the details for a representative portion of the structure.

Variations of equations (1.7) through (1.10) are useful in that they provide a systematic way of solving this problem. With a little rearrangement it can be shown that

$$T_{max} = H + mgd,$$

where T_{max} is the peak tension and d is the sag or dip of the catenary. Also

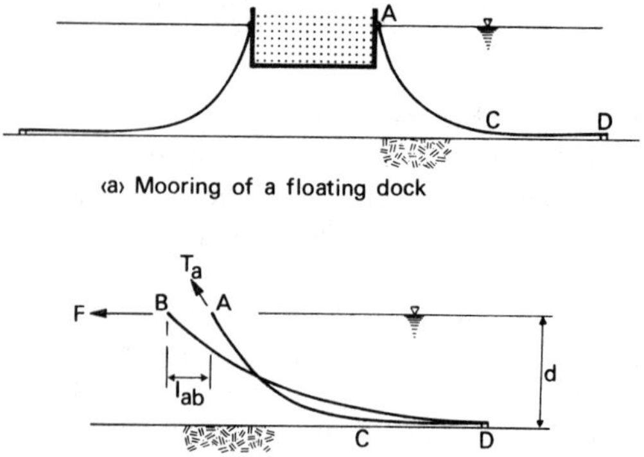

(a) Mooring of a floating dock

(b) Drift l_{ab} under the design load F

1.4

$$\frac{L}{2} = d\left(\frac{2T_{\max}}{mgd} - 1\right)^{1/2} = d\left(\frac{2H}{mgd} + 1\right)^{1/2}.$$

Initially the peak tension in the chain is T_A, and a portion L_{CD} lies on the bottom. If L_{AD} is the total length of the chain, then

$$L_{AD} = d\left(\frac{2T_A}{mgd} - 1\right)^{1/2} + L_{CD}.$$

The design drag force F is slightly greater than the horizontal component of tension induced in the windward chain, owing to the loss of tension in the leeward chain. But the correction is usually insignificant, and it is conservative to ignore it. On account of the inextensibility of the chain, $L_{BD} = L_{AD}$, and so

$$L_{AD} = d\left(\frac{2F}{mgd} + 1\right)^{1/2},$$

from which

$$L_{CD} = d\left\{\left(\frac{2F}{mgd} + 1\right)^{1/2} - \left(\frac{2T_A}{mgd} - 1\right)^{1/2}\right\}.$$

The drift is

$$l_{AB} = d\left[\frac{F}{mgd}\sinh^{-1}\left\{\frac{(2F/mgd + 1)^{1/2}}{F/mgd}\right\} - \left(\frac{T_A}{mgd} - 1\right)\sinh^{-1}\left\{\frac{(2T_A/mgd - 1)^{1/2}}{T_A/mgd - 1}\right\}\right] - L_{CD}.$$

This calculation of the drift is made easier by utilizing logarithmic representations of the inverse hyperbolic functions [1], so that

$$l_{AB} = d\ln\left[\frac{\{\alpha + (\alpha^2 + 1)^{1/2}\}^{\beta}}{\{\gamma + (\gamma^2 + 1)^{1/2}\}^{\delta}}\right] - L_{CD},$$

where

$$\alpha = \frac{(2F/mgd + 1)^{1/2}}{F/mgd}, \quad \beta = \frac{F}{mgd},$$

$$\gamma = \frac{(2T_A/mgd - 1)^{1/2}}{T_A/mgd - 1}, \quad \text{and} \quad \delta = \frac{T_A}{mgd} - 1.$$

Finally, the horizontal distance of the pile cap from the dock bollard is

$$l_{AD} = d \ln\{\gamma + (\gamma^2 + 1)^{1/2}\}^{\delta} + L_{CD}.$$

Consider now the following data which are broadly typical of the mooring requirements for floating docks that service large vessels: $d = 25$ m, $mg = 1$ kN/m, $T_A = 150$ kN, and $F = 1{,}500$ kN. The minimum chain length is

$$L_{AD} = 25\left(\frac{2 \times 1{,}500}{1 \times 25} + 1\right)^{1/2} = 275 \text{ m},$$

and the portion of this that lies on the ground is

$$L_{CD} = 275 - 25\left(\frac{2 \times 150}{1 \times 25} - 1\right)^{1/2} = 192 \text{ m}.$$

The drift under the design drag force is

$$l_{AB} = 25 \ln\left[\frac{\{11/60 + ((11/60)^2 + 1)^{1/2}\}^{60}}{\{(11)^{1/2}/5 + (11/25 + 1)^{1/2}\}^{5}}\right] - 192,$$

$$= 3.5 \text{ m},$$

and its magnitude illustrates a point common to most cable systems used for structural purposes—large, and possibly undesirable, changes in configuration may accompany the application of design live loads. The horizontal distance to the pile cap should be at least 270 m from the dock bollard, and a further measure of protection is assured if this distance is increased to, say, 280 m (with a consequent rise in the cable length to 285 m). Pretensioning of the chains is usually effected by winding them past the bollards until a marked length is reached. Although not considered here, tidal rise and fall usually needs to be included as well.

The Arch

Since one is the inverse of the other, it appears appropriate to pause at this point in this development of classical cable theory to delve into classical arch theory. The objectives are quite limited but the results are interesting, for we shall develop an exact profile for a common loading situation (in the sense that the line of thrust lies at the middepth of the ring of stones, or voussoirs, that provide the arch with its structural integrity), and we shall explore the stability against collapse when this arch has a different profile for its ring of voussoirs.

The theory of arches was many centuries in the making, but it received its greatest impetus when Hooke published his famous anagram in 1675.

Some thirty years later this was interpreted as, "as hangs a flexible cable so, inverted, stand the touching pieces of an arch." It has been pointed out, however, that Hooke's original tangled statement was more general than this translation implies, in that he referred to flexible cables under an arbitrary pattern of loads [15]. An early work on the statics of arches is the book *Traité de mécanique* by De la Hire, published in 1695. In the eighteenth and nineteenth centuries a good deal of effort was devoted to the study of arch theory, and many notable advances were recorded; a detailed review may be found in part of Heyman [12].

In the light of this it comes as somewhat of a surprise to observe that, although by 1854 Yvon Villarceau had developed general procedures for calculating lines of thrust which ensured that they lay at the middepth of the voussoirs, the specific problem of the arch ring coincident with the line of thrust that arises from backfill to a level upper surface appears to have been first given by Inglis [16]. This is in essence the problem of the stone bridge.

Therefore, following Inglis, and referring to figure 1.5, the differential equation governing the profile is

$$H\frac{d^2z}{dx^2} = -\gamma(c + d - z), \tag{1.11}$$

where H may be interpreted as the (constant) horizontal component of thrust in the arch, γ is the (uniform) specific weight of the arch materials, c is the depth of the overburden, d is the rise of the arch, and consequently $\gamma(c + d - z)$ is the intensity of the loading per unit span. After some manipulation the solution may be shown to be

$$z = d + c\left[1 - \cosh\left\{\left(1 - \frac{2x}{l}\right)\cosh^{-1}\left(\frac{c + d}{c}\right)\right\}\right], \tag{1.12}$$

with

$$H = \frac{\gamma l^2}{4}\frac{1}{\{\cosh^{-1}((c + d)/c)\}^2}.$$

Although there is just one profile for the problem as posed, variations

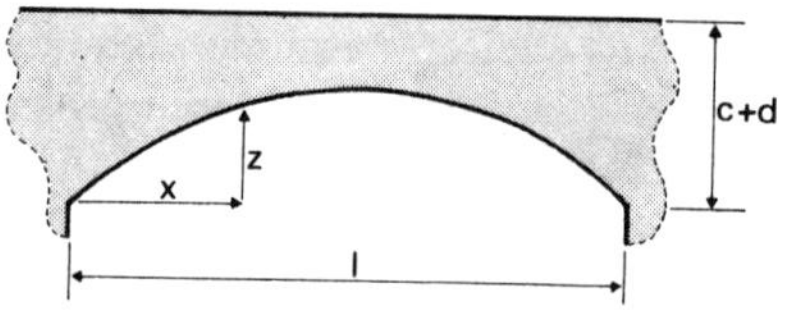

1.5 Definition diagram for an arch

can be (and have obviously had to be) used in practice. Many of the brick railway bridges built in Europe in the nineteenth century were fashioned with elliptical arch rings. This is certainly satisfactory if the voussoirs are sufficiently deep for the line of thrust to lie well within them. The workbooks of I. K. Brunel, one of the masters of arch bridge construction in that period, show abundant evidence of such concern [22]. If in addition the abutments are unyielding, there can then be no possibility of the joints opening up.

But this raises an obvious question: If the arch ring is not the true, or funicular, profile for the loading, how deep do the voussoirs which comprise the ring need to be to avoid collapse? We shall answer this for the classical Roman voussoir arch, but first the topic deserves a few comments of a historical nature.

The earliest theoretical studies of the stability against collapse of voussoir arches are probably due to Couplet in 1729 and 1730 [12, 15]. Recognizing that the line of thrust in an arch, consisting solely of a semicircular ring of voussoirs, is not itself a semicircle, Couplet was able to determine approximately the minimum voussoir depth necessary for the line of thrust to lie just within the arch ring. Somewhat similar principles were employed by Poleni in 1748, who experimented with a suitably weighted string of beads to show that the dome of St. Peters in Rome would not collapse, even though the meridional cracks appearing in it had caused some concern [7, 15]. The weighting of the beads increased steadily toward the supports to represent a lune of the dome. The influence of the hoop forces was ignored, the argument presumably being that a dome able to resist self-weight by calling on only meridional forces was surely also stable when the hoop forces (where they were compressive) were included.

If the seeds of the idea of the collapse mechanism were sown by Couplet (Leonardo da Vinci's sketch of this phenomenon was probably not widely known), it seems that the concept was first placed in a solid engineering context by Snell and by Barlow more than one hundred years later [22]. Much later still, the collapse of the voussoir arch was the subject of a detailed examination in Pippard and Baker [25], while Heyman's recent studies [11, 12, 13] have been largely concerned with understanding the stability of masonry structures from the point of view of plastic theory and its lower-bound theorem. These studies have been drawn on in performing a mathematical analysis of the stability of the Roman arch [18], some of the details of which are now outlined.

The Roman arch almost invariably consisted of a semicircular ring of voussoirs backfilled to provide a level upper surface. Whether used as bridges and/or as aqueducts, there are numerous fine examples extant, as figure 1.6 shows.

Methods of proportioning the arches were based largely on experience,

1.6 The Roman aqueduct at Segovia in Spain (reproduced from M. Hayden, 1976, *The Book of Bridges*, New York: Galahad Books)

with little recourse to the principles of mechanics. Indeed it appears that the Romans did not appreciate that a knowledge of both the velocity and the cross-sectional area was necessary to determine the discharge rate of an aqueduct [28]. Nonetheless the military engineers and master craftsmen of the time had a very good idea of the way an arch functioned, as can be seen from the books of Vitruvius [15], a Roman architect and engineer active in the first century BC who reported ancient building practices. Generally we assume that the properties of the arch ring are such that sliding cannot occur between adjacent voussoirs, the joints will open rather than transmit tension, and the voussoirs are both incompressible and of unlimited compressive strength.

The loading in figure 1.7 is taken to be vertical, and the structural resistance to this loading is regarded as being provided solely by the arch ring (clearly conservative assumptions). The intensity per unit span of this variable distributed loading is proportional to the length of the line *ab*. Although the arch ring is semicircular, the line of thrust is not semicircular because the loading is not the uniform hydrostatic pressure necessary for that purpose.

If the loading is assumed to be of uniform specific weight γ, and the mean radius of the arch ring R is given, the problem is to find the depth of the voussoirs t as a function of the overburden c, so that the line of

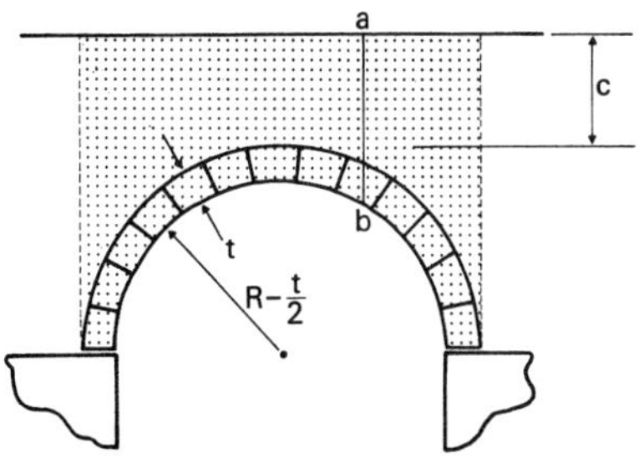

1.7 Definition diagram for the Roman arch

thrust is just able to lie within the voussoirs. This it may do by passing through the springings at A and E, touching the crown at C as well as touching the intrados at B and D (see figure 1.8). Points A through E may be regarded as hinges because relative rotations are possible between the faces of adjacent voussoirs. In this condition the arch is on the verge of collapse, and the voussoir depth is the minimum possible.

Simultaneously the analysis yields the hinge locations at B and D, and the horizontal component of thrust may be found also. There is no upper limit on the depth of the overburden, although a lower limit of zero is clearly necessary.

Because the full analysis is rather involved, we shall be content to present a simple analytical solution for one extreme and display the general results in tabular form. The method of attack is, however, quite general and could be used for a wide variety of such stability analyses.

Suppose therefore that the depth of the overburden is very great compared with the radius of the arch. The loading may then be assumed constant with span. Alternatively the case could be treated as arising with a weightless arch ring supporting a load uniformly distributed with span, although neither is realizable in practice. The line of thrust is then given by the parabola

$$z = \frac{W_0 l}{2H}\left\{\frac{x}{l} - \left(\frac{x}{l}\right)^2\right\}, \tag{1.13}$$

where W_0 is the total vertical load applied.

If this line of thrust is to lie just within the arch (in the sense outlined earlier), the span must be $l = 2R + t$, and the rise must be $R + t/2$ (see figures 1.7 and 1.8). Therefore

$$z = 4\left(R + \frac{t}{2}\right)\left\{\frac{x}{2R+t} - \left(\frac{x}{2R+t}\right)^2\right\},$$
$$H = \frac{W_0}{4}. \tag{1.14}$$

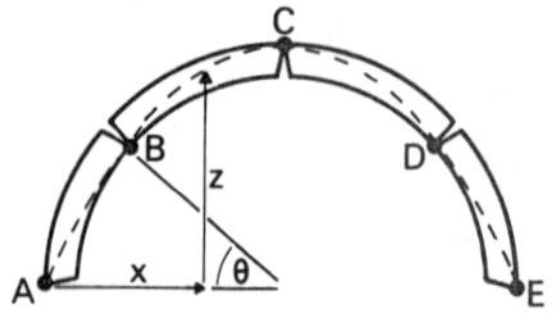

1.8 Assumed collapse mechanism showing hinge locations and line of thrust (dotted)

At point B (located by the angle θ in figure 1.8), where the line of thrust touches the intrados, two further geometric relations may be written. The first stipulates that the coordinates of B are $(R + t/2) - (R - t/2)\cos\theta$, $(R - t/2)\sin\theta$. Substitution of this into the first equation of (1.14) gives

$$\left(R - \frac{t}{2}\right)\sin\theta = \left\{\left(R + \frac{t}{2}\right) - \left(R - \frac{t}{2}\right)\cos\theta\right\}\left\{1 + \left(\frac{R - t/2}{R + t/2}\right)\cos\theta\right\}. \tag{1.15}$$

The second requirement is that at B the slope of the line of thrust must be tangential to the intrados. Therefore, differentiating the first equation of (1.14) and carrying out the required substitutions, we obtain

$$\sin\theta = \frac{1}{2}\left(\frac{R + t/2}{R - t/2}\right). \tag{1.16}$$

Manipulation of (1.15) and (1.16) leads to the results

$$\theta = \sin^{-1}\left(\frac{1}{\sqrt{3}}\right) = 35.3°,$$

$$\frac{t}{R} = 2(7 - 4\sqrt{3}) = 0.144, \tag{1.17}$$

or in terms of a ratio of span to depth $2R/t \simeq 14$.

These results have a simplicity uncharacteristic of this type of problem. For example, Heyman's results for a semicircular voussoir arch under self-weight which is assumed to be uniformly distributed around its centerline must be found numerically, for they involve the solution of the equation [12, 13]

$$\left(\frac{\pi}{2} - \theta\right)\tan\theta\left[\frac{(\pi - 2\theta)\sin\theta + \sin^2\theta\cos\theta + \cos\theta}{(\pi - 2\theta)\sin\theta + \sin^2\theta\cos\theta - \sin\theta}\right] = \frac{\pi}{2},$$

after which t/R may be found from

$$\frac{t}{R} = 2\frac{(\pi/2 - \theta - \cos\theta)(1 - \sin\theta)}{(\pi/2 - \theta)(1 + \sin\theta)}. \tag{1.18}$$

The results are $\theta = 31.2°$, $t/R = 0.106$, or again in terms of a ratio of span to depth $2R/t \simeq 19$. (Heyman also obtained results for circular arches in which the included angle was less than 180°.) Couplet, in assuming that $\theta = 45°$, obtained $t/R = 0.101$ [12]. Here the intensity of loading increases as the supports are approached, and a greater maximum span to voussoir depth may be achieved.

The general case of the Roman arch has no obviously simple solution, and, although the formulation is straightforward, numerical methods are necessary to effect a solution [18]. Table 1.1 contains the results.

It is clear that the analytical results of (1.17) are recovered as c/R becomes large. At the other extreme, when c/R is zero, and the crown of the arch lies in the upper surface, the voussoir depth necessary for stability is substantially less, indicating that the semicircle is not a bad approximation to the line of thrust when a high intensity of load is placed over the haunches of the arch. It is of note that a threefold increase in t/R is associated with only a 50 percent increase in θ. The location of the hinge at the intrados varies little even for widely differing forms of the arch [12].

1.3 The Suspended Elastic Cable under Concentrated Vertical Loads

This section treats the static response of a single cable suspended between two rigid supports that are not necessarily at the same level. The cable is of constant cross-sectional area when unloaded and is composed of a homogeneous material which is linearly elastic. Loading of the cable is provided by any number of concentrated vertical loads that may be arbitrarily placed along the cable's length. Expressions for the tension within the cable and the Cartesian coordinates describing the strained profile are derived as functions of a single independent variable which is conveniently taken as the Lagrangian coordinate associated with the unstrained profile—the unstrained length of cable between the support

Table 1.1 Values of the critical voussoir depth t and the angle locating the hinge at the intrados θ as a function of the overburden c. The mean radius of the arch ring is R.

c/R	θ	t/R	c/R	θ	t/R	c/R	θ	t/R
0	21.6°	0.047	0.2	32.0°	0.094	2.0	35.0°	0.133
0.001	21.7°	0.047	0.4	33.6°	0.104	4.0	35.1°	0.138
0.005	22.3°	0.049	0.6	34.2°	0.117	6.0	35.2°	0.140
0.01	23.0°	0.051	0.8	34.5°	0.122	10.0	35.2°	0.141
0.05	26.0°	0.063	1.0	34.7°	0.125	1000.0	35.3°	0.144
0.1	29.5°	0.078	1.5	34.9°	0.130	∞	35.3°	0.144

at the origin and some particle point. These expressions contain as unknowns the horizontal and vertical reactions of the support at the origin, quantities which can be determined by solving a transcendental equation simultaneously with an algebraic equation to give the complete solution.

The general problem is of some practical importance, quite apart from the fact that within the limitations of these assumptions it is a compact exact solution of which cable theory can boast very few. Although it is unnecessary to do so, we shall start with the elastic catenary and proceed to include response to a point load before treating the general case.

The Elastic Catenary

Even though the differential equations of equilibrium of a stretched cord were derived by the Bernoullis, it seems that the solution of the symmetrically suspended elastic catenary is due to Routh [29]. Feld [8] extended Routh's solution to the unsymmetric case, but unfortunately the coordinate system used in both analyses makes application awkward.

To get around this difficulty we shall adopt a Lagrangian approach, similar in concept to that presented in the chapters devoted to catenary problems in Ramsey [26] and Halstead et al. [9].

The cable shown in figure 1.9 is suspended between two fixed points A and B which have Cartesian coordinates $(0, 0)$ and (l, h), respectively. Thus the span of the cable is l, and the relative vertical displacement of the ends is h. The unstrained length of the cable is L_0, where L_0 is not necessarily greater than $(l^2 + h^2)^{1/2}$, although it obviously cannot be much less if Hooke's law is not to be violated. A point on the cable has Lagrangian coordinate s in the unstrained profile (the length of cable from the origin to that point is s when the cable is unloaded). Under self-weight of W $(= mgL_0)$ this point moves to occupy its new position in the strained profile described by Cartesian coordinates x and z and Lagrangian coordinate p.

The geometric constraint to be satisfied is

$$\left(\frac{dx}{dp}\right)^2 + \left(\frac{dz}{dp}\right)^2 = 1, \tag{1.19}$$

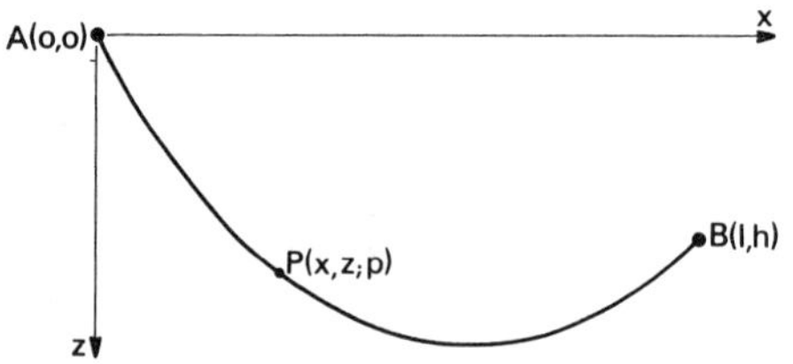

1.9 Coordinates for the elastic catenary

while with reference to figure 1.10 the balancing of horizontal and vertical forces yields

$$T\frac{dx}{dp} = H,$$
$$T\frac{dz}{dp} = V - W\frac{s}{L_0}, \tag{1.20}$$

where, owing to conservation of mass, the weight of that portion of the strained profile shown in the figure is simply Ws/L_0. The vertical reaction at the support is V, and as before H is the constant horizontal component of cable tension. Moment equilibrium for this segment of the cable yields an expression which equations (1.20) ensure is satisfied automatically (see problem 1.4).

A constitutive relation that is a mathematically consistent expression of Hooke's law is

$$T = EA_0\left(\frac{dp}{ds} - 1\right), \tag{1.21}$$

where E is Young's modulus, A_0 is the uniform cross-sectional area in the unstrained profile. Variations on this will be used throughout this book.

The end conditions at the cable supports A and B are

$$x = 0, \quad z = 0, \quad p = 0 \quad \text{at} \quad s = 0$$
$$x = l, \quad z = h, \quad p = L \quad \text{at} \quad s = L_0 \tag{1.22}$$

where L is the length of the cable in the strained profile.

We are now in a position to derive the parametric solution that describes the strained cable profile. In practice all that are of interest are the solutions for x, z, and T as functions of the independent variable s. Solutions for p and the strained cross-sectional area A are omitted but can be easily found.

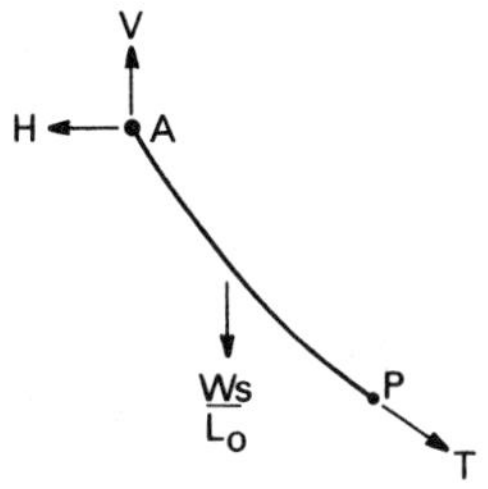

1.10 Forces on a segment of the strained cable profile

1. Solution for $T = T(s)$:
If equations (1.20) are squared, added, and substituted into the geometric constraint equation (1.19), we obtain

$$T(s) = \left\{H^2 + \left(V - W\frac{s}{L_0}\right)^2\right\}^{1/2}. \tag{1.23}$$

2. Solution for $x = x(s)$:
Note that $dx/ds = dx/dp\, dp/ds$ and that dx/dp is given as a function of T in the first equation of (1.20), while dp/ds may also be obtained as a function of T from (1.21). Hence substituting for T from (1.23) yields

$$\frac{dx}{ds} = \frac{H}{EA_0} + \frac{H}{[H^2 + \{V - Ws/L_0\}^2]^{1/2}}, \tag{1.24}$$

which integrates to

$$x(s) = \frac{Hs}{EA_0} + \frac{HL_0}{W}\left[\sinh^{-1}\left(\frac{V}{H}\right) - \sinh^{-1}\left\{\frac{V - Ws/L_0}{H}\right\}\right], \tag{1.25}$$

where the end condition $x = 0$ at $s = 0$ has been incorporated.

3. Solution for $z = z(s)$:
Following a procedure analogous to that employed for x yields

$$z(s) = \frac{Ws}{EA_0}\left(\frac{V}{W} - \frac{s}{2L_0}\right) + \frac{HL_0}{W}\left[\left\{1 + \left(\frac{V}{H}\right)^2\right\}^{1/2} - \left\{1 + \left(\frac{V - Ws/L_0}{H}\right)^2\right\}^{1/2}\right]. \tag{1.26}$$

4. Solutions for H and V:
In deriving the solutions for x and z, only the end conditions at $s = 0$ in (1.22) were used. By satisfying the other end conditions for x and z, we obtain

$$l = \frac{HL_0}{EA_0} + \frac{HL_0}{W}\left\{\sinh^{-1}\left(\frac{V}{H}\right) - \sinh^{-1}\left(\frac{V - W}{H}\right)\right\}, \tag{1.27}$$

which is a transcendental equation in H and V. The second is an algebraic equation in these two quantities, namely,

$$h = \frac{WL_0}{EA_0}\left(\frac{V}{W} - \frac{1}{2}\right) + \frac{HL_0}{W}\left[\left\{1 + \left(\frac{V}{H}\right)^2\right\}^{1/2} - \left\{1 + \left(\frac{V - W}{H}\right)^2\right\}^{1/2}\right], \tag{1.28}$$

The simultaneous solution of these equations for H and V then allows the expressions for x, z, and T to be used. In general numerical methods are necessary, and it has been found that techniques such as a modified two-dimensional Newton's method are straightforward to implement, and with care convergence is fast [17]. This method is mentioned here because the general problem of response to many point loads is conceptually no more difficult to solve, and a two-dimensional Newton's method, or a similar technique, suffices to determine H and V (see equations 1.44 and 4.33).

When the supports are level, $h = 0$ and equation (1.28) supplies the expected result of $V = W/2$. Equation (1.27) can now be written in terms of the single dependent variable H:

$$\sinh\left(\frac{Wl}{2HL_0} - \frac{W}{2EA_0}\right) = \frac{W}{2H}, \tag{1.29}$$

and we may profitably spend some time exploring some of the properties of the symmetric elastic catenary as various limits are taken.

If the cable is assumed inextensible, (1.29) reduces to (1.9), and in all other respects the results for the classical catenary are recovered. A graphical solution of (1.29) quickly establishes the upper bound

$$H < EA_0\frac{l}{L_0}, \tag{1.30}$$

which will clearly be excessively conservative when cables composed of normal structural materials such as steel are used for structural purposes. It is of interest to note, however, that, if the cable is composed of a really extensible material, $H \to EA_0 l/L_0$, and the profile is parabolic [26].

When $l \leq L_0$, a lower bound for H is simply $H > 0$. But, when $l > L_0$, this lower bound is

$$H > EA_0\left(\frac{l}{L_0} - 1\right), \tag{1.31}$$

which implies that, because the cable's self-weight must be supported, the horizontal component of tension is always greater than the force necessary just to stretch the cable into position between the supports.

When l is little different from L_0, and the cable is relatively inextensible, a situation common in practice, equations (1.25), (1.26), and (1.29) may be shown to reduce to the approximate results

$$z(s) = \frac{1}{2}\left(1 + \frac{H}{EA_0}\right)\frac{W}{HL_0}s(L_0 - s), \tag{1.32}$$

where H is found from

$$H^3 + \frac{EA_0}{L_0}(L_0 - l)H^2 - \frac{EA_0 W^2}{24} = 0. \tag{1.33}$$

From Descartes' rule of signs [31] there is just one positive root of this cubic, regardless of the sign of $(L_0 - l)$, which is the required value of H. Equations (1.32) and (1.33) constitute what may be called the elastic parabola, and we may observe that the well-known parabolic approximation is obtained if EA_0 is assumed infinite and L_0 is only slightly greater than l. Then $H = [W^2/\{24(1 - l/L_0)\}]^{1/2}$. The effect of small but finite extensibility is thus to reduce this value of H when $L_0 > l$, as can be seen by substituting it in (1.33).

Response to a Point Load

Suppose that in addition to cable self-weight a concentrated vertical load F_1 is hung from the cable at a point s_1 along the cable from the support. The properties of this new profile can be found with little extra effort. The approach is as follows [17].

The geometric constraint equation (1.19) holds as before, except at the point of application of the concentrated load where the slope has to be discontinuous. The first of the equilibrium equations (see 1.20) is unaltered, while the second reads for $0 \leq s < s_1$

$$T\frac{dz}{dp} = V - W\frac{s}{L_0}$$

and for $s_1 < s \leq L_0$

$$T\frac{dz}{dp} = V - F_1 - W\frac{s}{L_0}. \tag{1.34}$$

The constitutive relation (1.21) is unaltered, except that it too cannot hold right at the point load. The end conditions are the same as given by (1.22). In addition we stipulate the following matching conditions, which ensure continuity of the cable at s_1:

$$x_1^- = x_1^+, \quad z_1^- = z_1^+, \quad p_1^- = p_1^+ \quad \text{at } s = s_1, \tag{1.35}$$

where $x_1^- = \lim\limits_{e \to 0} x(s_1 - e), e > 0$, with analogous definitions for the rest.

The results are

1. Solutions for $T = T(s)$:

$$\begin{aligned} T(s) &= \left\{H^2 + \left(V - W\frac{s}{L_0}\right)^2\right\}^{1/2}, \quad 0 \leq s < s_1, \\ T(s) &= \left\{H^2 + \left(V - F_1 - W\frac{s}{L_0}\right)^2\right\}^{1/2}, \quad s_1 < s \leq L_0. \end{aligned} \tag{1.36}$$

2. Solutions for $x = x(s)$:

$$x(s) = \frac{Hs}{EA_0} + \frac{HL_0}{W}\left[\sinh^{-1}\left(\frac{V}{H}\right) - \sinh^{-1}\left\{\frac{V - Ws/L_0}{H}\right\}\right],$$

$$0 \le s \le s_1,$$

$$x(s) = \frac{Hs}{EA_0} + \frac{HL_0}{W}\left[\sinh^{-1}\left(\frac{V}{H}\right) - \sinh^{-1}\left\{\frac{V - F_1 - Ws/L_0}{H}\right\}\right.$$

$$\left. + \sinh^{-1}\left\{\frac{V - F_1 - Ws_1/L_0}{H}\right\} - \sinh^{-1}\left\{\frac{V - Ws_1/L_0}{H}\right\}\right],$$

$$s_1 \le s \le L_0. \tag{1.37}$$

3. Solutions for $z = z(s)$:

$$z(s) = \frac{Ws}{EA_0}\left(\frac{V}{W} - \frac{s}{2L_0}\right) + \frac{HL_0}{W}\left[\left\{1 + \left(\frac{V}{H}\right)^2\right\}^{1/2}\right.$$

$$\left. - \left\{1 + \left(\frac{V - Ws/L_0}{H}\right)^2\right\}^{1/2}\right], \quad 0 \le s \le s_1,$$

$$z(s) = \frac{Ws}{EA_0}\left(\frac{V}{W} - \frac{s}{2L_0}\right) + \frac{HL_0}{W}\left[\left\{1 + \left(\frac{V}{H}\right)^2\right\}^{1/2}\right.$$

$$- \left\{1 + \left(\frac{V - F_1 - Ws/L_0}{H}\right)^2\right\}^{1/2} + \frac{F_1}{H}\frac{W}{EA_0}\left(\frac{s_1}{L_0} - \frac{s}{L_0}\right)$$

$$+ \left\{1 + \left(\frac{V - F_1 - Ws_1/L_0}{H}\right)^2\right\}^{1/2}$$

$$\left. - \left\{1 + \left(\frac{V - Ws_1/L_0}{H}\right)^2\right\}^{1/2}\right], \quad s_1 \le s \le L_0. \tag{1.38}$$

Finally, the equations by which H and V may be found are obtained by setting $x = l$ and $z = h$, at $s = L_0$, in the second equations of (1.37) and (1.38).

Writing out of solutions will get progressively more tedious as the number of point loads applied increases. But a remedy exists, and this is presented with the general results in section 1.3. First, however, it is profitable to discuss some experimental results.

The specific problem analyzed involved a cable of deep profile with supports at the same level. Initially the cable hung under its own weight (in reality a series of twenty weights placed uniformly along its length), and subsequently the cable was loaded in two equal increments at a point off-center. The experimental results are well illustrated by figure 1.11. Further details, which are not needed here, may be found in the report

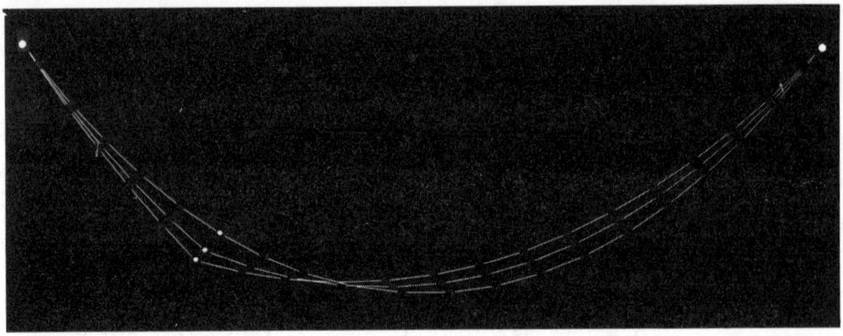

1.11 Multiple exposure photograph illustrating geometrically nonlinear response of a suspended cable

on the work; the excellent agreement found between theory and experiment can, perhaps, best be summarized by noting that the discrepancies in measured and calculated displacements were at most 4 percent [17].

The geometric nonlinearity of the response is clearly noticeable, as the displacement of the load point is markedly reduced with the application of the second increment of the point load. This stiffening effect characterizes cable response when the additional loads are appreciable. This nonlinearity is solely geometric, and in fact its origin may be traced to equation (1.19). The horizontal and vertical displacements of the load point are of the same magnitude because of the relatively deep profile and the positioning of the load point near the quarter-span point. In this particular example the effects of cable elasticity are practically negligible.

Response to Many Point Loads

Conceptually the results for one point load can be extended to include the response when many such loads are hung from the cable. A systematic way of writing and applying the solutions is an approach developed by Sinclair [17] which is also admirably compact.

In figure 1.12, the statement of vertical equilibrium at a point P on the strained profile is

$$T\frac{dz}{dp} = V - \sum_{i=0}^{n} F_i - W\frac{s}{L_0}, \tag{1.39}$$

provided that n is chosen so that P lies between P_n and P_{n+1} for $n = 0, 1, \ldots, N$. There are N point loads applied, and F_0 is defined as $F_0 = 0$, so that the general results will include the case of the cable under self-weight alone (when $N = 0$, the elastic catenary). The statement of horizontal equilibrium is the same as before. The ends of the cable are now conveniently labeled P_0 and P_{N+1}, and the end conditions are as given

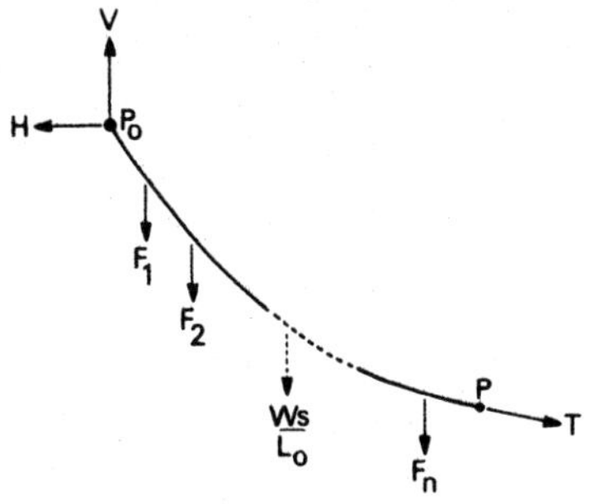

1.12 Representative segment when many concentrated vertical loads act in addition to cable self-weight

by (1.22). The matching conditions at the points of concentrated load application are

$$x_n^- = x_n^+, \quad z_n^- = z_n^+, \quad p_n^- = p_n^+ \quad \text{at } s = s_n, \tag{1.40}$$

for $n = 1, 2, \ldots, N$. Here $x_n^- = \lim_{e \to 0} x(s_n - e)$, $e > 0$, and so on.

The remaining details of the solution are mostly straightforward, especially once a recurrence relation is found for the constants of integration that arise by satisfying the matching conditions at each load point [17]. With a view toward a concise solution, the following dimensionless quantities are introduced (their use is confined just to section 1.3):

$\xi = x/L_0$, horizontal coordinate of the strained profile,
$\eta = z/L_0$, vertical coordinate of the strained profile,
$\sigma = s/L_0$, Lagrangian coordinate of the unstrained profile,
$\sigma_n = s_n/L_0$, Lagrangian load coordinate ($n = 1, 2, \ldots, N$),
$\delta = h/L_0$, relative vertical displacement of the end points,
$\gamma = l/L_0$, cable aspect ratio,
$\tau = T/W$, cable tension,
$\chi = H/W$, horizontal reaction at supports P_0, P_{N+1},
$\phi = V/W$, vertical reaction at support P_0,
$\psi_n = F_n/W$, applied concentrated vertical loads ($n = 1, 2, \ldots, N$),
$\beta = W/EA_0$, flexibility factor.

We define $\sigma_0 = 0$, $\psi_{-1} = 0$, and $\psi_0 = 0$, and we use the partial sum

$$\Psi_n = \sum_{j=-1}^{n} \psi_j,$$

so that $\Psi_{-1} = 0$, as does Ψ_0. We therefore have

1. Solution for cable tension $\tau = \tau(\sigma)$:

$$\tau(\sigma) = \{\chi^2 + (\phi - \Psi_n - \sigma)^2\}^{1/2}, \tag{1.41}$$

which holds on $\sigma_n < \sigma < \sigma_{n+1}$, for $n = 0, 1, \ldots, N$.

2. Solution for the horizontal coordinate of the strained profile $\xi = \xi(\sigma)$:

$$\xi(\sigma) = \chi\left[\beta\alpha + \sinh^{-1}\left(\frac{\phi}{\chi}\right) - \sinh^{-1}\left(\frac{\phi - \Psi_n - \sigma}{\chi}\right)\right. \tag{1.42}$$
$$\left. + \sum_{i=0}^{n}\left\{\sinh^{-1}\left(\frac{\phi - \Psi_i - \sigma_i}{\chi}\right) - \sinh^{-1}\left(\frac{\phi - \Psi_{i-1} - \sigma_i}{\chi}\right)\right\}\right],$$

which holds on $\sigma_n \leq \sigma \leq \sigma_{n+1}$, for $n = 0, 1, \ldots, N$.

3. Solution for the vertical coordinate of the strained profile $\eta = \eta(\sigma)$:

$$\eta(\sigma) = \beta\sigma\left(\phi - \frac{\sigma}{2}\right) + \{\chi^2 + \phi^2\}^{1/2} - \{\chi^2 + (\phi - \Psi_n - \sigma)^2\}^{1/2}$$
$$+ \sum_{i=0}^{n}[\beta\psi_i(\sigma_i - \sigma) + \{\chi^2 + (\phi - \Psi_i - \sigma_i)^2\}^{1/2} \tag{1.43}$$
$$- \{\chi^2 + (\phi - \Psi_{i-1} - \sigma_i)^2\}^{1/2}],$$

which also holds on $\sigma_n \leq \sigma \leq \sigma_{n+1}$, for $n = 0, 1, \ldots, N$.

The equations from which χ and ϕ are found are

$$\frac{\gamma}{\chi} - \beta = \sinh^{-1}\left(\frac{\phi}{\chi}\right) - \sinh^{-1}\left(\frac{\phi - \Psi_N - 1}{\chi}\right)$$
$$+ \sum_{i=0}^{N}\left\{\sinh^{-1}\left(\frac{\phi - \Psi_i - \sigma_i}{\chi}\right) - \sinh^{-1}\left(\frac{\phi - \Psi_{i-1} - \sigma_i}{\chi}\right)\right\},$$

$$\delta = \beta\left(\phi - \frac{1}{2}\right) + (\chi^2 + \phi^2)^{1/2} - \{\chi^2 + (\phi - \Psi_N - 1)^2\}^{1/2}$$
$$+ \sum_{i=0}^{N}[\beta\psi_i(\sigma_i - 1) + \{\chi^2 + (\phi - \Psi_i - \sigma_i)^2\}^{1/2}$$
$$- \{\chi^2 + (\phi - \Psi_{i-1} - \sigma_i)^2\}^{1/2}]. \tag{1.44}$$

The values of these two parameters may be determined via a modified two-dimensional Newton's method, but other techniques such as Brown's method for simultaneous nonlinear equations may be useful [6].

Equations (1.41) through (1.44) constitute an exact solution for the static response of an elastic cable under self-weight and N concentrated vertical loads. By setting $N = 0, 1$, the earlier results of this section are recovered, as expected. The present problem can be generalized to the case of nonvertical loads (although these often cannot be as easily applied) and to distributed loadings (or even loaded networks in which case the mushrooming of the algebra effectively precludes their study by this method, except for very simple networks [14]). For distributed loadings such as fluid pressure a different formulation is preferable, especially because usually the cable (now obviously a two-dimensional membrane) self-weight may be ignored.

1.4 An Alternative Formulation for Distributed Loadings

Ignoring cable self-weight and extensibility for the present, we concentrate on the equations of equilibrium of a cable element when they are expressed in terms of coordinates directed along and perpendicular to the element. For coordinates parallel to a pair of fixed directions, such as the x and z axes in figure 1.13a, the equations are simply

$$\begin{aligned} \frac{d}{ds}\left(T\frac{dx}{ds}\right) - e\frac{dz}{ds} &= 0, \\ \frac{d}{ds}\left(T\frac{dz}{ds}\right) + f\frac{dx}{ds} &= 0, \end{aligned} \tag{1.45}$$

where e is the intensity per unit vertical length of the distributed load that acts in the negative horizontal direction, while f is the intensity per unit horizontal length of the load acting in the vertical direction.

In figure 1.13b a pair of mutually perpendicular quantities (X, Z) is rotated through an angle ψ to form the pair (X', Z'). It is easily shown that

$$\begin{Bmatrix} X' \\ Z' \end{Bmatrix} = \begin{bmatrix} \cos\psi & \sin\psi \\ -\sin\psi & \cos\psi \end{bmatrix} \begin{Bmatrix} X \\ Z \end{Bmatrix}. \tag{1.46}$$

Applying this transformation to (1.45), we obtain

$$\begin{aligned} \cos\psi\frac{d}{ds}\left(T\frac{dx}{ds}\right) - \cos\psi\, e\frac{dz}{ds} + \sin\psi\frac{d}{ds}\left(T\frac{dz}{ds}\right) + \sin\psi\, f\frac{dx}{ds} &= 0, \\ -\sin\psi\frac{d}{ds}\left(T\frac{dx}{ds}\right) + \sin\psi\, e\frac{dz}{ds} + \cos\psi\frac{d}{ds}\left(T\frac{dz}{ds}\right) + \cos\psi\, f\frac{dx}{ds} &= 0. \end{aligned} \tag{1.47}$$

In figure 1.13c, we observe that $dx/ds = \cos\psi$ and $dz/ds = \sin\psi$, so that equations (1.47) simplify to

$$\begin{aligned} \frac{dT}{ds} &= (e - f)\sin\psi\cos\psi, \\ T\frac{d\psi}{ds} &= -e\sin^2\psi - f\cos^2\psi, \end{aligned} \tag{1.48}$$

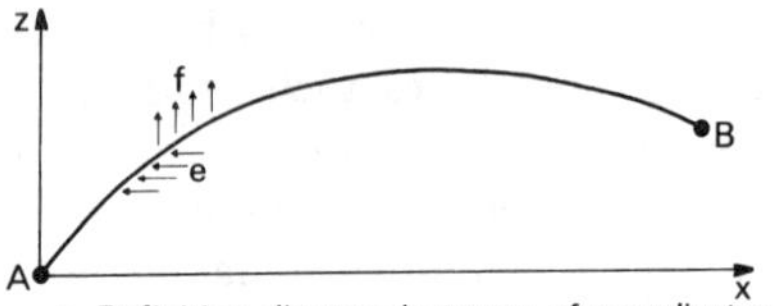

(a) Definition diagram in terms of coordinates parallel to a pair of fixed directions

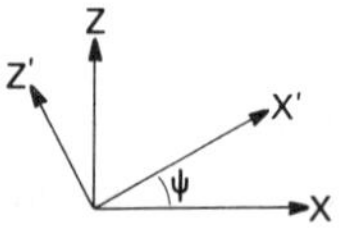

(b) Coordinate transformation

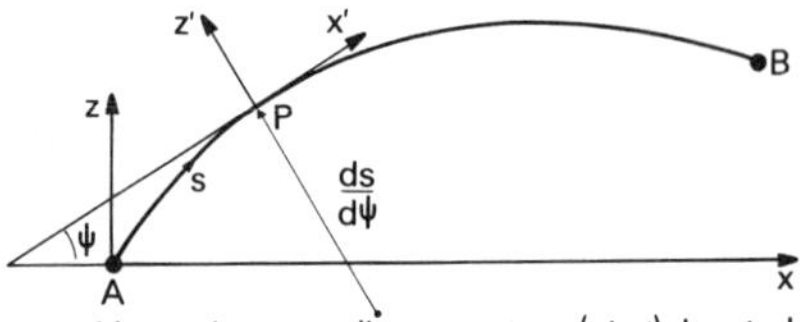

(c) Alternative coordinate system (x', z') located at the point P

1.13

where $ds/d\psi$ is recognized as the radius of curvature of the element.

Suppose for argument's sake that the applied loading is hydrostatic, namely, $e = f$. Since there is now no loading applied tangentially to the cable, the tension is everywhere constant, $T = T_0$. The second equation of (1.48) reduces to

$$T_0 \frac{d\psi}{ds} = -f, \tag{1.49}$$

and together with the geometric relations

$$\begin{aligned} \frac{dx}{ds} &= \cos\psi, \\ \frac{dz}{ds} &= \sin\psi, \end{aligned} \tag{1.50}$$

and the pertinent set of end conditions, this facilitates the solution of problems in which hydrostatic pressure is the dominant loading. The inclusion of elastic effects is particularly straightforward on account of the constancy of the tension.

As was mentioned earlier, we are in reality discussing a representative slice of a two-dimensional membrane, but we are not including the influence of end effects in the other direction. Because of their complexity, such problems are often considered respectable by theoretical elasticians, so treatment of them lies well outside the scope of this work.

This formulation of the equations of equilibrium in which the angle ψ is the independent parameter and x, z, and s are all dependent on it provides direct solutions in several instances.

1.5 Two Applications

We treat in detail two practical applications of the exact theory. The first concerns the shape of an enveloping cable used to contain and move a loose raft of logs. The second treats elements of the theory and behavior of inflatable dams.

The Shape of a Towed Boom of Logs

To set the scene, we may quote verbatim from Newman [21], whose work we refer to extensively in this problem.

> In Canada, and in parts of Europe, wood is sometimes transported as a loose floating raft of logs surrounded by a boom of larger timbers. In Canada the wood is usually pulp wood consisting of short logs 4 ft long and less than 1 ft in diameter. The booms consist of larger timbers, typically 20–30 ft long, and when large-enough diameters are not available, each "boomstick" may consist of three or four timbers bolted together in an arrangement peculiar to the locality. The complete boom consists of several hundred such sticks coupled together with chains. . . .
>
> A rafting boom is usually made up in the following manner. The wood floats freely down a river during the spring and early summer and is stopped by a holding boom where the river widens or enters a lake. Here the rafting boom is stretched from shore to shore downstream of the holding boom, which is then opened, permitting the wood to flow in. The rafting boom is then closed to form a circular raft of floating wood. When one of the boom chains is joined by a cable to the towing tug, the wood moves rearwards, and the boom assumes a characteristic oval shape with open water at the front and logs piled three or four deep at the rear, being held there by the greater depth of the boom. As the tow proceeds, the wood continues to move rearwards, particularly if strong head winds are encountered, and a small amount of wood (usually less than 2 percent) is lost. Modern tow boats are equipped with diesel engines of 500–1,000 hp and tow the raft at about 1 mi/hr. The rafts hold 10^5–10^6 logs and are very large, extending over areas which are typically 10–40 acres in extent.

In developing the theory for the properties of such a boom, several assumptions are necessary. For example, aerodynamic effects and the wave-making ability of the raft are considered negligible on account of its low speed. The shaping force is thus skin friction from the highly turbulent passage of water around the logs which, because they are so numerous, are best treated as a continuum. In view of the ever-present water, this continuum is essentially free of shear stresses. The skin friction is judged constant everywhere, and the logs in the raft experience a hydrostatic state of compressive stress which therefore increases in magnitude directly as the rear of the boom is approached. The interconnected line of boomsticks has more in common which a giant cable than anything else.

In a tow under constant velocity in which the ordinate x marks the forward position of the logs, as is evident from figure 1.14, the equation of equilibrium is

$$T_0 \frac{d\psi}{ds} = -\tau z, \tag{1.51}$$

where T_0 is the constant tension in the boomsticks, ψ is the angle between the tangent to the curve and the x axis, and τ is the skin friction.

The geometric relations are given by (1.50). The problem is identical to that of determining the shape of a two-dimensional droplet or the elastica—a beam under a pair of equal and opposite end loads—for which the solutions are well known.

The second of equations (1.50), together with (1.51) yields on integration

$$\tau z^2 = 2T_0 \cos \psi + C, \tag{1.52}$$

where the constant C is found from the condition that $\psi = \pi/2 + \alpha$ at $z = 0$ (the angle 2α is the angle between the boomsticks at their junction with the tug's hawser). Therefore

$$\tau z^2 = 2T_0 (\cos \psi + \sin \alpha). \tag{1.53}$$

Substituting this into (1.51) provides

$$s = \left(\frac{T_0}{2\tau}\right)^{1/2} \int_{\psi}^{\pi/2+\alpha} \frac{d\psi}{(\cos \psi + \sin \alpha)^{1/2}}. \tag{1.54}$$

Let $\psi = \pi/2 + \theta$, and set $\sin \phi = \{(1 + \sin \theta)/(1 + \sin \alpha)\}^{1/2}$. Since $-\pi/2 \le \theta \le \alpha$, it follows that $0 \le \sin \phi \le 1$. The angle θ is that between the tangent to the profile and the z axis. After some manipulation we obtain

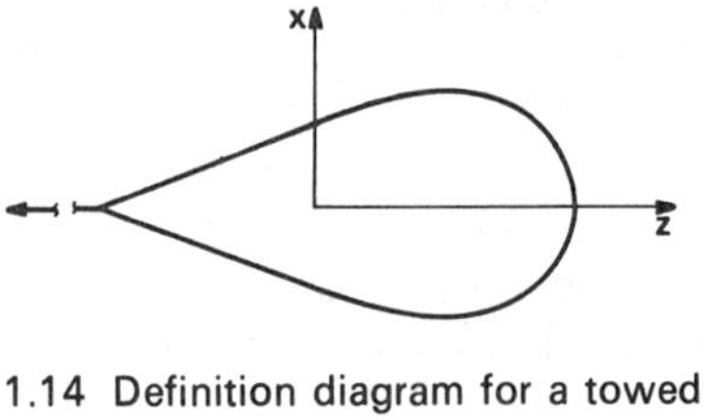

1.14 Definition diagram for a towed raft of logs

the solution in a standard form for an elliptic integral of the first kind:

$$s = \left(\frac{T_0}{\tau}\right)^{1/2} \int_{\phi}^{\pi/2} \frac{d\phi}{(1 - m \sin^2 \phi)^{1/2}}, \tag{1.55}$$

where $m = (1 + \sin\alpha)/2$. In the present problem the bounds on m are $1/2 < m < 1$. The lower bound holds because α can approach but never reach zero; the weak upper bound holds because α has a peak value of about 41°. So, adopting standard terminology [1], the length of the profile reads

$$s(\phi) = \left(\frac{T_0}{\tau}\right)^{1/2} \left\{F\left(\frac{\pi}{2}\middle| m\right) - F(\phi|m)\right\}. \tag{1.56}$$

With a similar approach we also obtain

$$x = \left(\frac{T_0}{\tau}\right)^{1/2} \left\{2\int_0^{\phi} (1 - m \sin^2 \phi)^{1/2}\, d\phi - \int_0^{\phi} \frac{d\phi}{(1 - m \sin^2 \phi)^{1/2}}\right\}, \tag{1.57}$$

where the lower limit applies since $x = 0$ at $\phi = 0$. This is written

$$x(\phi) = \left(\frac{T_0}{\tau}\right)^{1/2} \{2E(\phi|m) - F(\phi|m)\}, \tag{1.58}$$

where $E(\ )$ and $F(\ )$ are elliptic integrals of the second and first kinds, respectively. When $\theta = \alpha$, $\phi = \pi/2$, and the peak value of α occurs when $x(\pi/2) = 0$. This requires $2E(\pi/2|m) = F(\pi/2|m)$, from which $\alpha_{max} \simeq 41°$ (see figure 1.15): the boom cannot be made fuller than this.

The final theoretical aspect concerns the overall equilibrium which, if A is the total area in plan of the logs, is expressed by

$$2T_0 \cos\alpha = A\tau,$$

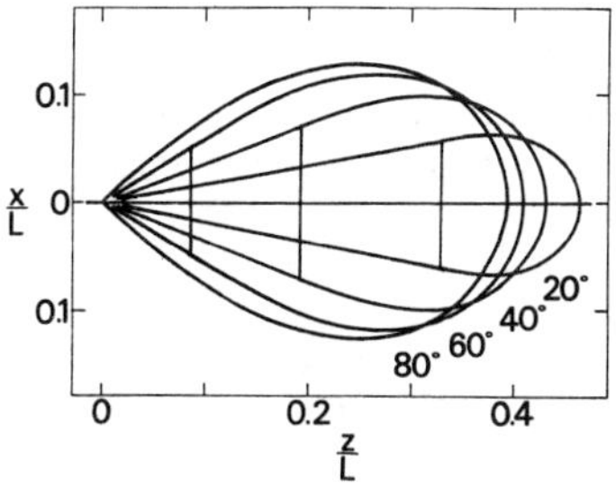

1.15 Raft profiles for varying values of the included angle at the towpoint, scaled with respect to profile length (courtesy B. G. Newman)

or

$$A = 4\left(\frac{T_0}{\tau}\right) m^{1/2}(1 - m)^{1/2}. \tag{1.59}$$

Hence our equations may be restated as

$$\begin{aligned}
\frac{z}{A^{1/2}} &= \frac{m^{1/2}\cos\phi}{(m - m^2)^{1/4}},\\
\frac{s}{A^{1/2}} &= \frac{1}{2(m - m^2)^{1/4}}\left\{F\left(\frac{\pi}{2}\middle| m\right) - F(\phi|m)\right\},\\
\frac{x}{A^{1/2}} &= \frac{1}{2(m - m^2)^{1/4}}\{2E(\phi|m) - F(\phi|m)\}.
\end{aligned} \tag{1.60}$$

A little further manipulation gives the total length of the boom as

$$\frac{L}{A^{1/2}} = \frac{2}{(2m - 1)(m - m^2)^{1/4}}\left\{E\left(\frac{\pi}{2}\middle| m\right) - (1 - m)F\left(\frac{\pi}{2}\middle| m\right)\right\}. \tag{1.61}$$

If a particular boom length is chosen, the preceding equations may be used to illustrate the chain of events as a raft is taken in tow and the logs move rearward (see figure 1.15).

Newman reported several attempts at experimental verification of the theory, the outcome being somewhat inconclusive on account of the very real difficulty of achieving satisfactory scaling of the physical effects. However, with the aid of aerial photography, of which figure 1.16 is a good example, he was able to demonstrate the broad validity of his assumptions. In figure 1.16, the included angle at the apex is around 40°, which is typical of a fully developed tow. Figure 1.17, on the other hand, depicts conditions near the start of a tow, and the included angle is somewhat greater.

We began this example with a quotation from Newman's work, and we may close it by again drawing directly on his work:

The analyses may prove useful in a number of ways. For example, the total area of the wood under tow can be determined from simple measure-

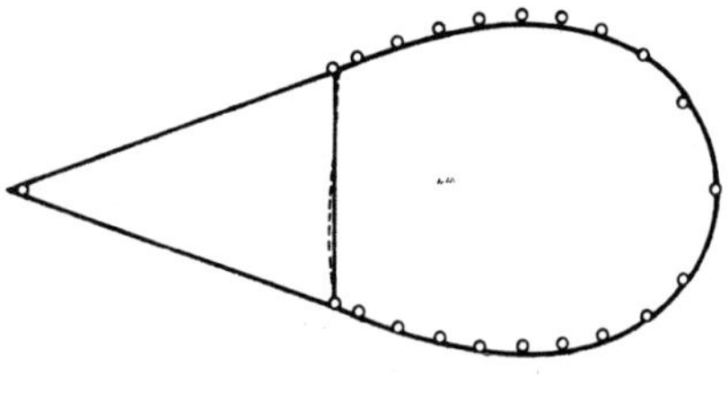

1.16 Comparison of theory with a surveyed boom under tow (courtesy B. G. Newman)

1.17 Aerial photograph of a tow in progress (courtesy B. G. Newman)

ments of boom geometry. The power required to tow the boom at various speeds may be estimated. It would also be possible to determine the best point to cinch a boom to clear a narrow passageway subject perhaps to the requirement that the tension in the various portions of the boom should not be increased beyond acceptable limits.

Newman's work is a case in point of how a little analysis and sound engineering judgment can be usefully combined.

Inflatable Dams

Inflatable dams have been used in several parts of the world to impound water. Since concerns about long-term fabric durability and performance naturally prevail, their use has been confined mostly to temporary works. An interesting discussion of the behavior of one such dam, used on the Mangla project in Pakistan, has been published by Binnie et al. [5], and experiments on models have been reported by Baker et al. [3] and Anwar [2].

The theory of inflatable dams has been treated to varying degrees of completeness by Anwar [2], Binnie [4], and Parberry [23]. The treatment given here is based on these studies.

The properties of a dam under inflation are used to help construct an exact solution for the inflated dam, impounding water to its crest.

By specifying independent variables such as would be used in practice, the dependent variables of this problem may be found. However, if tables of the elliptic functions are not readily available, the same results may be found by using formulas in which uniformly valid asymptotic expansions replace the elliptic integrals. These approximate solutions involve simple trigonometric functions and are sufficiently accurate for practical purposes.

Properties of the Inflated Dam

The inflatable dam consists of a flexible material loaded by hydrostatic pressure and founded on a horizontal base. The dam is assumed to be infinitely long, so that its generators are straight (in practice a length ten times the crest height would qualify). The cross section shown in figure 1.18 is typical of such a dam.

Ignoring the material self-weight and elasticity (which in fact is typically of little consequence) and introducing dimensionless variables,

$$\mathbf{x} = \frac{x}{l}, \quad \mathbf{z} = \frac{z}{l}, \quad \mathbf{s} = \frac{s}{l},$$

and

$$\mathbf{L} = \frac{L_0}{l}, \quad \mathbf{h} = \frac{h}{l}, \quad \mathbf{T} = \frac{T_0}{\rho g l^2},$$

where L_0 is the profile length, l is the width of the base, h is the total head on the base of the dam, T_0 is the tension, and ρg is the specific weight of the inflating fluid (usually water), (1.49) and (1.50) become

$$\frac{d\psi}{d\mathbf{s}} = -\frac{1}{\mathbf{T}^2}(\mathbf{h} - \mathbf{z}), \tag{1.62}$$

and

$$\begin{aligned} \frac{d\mathbf{x}}{d\mathbf{s}} &= \cos \psi, \\ \frac{d\mathbf{z}}{d\mathbf{s}} &= \sin \psi. \end{aligned} \tag{1.63}$$

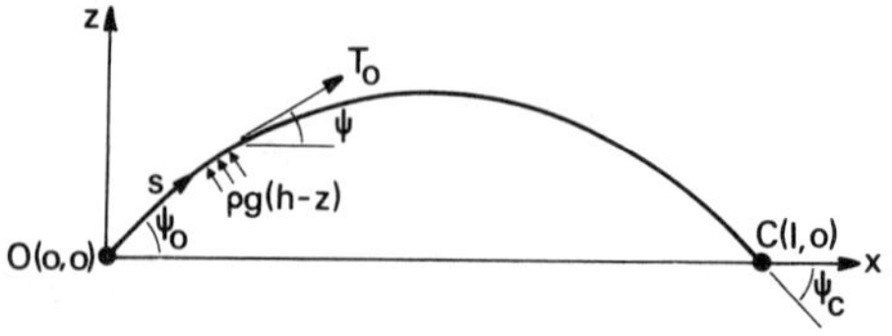

1.18 Definition diagram for a dam under inflation alone

Combining the first and the third of these, and then integrating and using the boundary condition of $\psi = \psi_0$ and $\mathbf{z} = 0$, gives

$$\cos\psi - \cos\psi_0 = \frac{1}{2\mathbf{T}^2}(2\mathbf{hz} - \mathbf{z}^2), \tag{1.64}$$

from which it follows that $\psi_c = -\psi_0$, as expected. Therefore

$$\mathbf{z} = \mathbf{h}\left[1 - \left\{1 - \frac{\varepsilon^2}{2}(\cos\psi - \cos\psi_0)\right\}^{1/2}\right], \tag{1.65}$$

where for convenience $\varepsilon = 2\mathbf{T}/\mathbf{h}$. For (1.65) to yield a real solution, it is necessary that $\varepsilon^2 \leq 2/(\cos\psi - \cos\psi_0)$, and a least upper bound for ε^2, corresponding to the crest $\psi = 0$, is $\varepsilon^2 \leq 2/(1 - \cos\psi_0)$. The maximum value of $\mathbf{z}$ occurs at $\psi = 0$, so that

$$\mathbf{z}_{\mathbf{max}} = \mathbf{h}\left[1 - \left\{1 - \frac{\varepsilon^2}{2}(1 - \cos\psi_0)\right\}^{1/2}\right], \tag{1.66}$$

and the least upper bound ensures that $\mathbf{z}_{\mathbf{max}} \leq \mathbf{h}$. Therefore full inflation always occurs.

Introduce $\sin\beta$ defined by

$$\sin^2\beta = \frac{\varepsilon^2}{\{1 + \varepsilon^2\cos^2(\psi_0/2)\}}, \tag{1.67}$$

where, again with the aid of the least upper bound, we know that $\sin^2\beta \leq 1$. Equation (1.65) may be written in the alternative form

$$\mathbf{z} = \mathbf{h}\left[1 - \frac{\varepsilon}{\sin\beta}\{1 - \sin^2\beta\cos^2(\psi/2)\}^{1/2}\right], \tag{1.68}$$

and so

$$\mathbf{z}_{\mathbf{max}} = \mathbf{h}(1 - \varepsilon\cot\beta). \tag{1.69}$$

As an interim measure for finding $\mathbf{s} = \mathbf{s}(\psi)$, it is noted that (1.62) and (1.68) between them yield

$$\frac{d\mathbf{s}}{d\psi} = -\frac{\mathbf{T}}{2}\sin\beta\frac{1}{\{1 - \sin^2\beta\cos^2(\psi/2)\}^{1/2}}, \tag{1.70}$$

which upon integration, using the substitution $\phi = (\pi - \psi)/2$ and the end condition $\mathbf{s} = 0$ at $\psi = \psi_0$, gives

$$\mathbf{s} = \mathbf{T}\sin\beta\int_{\pi/2-\psi_0/2}^{\pi/2-\psi/2}\frac{d\phi}{(1 - \sin^2\beta\sin^2\phi)^{1/2}}, \tag{1.71}$$

so that in terms of the elliptic integral of the first kind [1]

$$\mathbf{s} = \mathbf{T}\sin\beta\left\{F\left(\frac{\pi}{2} - \frac{\psi}{2}\backslash\beta\right) - F\left(\frac{\pi}{2} - \frac{\psi_0}{2}\backslash\beta\right)\right\}. \tag{1.72}$$

At $\psi = 0$, $\mathbf{s} = \mathbf{L}/2$, so that

$$\mathbf{L} = 2\mathbf{T}\sin\beta\left\{F\left(\frac{\pi}{2}\backslash\beta\right) - F\left(\frac{\pi}{2} - \frac{\psi_0}{2}\backslash\beta\right)\right\}. \tag{1.73}$$

Again we note in the interim that

$$\frac{d\mathbf{x}}{d\psi} = -\frac{\mathbf{T}}{2}\sin\beta\frac{\cos\psi}{\{1 - \sin^2\beta\cos^2(\psi/2)\}^{1/2}}, \tag{1.74}$$

which leads to

$$\mathbf{x} = \frac{\mathbf{T}}{\sin\beta}(2 - \sin^2\beta)\int_{\pi/2-\psi_0/2}^{\pi/2-\psi/2}\frac{d\phi}{(1 - \sin^2\beta\sin^2\phi)^{1/2}} - \frac{2\mathbf{T}}{\sin\beta}\int_{\pi/2-\psi_0/2}^{\pi/2-\psi/2}(1 - \sin^2\beta\sin^2\phi)^{1/2}\,d\phi. \tag{1.75}$$

Now, in terms of elliptic integrals of both the first and second kinds,

$$\mathbf{x} = \frac{\mathbf{T}}{\sin\beta}(2 - \sin^2\beta)\left\{F\left(\frac{\pi}{2} - \frac{\psi}{2}\backslash\beta\right) - F\left(\frac{\pi}{2} - \frac{\psi_0}{2}\backslash\beta\right)\right\} - \frac{2\mathbf{T}}{\sin\beta}\left\{E\left(\frac{\pi}{2} - \frac{\psi}{2}\backslash\beta\right) - E\left(\frac{\pi}{2} - \frac{\psi_0}{2}\backslash\beta\right)\right\}. \tag{1.76}$$

At $\psi = 0$, $\mathbf{x} = 1/2$, with the result

$$\frac{1}{2} = \frac{\mathbf{T}}{\sin\beta}(2 - \sin^2\beta)\left\{F\left(\frac{\pi}{2}\backslash\beta\right) - F\left(\frac{\pi}{2} - \frac{\psi_0}{2}\backslash\beta\right)\right\} - \frac{2\mathbf{T}}{\sin\beta}\left\{E\left(\frac{\pi}{2}\backslash\beta\right) - E\left(\frac{\pi}{2} - \frac{\psi_0}{2}\backslash\beta\right)\right\}. \tag{1.77}$$

The cross-sectional area of the profile is

$$\mathbf{A} = \int_0^1 \mathbf{z}\,d\mathbf{x} = \int_{\psi_0}^{-\psi_0}\mathbf{z}\frac{d\mathbf{x}}{d\psi}\,d\psi = \mathbf{h} - 2\mathbf{T}^2\sin\psi_0. \tag{1.78}$$

where $\mathbf{A} = A/l^2$.

Between (1.73) and (1.77) we can find expressions for **T** and **L** that may be written solely in terms of β and ψ_0. Then using series expansions for the elliptic functions, we obtain

$$\mathbf{T} \simeq \frac{1}{\sin\beta}\left\{\frac{1}{\sin\psi_0} - \sin^2\beta\frac{(3\psi_0 + 4\sin\psi_0 + \sin\psi_0\cos\psi_0)}{8\sin\psi_0(\psi_0 + \sin\psi_0)}\right\},$$

$$\mathbf{L} \simeq \frac{\psi_0}{\sin\psi_0} - \sin^2\beta\frac{(\psi_0^2 - 2\sin^2\psi_0 + \psi_0\sin\psi_0\cos\psi_0)}{8\sin\psi_0(\psi_0 + \sin\psi_0)}. \quad (1.79)$$

(These results are due to Michael J. O'Sullivan.)

Inflation Plus Impoundment

A loading condition often worthwhile checking out is the extreme case of no freeboard: the dam impounds water to its crest on the upstream side, and there is no tailwater. Figure 1.19 shows two regions that require consideration. The first, the downstream profile, is an adaptation of the results just presented, and the other, the upstream face, is simple to describe from first principles since the uniform pressure on all of this face means that it is circular in profile.

One way to proceed to the solution is as follows. The angle that the profile subtends to the horizontal at the downstream toe ψ_1 is specified in advance, as are the head on the base of the dam h and crest height $z_{max}(z_{max} \leq h)$, which is also equal to the depth of the impounded water. Thus from a rearrangement of (1.67) and (1.69) we obtain

$$\varepsilon = \frac{\{1 - (1 - z_{max}/h)^2\}^{1/2}}{\sin\psi_1},$$

$$\sin\beta = \frac{\varepsilon}{\{\varepsilon^2 + (1 - z_{max}/h)^2\}^{1/2}}. \quad (1.80)$$

Since $\varepsilon = 2\mathbf{T}/\mathbf{h}$, the tension is

$$T_0 = \frac{\varepsilon^2 \rho g h^2}{4}. \quad (1.81)$$

Substitution of this into (1.77) yields for the base width of the downstream portion of the dam

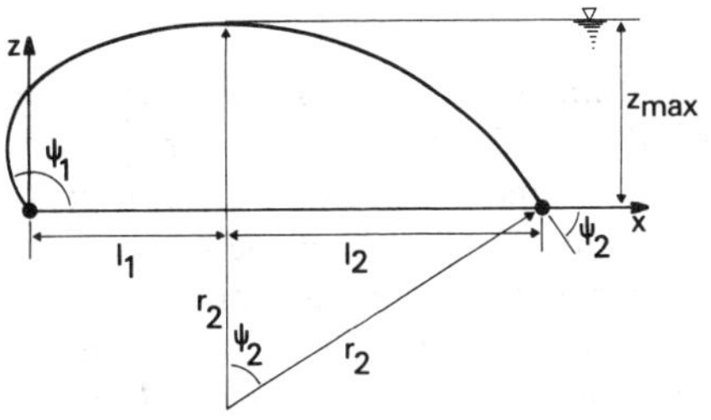

1.19 Definition diagram for a dam impounding water to its crest with no tailwater

$$l_1 = \frac{\varepsilon h}{2 \sin \beta}\left[(2 - \sin^2 \beta)\left\{F\left(\frac{\pi}{2}\backslash\beta\right) - F\left(\frac{\pi}{2} - \frac{\psi_1}{2}\backslash\beta\right)\right\}\right.$$
$$\left. - 2\left\{E\left(\frac{\pi}{2}\backslash\beta\right) - E\left(\frac{\pi}{2} - \frac{\psi_1}{2}\backslash\beta\right)\right\}\right], \tag{1.82}$$

and from (1.72) the length of the downstream face of the profile is

$$L_1 = \frac{\varepsilon h}{2} \sin \beta \left\{F\left(\frac{\pi}{2}\backslash\beta\right) - F\left(\frac{\pi}{2} - \frac{\psi_1}{2}\backslash\beta\right)\right\}. \tag{1.83}$$

But if instead of the elliptic functions their series expansions are used (see 1.79), the results will not be too different.

Turning now to the upstream face, and following Binnie [4], we note that the radius of the circular arc is

$$r_2 = \frac{T_0}{\rho g(h - z_{\max})}, \tag{1.84}$$

so that this part of the profile is quickly sketched in. In particular the angle at the upstream toe is given by

$$\cos \psi_2 = \frac{1 - z_{\max}}{r_2},$$

so that

$$l_2 = r_2(1 - \cos^2 \psi_2)^{1/2}, \tag{1.85}$$

and

$$L_2 = r_2 \psi_2.$$

These last two results allow the basewidth $l(= l_1 + l_2)$ to be determined and also the length of the profile $L(= L_1 + L_2)$. The total cross-sectional area is

$$A = hl_1 - \frac{T_0}{\rho g} \sin \psi_1 + \frac{1}{2}\left\{\frac{\psi_2 - \frac{1}{2}(\sin 2\psi_2)}{2}\right\} r_2^2. \tag{1.86}$$

Numerous other problems readily come to mind, such as determining properties of the same dam under inflation alone, or when, as frequently occurs, impoundment is not to the crest and some tailwater is present. In theory all these problems may be solved along the same lines as the relatively straightforward one that we have chosen to consider. But, because many of the calculations are tedious and involve the implicit solution of rather involved equations, we shall let the matter rest. Further guidance may be sought in the literature already cited. However, a word of caution is in order: in trying to determine new properties for the same dam under different external loading conditions, one too many unknowns will exist unless the extra condition of conservation of mass is stipulated. Incompressibility for the contained water requires the cross-sectional areas to be identical in both instances. The example that follows illustrates the application of the results.

Example 1.2 An Inflatable Dam for a River Diversion

To help reduce flooding in a low-lying area, a cut is proposed to allow floodwater to bypass a loop in the river. Under normal flows the cut is dry, and inflatable dams are planned at each end to hold back the river. Under high-flood conditions the dams are deflated. The critical design condition is expected to exist under impoundment to a crest height of 3 m. The head on the base of the dam is set at 5 m, and it is considered prudent not to let the downstream toe of the dam get closer than 135° to the horizontal. We need then to know what tension will be induced and what fabric length and basewidth are necessary. The specific weight of water is 10 kN/m^3.

Systematic application of (1.80) through (1.86) yields

$$\varepsilon = \frac{\{1 - (1 - 3/5)^2\}^{1/2}}{\sin 3\pi/4} = 1.30.$$

Therefore

$$T_0 = \frac{1.30^2 \times 10 \times 25}{4} = 106 \text{ kN/m},$$

and this tension corresponds to a stress of approximately 0.22 MPa for a fabric of thickness 0.005 m, which is low. Proceeding, we have

$$\sin \beta = \frac{1.30}{(1.30^2 + 4/25)^{1/2}} = 0.956,$$

and the partial basewidth is

$$l_1 = \frac{1.30 \times 5}{2 \times 0.956}\{(2 - 0.956^2)(2.65 - 0.40) - 2(1.09 - 0.38)\}$$

$$= 3.5 \text{ m}.$$

(If tables of elliptic functions are not readily available, equation 1.79 may be used instead.)

The length of the downstream face is

$$L_1 = \frac{1.30 \times 5}{2} \times 0.956(2.65 - 0.40)$$

$$= 7.0 \text{ m}.$$

The radius of curvature of the upstream face is

$$r_2 = \frac{106}{10 \times 2} = 5.3 \text{ m},$$

and the angle at the base is

$$\psi_2 = \cos^{-1}\left(\frac{5.3 - 3}{5.3}\right) = 64°.$$

The portions of the base and the profile length associated with the upstream face are, respectively,

$$l_2 = 5.3\left\{1 - \left(\frac{2.3}{5.3}\right)^2\right\}^{1/2} = 4.8 \text{ m},$$

and

$$L_2 = 5.9 \text{ m}.$$

The required basewidth is $l = 8.3$ m, and the profile length is $L = 12.9$ m; the cross-sectional area is $A = 20.2$ m^2. A typical elevation is shown in figure 1.20.

Exercises

1.1 Show that a cable can never be straight unless it is vertical. This was a theorem known to both Leonardo da Vinci and Galileo. An elegant proof is given by Routh [29].

1.2 Prove Pardies' theorem (1673): the point of intersection of any two tangents to the cable profile lies on the vertical through the center of gravity of the included portion [30].

1.3 With reference to figure 1.21 show from statics alone that the difference equation for the slopes of the links is

$$\tan\theta_{k+2} - 2\tan\theta_{k+1} + \tan\theta_k = 0,$$

from which

$$\tan\theta_k = Ak + B,$$

where A and B are constants. This constitutes a proof of Huygen's theorem (1690) that the slopes of the weightless links interconnecting a series of equal concentrated loads increase in arithmetic progression [30]. If in addition the horizontal projections of the links are all equal, it may be

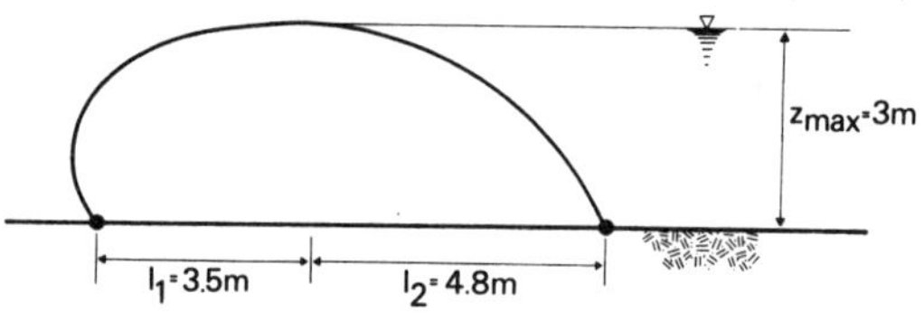

1.20 Typical cross section of an inflatable dam for a river diversion

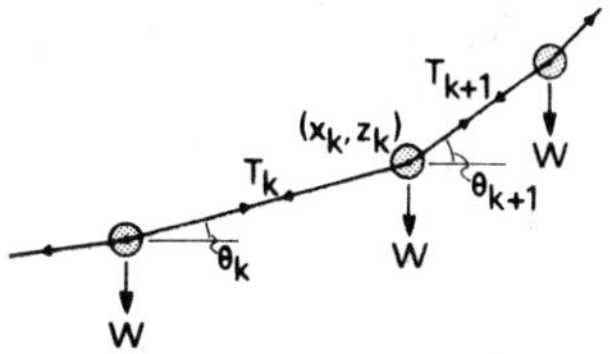

1.21

shown that the points (x_k, z_k) lie on a parabola. If the links themselves are all equal, it may be shown with somewhat more difficulty that the curve through (x_k, z_k) is the catenary.

1.4 The equilibrium equations for figure 1.22 are

$$T\frac{dz}{dp} = V - mgs - \int_0^x f(\xi)\,d\xi$$

and

$$T\frac{dx}{dp} = H - \int_0^z e(\xi)\,d\xi$$

for the vertical and horizontal forces, with moment equilibrium about the support requiring

$$\left(T\frac{dz}{dp}\right)x - \left(T\frac{dx}{dp}\right)z = \int_0^z e(\xi)\xi d\xi - \int_0^s mgx(\xi)d\xi - \int_0^x f(\xi)\xi d\xi,$$

where e and f are distributed loads per unit vertical and horizontal length, respectively, and ξ is a dummy variable of integration. Manipulate these three equations to prove the statement given in section 1.3 that moment equilibrium supplied no additional information to that already inherent in the equations of horizontal and vertical force equilibrium for a cable segment.

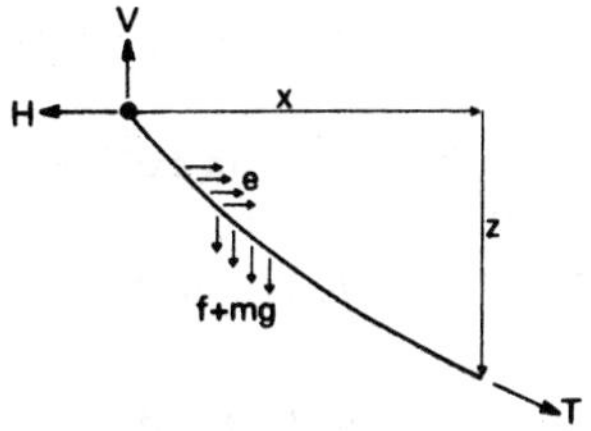

1.22

1.5 The peak tension in a symmetrically suspended catenary will be high if the sag is large compared to the span, and it will again be large as the sag becomes a small fraction of the span. This suggests a minimum value for the peak tension at some intermediate value of the ratio of sag to span. Show that this occurs when the ratio is

$$\frac{\{\cosh (mgl/2H) - 1\}}{2(mgl/2H)},$$

where the required value $mgl/2H$ is found from the root of $\tanh (mgl/2H) = 1/(mgl/2H)$. In fact this root is about unity, and the ratio is then about 1:4, which is too high for structural efficiency.

1.6 Show that the ellipse is the funicular profile for a loading consisting of components with constant but different intensities per unit vertical and horizontal length. This type of loading is known to exist in some undisturbed soil deposits where for geological reasons lateral pressures may coexist with markedly different overburden pressures. It might be thought then that a suitably scaled ellipse would provide the best lining profile for a tunnel driven through this type of material. However, site measurements indicate that the loading a disturbed soil eventually imposes on a lining is frequently hydrostatic, or nearly so, so that the circle is the better shape.

References

1. Abramowitz, M., and Stegun, I. A. 1965. *Handbook of Mathematical Functions.* New York: Dover.

2. Anwar, H. D. 1967. *J. Hydr. Div., Proc. ASCE.* 93:99–119.

3. Baker, P. J., et al. 1965. *Brit. Hydromech. Res. Assoc., Res. Reps.* 803, 827.

4. Binnie, A. M. 1973. *J. Hydr. Res.,* 11:61–68.

5. Binnie, G. M., et al. 1968. *Proc. Instn. Civ. Engrs.,* 41:197.

6. Brown, K. M. 1969. *SIAM J. Num. Anal.,* 6:560–569.

7. Cowan, H. J. 1977. *Proc. Sixth Australasian Conf. Mech. Struct. Mats.* (Christchurch, New Zealand), 21–29.

8. Feld, J. 1930. *J. Franklin Inst.,* 209:83–108 (this work contains an extensive bibliography of work done in the early part of the century).

9. Halstead, H. J., et al. 1969. *A Course in Pure and Applied Mathematics.* 2nd ed. Melbourne, Australia: Macmillan, pp. 487–499.

10. Hauri, H. H., and Peters, T. F. 1979. *Ann. Conv. Am. Soc. Civ. Engrs.,* Boston, Preprint No. 3590.

11. Heyman, J. 1966. *Int. J. Sol. Struct.,* 2:249–279.

12. Heyman, J. 1969. *Int. J. Mech. Sci.,* 11:363–384.

13. Heyman, J. 1977. *Equilibrium of Shell Structures*. Oxford: Oxford University Press, ch. 4.

14. Hodder, S. B. 1976. Master of Engineering thesis, University of Auckland, Auckland, New Zealand.

15. Hopkins, H. J. 1970. *A Span of Bridges*. Newton Abbot, England: David and Charles, ch. 1.

16. Inglis, Sir C. 1963. *Applied Mechanics for Engineers*. New York: Dover, p. 51.

17. Irvine, H. M., and Sinclair, G. B. 1976. *Int. J. Sol. Struct.*, 12:309–317.

18. Irvine, H. M. 1979. *Int. J. Mech. Sci.* 21:467–475.

19. Lamb, H. 1925. *The Dynamical Theory of Sound*. 2nd ed. London: Edward Arnold, ch. 5.

20. Needham, J. 1954ff. *Science and Civilisation in China*, 8 vols (to date). Cambridge: Cambridge University Press, see vol. 4, pt. 3, pp. 184ff.

21. Newman, B. G. 1975. *Proc. Roy. Soc.* (Lond.), A346:329–348.

22. Owen, J. B. B. 1976. *The Works of Isambard Kingdom Brunel*. Ed. Sir A. G. Pugsley. Bristol: Inst. Civ. Engrs., Bristol University, ch. 5.

23. Parberry, R. D. 1978. *Proc. Instn. Civ. Engrs.* 65(pt. 2):645–654.

24. Parsons, W. B. 1976. *Engineers and Engineering in the Renaissance*, Cambridge, Massachusetts: The MIT Press, pp. 71–72.

25. Pippard, A. J. S., and Baker, Sir J. F. 1968. *The Analysis of Engineering Structures*. 4th ed. New York: American Elsevier, ch. 16.

26. Ramsey, A. S. 1946. *Statics*. 2nd ed. Cambridge: Cambridge University Press, ch. 12.

27. Rayleigh, Lord. 1945. *The Theory of Sound*. Vol. 1. 2nd ed. New York: Dover, chs. 6, 9.

28. Rouse, H., and Ince, S. 1963. *History of Hydraulics*. New York: Dover, ch. 3.

29. Routh, E. J. 1891. *Analytical Statics*. Vol. 1. Cambridge: Cambridge University Press, ch. 10.

30. Truesdell, C. 1960. *The Rational Mechanics of Flexible or Elastic Bodies 1638–1788*, Turici: Orell Füssli.

31. Uspensky, J. V. 1948. *Theory of Equations*. New York: McGraw-Hill, ch. 6.

32. Watson, G. N. 1966. *Theory of Bessel Functions*. 2nd ed. Cambridge: Cambridge University Press, pp. 3–6.

33. Whittaker, E. T. 1970. *Analytical Dynamics*. 4th ed. Cambridge: Cambridge University Press, pp. 34–177.

2

Statics of a Suspended Cable

2.1 Introduction

By the close of the first chapter we found that the exact analysis of simple suspended cable problems was somewhat restricted because the solution methods tend to be cumbersome. Simplifications can be made when the profile of the cable is flat, and, since this often corresponds to the situation where cables with relatively low sag are used for structural purposes, we investigate here in some detail the engineering theory of the suspended cable. In its most general form [8, 9] this approximate theory provides explicit, consistent methods for finding static response to applied loads accurate to the third order of small quantities.

2.2 The Parabolic Profile and the Influence of Flexural Rigidity

Consider the profile of a uniform cable hanging under its own weight between two supports at the same level and the properties associated with it. If the profile is flat, so that the ratio of sag to span is 1:8 or less, the differential equation governing vertical equilibrium of an element is accurately specified by

$$H\frac{d^2z}{dx^2} = -mg, \tag{2.1}$$

for which the pertinent solution is

$$\mathbf{z} = \frac{1}{2}\mathbf{x}(1 - \mathbf{x}), \tag{2.2}$$

where we have nondimensionalized the variables: $\mathbf{x} = x/l$, in which l is the span of the cable, and $\mathbf{z} = z/(mgl^2/H)$, in which mg is the cable weight per unit length and H is the horizontal component of cable tension. (The exact equation is $H(d^2z/dx^2) = -mg\{1 + (dz/dx)^2\}^{1/2}$, so, in assuming the profile is flat, we are choosing to ignore $(dz/dx)^2$ in comparison to unity. Nevertheless, the cutoff in the ratio of sag to span at 1:8 is largely one of convenience, for then $H \geq mgl$. In fact independent finite element analyses have shown that the theory of this chapter retains its accuracy even with profiles as deep as 1:5 or 1:4.) We shall frequently resort to suitable dimensionless forms for subsequent results. The horizontal component of cable tension is

$$H = \frac{mgl^2}{8d}, \tag{2.3}$$

where d is the sag, and the cable length is

$$\begin{aligned} L &= \int_0^l \left\{1 + \left(\frac{dz}{dx}\right)^2\right\}^{1/2} dx \\ &= l\left\{1 + \frac{8}{3}\left(\frac{d}{l}\right)^2 - \frac{32}{5}\left(\frac{d}{l}\right)^4 + \dots\right\}. \end{aligned} \tag{2.4}$$

In these calculations the effects of cable stretch can be assumed to have been accounted for, but it is necessary to investigate at least in part the probable influence on response of the finite flexural rigidity of the cable. Before that, however, an example is presented that indicates how to find the profile of a cable when the supports are not at the same level.

Example 2.1 The Profile of a Guywire

Figure 2.1 shows a guywire stretched between point A and a point B some distance below A. We are interested in obtaining an approximate solution for the profile of this cable when it hugs closely to the chord AB which itself may be inclined at some angle θ to the horizontal, where $0 \leq \theta < \pi/2$.

If z here measures the dip of the profile below the chord, vertical equilibrium is satisfied by

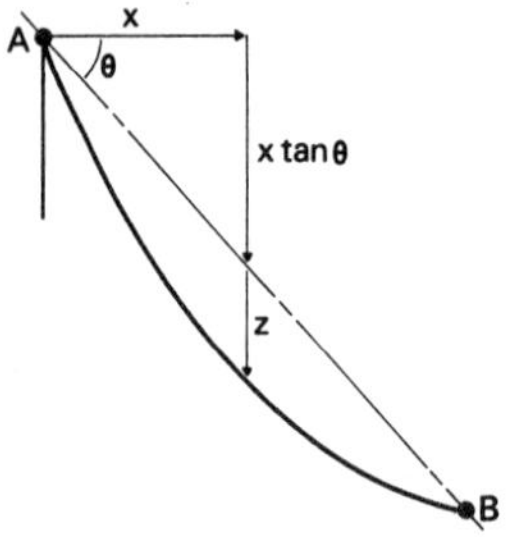

2.1 Definition diagram for an inclined cable

$$H\frac{d^2z}{dx^2} = -mg\left\{1 + \left(\tan\theta + \frac{dz}{dx}\right)^2\right\}^{1/2}.$$

If dz/dx is considered sufficiently small for its square to be ignored, the equation reduces to

$$\frac{d^2\mathbf{z}}{d\mathbf{x}^2} + \varepsilon\frac{d\mathbf{z}}{d\mathbf{x}} = -1,$$

where $\mathbf{z} = z/(mg \sec\theta l^2/H)$, $\mathbf{x} = x/l$, and $\varepsilon = mgl \sin\theta/H$. The parameter ε is small because, for the cable to lie close to the chord, mgl must be a small fraction of H.

This linear second-order ordinary differential equation can be integrated directly to give a first-order equation, and a solution may then be found using the theory of linear first-order equations. But there is no point in this because we are neglecting powers of $d\mathbf{z}/d\mathbf{x}$ higher than the first power. On the other hand, if we let

$$\mathbf{z} = \mathbf{z}_0 + \varepsilon\mathbf{z}_1 + \ldots,$$

substitute this in the differential equation, and collect like terms, we obtain

$$\frac{d^2\mathbf{z}_0}{d\mathbf{x}^2} = -1,$$

and

$$\frac{d^2\mathbf{z}_1}{d\mathbf{x}^2} = -\frac{d\mathbf{z}_0}{d\mathbf{x}},$$

whose solutions are required to satisfy the zero displacement at A and B.

These solutions are $\mathbf{z}_0 = \mathbf{x}(1 - \mathbf{x})/2$ and $\mathbf{z}_1 = \mathbf{x}(1 - \mathbf{x})(1 - 2\mathbf{x})/12$. Consequently

$$\mathbf{z} = \frac{1}{2}\mathbf{x}(1 - \mathbf{x})\left\{1 + \frac{\varepsilon}{6}(1 - 2\mathbf{x})\right\},$$

and this perturbation solution is accurate to the first power of ε, which is all that is claimed of the differential equation that generated it.

The correction is often small and can be ignored when the slope of the chord is small. Of course the correction disappears when the supports are horizontal. The matter is taken up again in example 2.2 and in a different way in example 3.1, in the context of free vibrations of an inclined cable.

Returning to the question of the influence of a cable's flexural rigidity, the statement of shear force equilibrium of a uniform beam under self-weight and axial tension is

$$-EI\frac{d^3z}{dx^3} + H\frac{dz}{dx} = mg\left(\frac{l}{2} - x\right), \tag{2.5}$$

where E is Young's modulus and I is the second moment of area of the cross section. In the case of a cable composed of many separate strands, the calculations involved in finding the second moment of area based on the gross area will be somewhat conservative.

The solution we are interested in is that which satisfies conditions of zero displacement and zero bending moment (zero second derivatives of the displacement) at each end. In dimensionless form this solution reads

$$\mathbf{z} = \frac{1}{2}\mathbf{x}(1 - \mathbf{x}) - \frac{1}{\gamma^2}\left\{1 + \tanh\frac{\gamma}{2}\sinh\gamma\mathbf{x} - \cosh\gamma\mathbf{x}\right\}, \tag{2.6}$$

where the only new quantity is $\gamma = (Hl^2/EI)^{1/2}$, which indicates the relative importance of cable and beam actions. At midspan the sag is

$$d = \frac{mgl^2}{8H}\left\{1 - \frac{8}{\gamma^2}\left(1 - \operatorname{sech}\frac{\gamma}{2}\right)\right\} < \frac{mgl^2}{8H}. \tag{2.7}$$

When γ is very small, indicating that beam action predominates,

$$\left\{1 - \frac{8}{\gamma^2}\left(1 - \operatorname{sech}\frac{\gamma}{2}\right)\right\} \to \frac{5}{12}\left(\frac{\gamma}{2}\right)^2, \tag{2.8}$$

so that

$$d \to \frac{5}{384}\frac{mgl^4}{EI}. \tag{2.9}$$

This is the result for the central deflection of a uniform beam simply supported at each end. Conversely, when γ is large, cable action is of primary importance, and $d \to mgl^2/8H$.

In the vast majority of cable problems γ is large, often of order 10^3 or more. The effects of flexural rigidity are then quite unimportant and may be ignored. However, when rapid changes in curvature are unavoidable, as in the vicinity of point loads or, for example, at the saddles of the towers of suspension bridges, bending effects may be locally important. This suggests the existence of a boundary layer close to the load in which beam action is significant, while elsewhere cable action predominates. In fact the analogy with the traditional boundary layer of fluid mechanics is complete. Such boundary layer concepts are important in perturbation solution methods, and indeed Cole [4] uses an example drawn from suspension bridge theory to illustrate one section of his work.

There are two other aspects of flexural rigidity that fit in with our discussion here. The first concerns the bending stresses in an eye bar chain. Flexural effects must be considered because each bar acts as a beam of rectangular cross section simply supported between its ends, as may be seen from figure 2.2.

The funicular polygon does not lie along the centerlines of the inter-

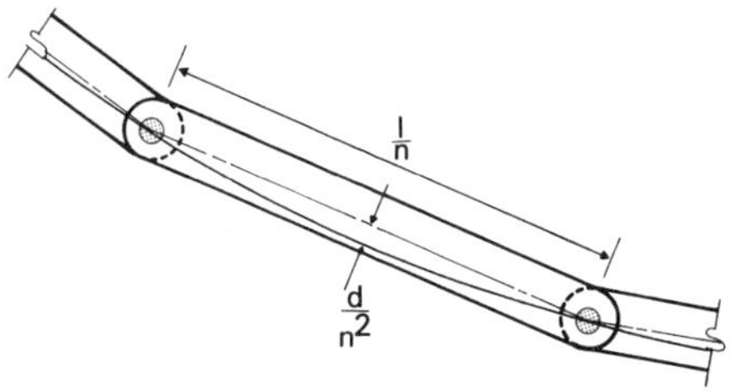

2.2 A typical eye bar in an eye bar chain showing the offset of the funicular curve from the centerline

connected eye bars; rather it is curved between the eyes, the offset being d/n^2, where d is the sag of the whole system and n the number of links that comprise it. It is not difficult to show that the nominal direct stress is increased on the convex face by a fractional amount $6d/(n^2b)$, where b is the depth of the eye bar, and reduced by the same amount on the concave or upper face.

By way of illustration we may consider Telford's Menai Straits suspension bridge. When it was built in 1826, the bridge had massive eye bar chains and a relatively light deck. Thus it may be conjectured from the data [12], $l = 177$ m, $d = 177/13.5 = 13.1$ m, $n = 58$, and $b = 0.083$ m, that bending effects added $\pm$ 28 percent to the direct dead load stresses.

On the other hand, a true chain composed of many small links might be expected to behave as a perfect cable. Yet this too is deceptive, for local bending stresses in the individual links are exceedingly important. For this reason the factor of safety based on nominal direct stresses is high, with the end result of chains often looking too heavy for the task at hand. Simple methods of structural analysis provide a means of assessing these effects, as shown in problem 2.1.

2.3 Response to a Point Load

Around the middle of the nineteenth century several papers appeared in connection with suspension bridge design, proposing approximate analyses for the behavior of a heavy parabolic cable under various types of applied loading. Theories were developed partly by Rankine in 1858 but mainly by an anonymous writer in 1860 and 1862 [13]. It was realized at this time that the response of the cable was nonlinear: successive equal increments of load were seen to cause successive increments in the corresponding deflection, each smaller than the last. This nonlinear stiffening has been discussed at length by Pugsley, in whose book an account is also given of the analysis [14]. Timoshenko also discusses static cable response from a practical approximate point of view [16, 17].

The treatment in this chapter is broadly applicable to simple suspended cable problems [8, 9]. There is an extensive literature on numerical methods that may be consulted in cases not amenable to a hand analysis.

The equations of equilibrium are solved here in a straightforward manner, and compatibility of displacements satisfied by a cable equation in which all important terms are retained. Simplifications can then be made to the general results. The solutions may be linearized or adapted to apply to cables that are initially taut and flat. In either case the results are considerably simplified.

Consider, for example, a point load P acting at a distance x_1 from the left hand support (see figure 2.3). Provided that the additional movements of the cable are small, so that the profile remains relatively shallow, vertical equilibrium at a cross section of the cable requires that

$$(H + h)\frac{d}{dx}(z + w) = P\left(1 - \frac{x_1}{l}\right) + \frac{mgl}{2}\left(1 - \frac{2x}{l}\right), \tag{2.10}$$

for $0 \leq x < x_1$, where w is the additional vertical cable deflection and h is the increment in the horizontal component of cable tension, owing to the point load. The righthand side of (2.10) is analogous to the shear force in a simply supported beam of uniform weight under the action of a point load. Expanding (2. 10), and removing the self-weight terms (which cancel identically), we obtain

$$(H + h)\frac{dw}{dx} = P\left(1 - \frac{x_1}{l}\right) - h\frac{dz}{dx}. \tag{2.11}$$

Similarly it is found that

$$(H + h)\frac{dw}{dx} = -\frac{Px_1}{l} - h\frac{dz}{dx}, \tag{2.12}$$

for $x_1 < x \leq l$, Equations (2.11) and (2.12) may be integrated directly, and, after the boundary conditions have been satisfied, the dimensionless equations for the additional vertical deflection are

$$\mathbf{w} = \frac{1}{(1 + \mathbf{h})}\left\{(1 - \mathbf{x}_1)\mathbf{x} - \frac{\mathbf{h}}{2\mathbf{P}}\mathbf{x}(1 - \mathbf{x})\right\}, \tag{2.13}$$

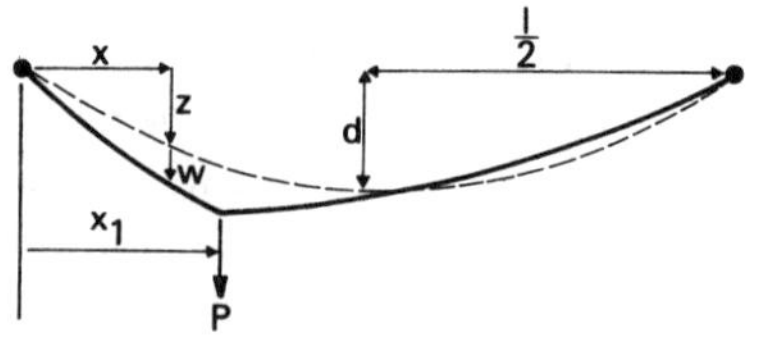

2.3 Definition diagram for a point load on a cable

for $0 \le \mathbf{x} \le \mathbf{x}_1$, and

$$\mathbf{w} = \frac{1}{(1 + \mathbf{h})}\left\{\mathbf{x}_1(1 - \mathbf{x}) - \frac{\mathbf{h}}{2\mathbf{P}}\mathbf{x}(1 - \mathbf{x})\right\}, \tag{2.14}$$

for $\mathbf{x}_1 \le \mathbf{x} \le 1$, where $\mathbf{w} = w/(Pl/H)$, $\mathbf{h} = h/H$, and $\mathbf{P} = P/mgl$.

To complete the solution, $\mathbf{h}$ must now be evaluated. Use is made of a cable equation that incorporates Hooke's law to provide a closure condition relating the changes in cable tension to the changes in cable geometry when the cable is displaced from its original equilibrium profile. The geometry of this displacement is shown in figure 2.4. If ds is the original length of the element, and ds' is its new length, then

$$\begin{aligned} ds^2 &= dx^2 + dz^2, \\ ds'^2 &= (dx + du)^2 + (dz + dw)^2, \end{aligned} \tag{2.15}$$

where u and w are the longitudinal and vertical components of the displacement, respectively. For flat-sag cables the fractional change in length, correct to the second order of small quantities, is

$$\frac{ds' - ds}{ds} = \frac{du}{ds}\frac{dx}{ds} + \frac{dz}{ds}\frac{dw}{ds} + \frac{1}{2}\left(\frac{dw}{ds}\right)^2, \tag{2.16}$$

while Hooke's law stipulates that

$$\frac{\tau}{EA} = \frac{ds' - ds}{ds}, \tag{2.17}$$

where τ is the increment in tension exerted on the element. But, to second order, $\tau = h\,ds/dx$, so that the cable equation for the element reads

$$\frac{h(ds/dx)^3}{EA} = \frac{du}{dx} + \frac{dz}{dx}\frac{dw}{dx} + \frac{1}{2}\left(\frac{dw}{dx}\right)^2. \tag{2.18}$$

It is convenient to use the cable equation in the integrated form, for in this

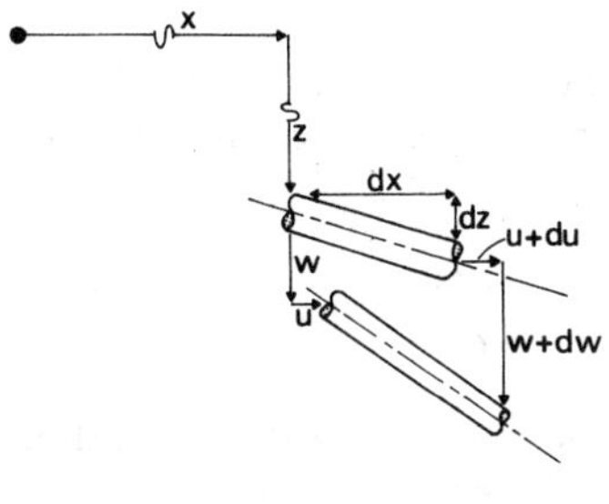

2.4 Displacements of an element of the cable

way we to ensure that h is constant, as it must be because no longitudinal loads are acting. We have

$$\frac{hL_e}{EA} = u(l) - u(0) + \int_0^l \frac{dz}{dx}\frac{dw}{dx} + \frac{1}{2}\int_0^l \left(\frac{dw}{dx}\right)^2, \tag{2.19}$$

where $L_e = \int_0^l (ds/dx)^3\,dx \simeq l(1 + 8(d/l)^2)$, a quantity usually only a little greater than the span itself, and $u(l)$ and $u(0)$ are the longitudinal movements of the supports. (If the effects of a uniform temperature rise of ΔT need to be incorporated, we add a term $-\alpha\Delta T L_t$ to the righthand side of (2.19), where α is the coefficient of expansion and $L_t = \int_0^l (ds/dx)^2\,dx$.) If the longitudinal movement is due to the additional tension and is linear in it (a common occurrence with cable-supported roofs where the cables may be anchored to a supporting frame), the reduction in additional tension due to this support flexibility can be quite marked: If horizontal support flexibilities of f_1 and f_2 occur at each end, respectively, one may replace the axial stiffness of the cable by $EA/\{1 + (f_1 + f_2)EA/L_e\}$, and then proceed as if the supports were unyielding. If u and w are both considered to vanish at the supports then, since dz/dx is continuous along the span, (2.19) integrates to

$$\frac{hL_e}{EA} = \frac{mg}{H}\int_0^l w\,dx + \frac{1}{2}\int_0^l \left(\frac{dw}{dx}\right)^2 dx, \tag{2.20}$$

which is a convenient final form of the cable equation. Longitudinal movement of the cable may be estimated by letting the integrals remain in their indefinite form, so that

$$u(x) = \frac{hL_x}{EA} - \frac{mg}{H}\int_0^x w\,dx - \frac{1}{2}\int_0^x \left(\frac{dw}{dx}\right)^2 dx,$$

where $L_x = \int_0^x (ds/dx)^3\,dx$.

Under point loading dw/dz is discontinuous at the position of load application and the last integral in equation (2.20). When integrated by parts, this gives

$$\frac{1}{2}\int_0^l \left(\frac{dw}{dx}\right)^2 = -\frac{1}{2}\left\{\frac{dw}{dx}w\Big|_{x_1^-}^{x_1^+} + \frac{d^2w}{dx^2}\int_0^{x_1} w\,dx + \frac{d^2w}{dx^2}\int_{x_1}^l w\,dx\right\}. \tag{2.21}$$

Substituting equation (2.21) into (2.20), and performing the integration using (2.13) and (2.14), yields the following dimensionless cubic equation for **h**:

$$\mathbf{h}^3 + (2 + \lambda^2/24)\mathbf{h}^2 + (1 + \lambda^2/12)\mathbf{h} - \lambda^2\mathbf{x}_1(1 - \mathbf{x}_1)\mathbf{P}(1 + \mathbf{P})/2 = 0, \tag{2.22}$$

where $\lambda^2 = (mgl/H)^2 l/(HL_e/EA)$.

The independent parameter λ^2 is of fundamental importance in the static (and also dynamic) response of suspended cables and will arise repeatedly in this book. Basically the parameter accounts for geometric and elastic effects. One way of thinking of it is as resembling the stiffness of two springs in series. Consider a sagging cable in which the ends are being stretched apart. Some of the resistance supplied is geometric because the sag is being reduced: this stiffness is $12H/\{l(mgl/H)^2\}$. The remaining stiffness is axial and quantified by a term like EA/l.

A combination of these two effects is apparent in λ^2. But generally the profile description mgl/H determines the size of λ^2. For metal cables such as steel, the strains, measured by H/EA, are small, but $(mgl/H)^2$ is even smaller for a taut flat cable. Consequently λ^2 is small and becomes smaller as the cable material becomes more extensible. Such a cable must stretch to resist applied load, but this stretching is of the second order in the additional deflection, so that first-order changes in tension are absent. On the other hand, when mgl/H is more appreciable, say, approaching unity as it will in a suspension bridge cable, λ^2 is typically large, and then the relative inextensibility typical of metal cable manifests itself. Additional tension can be generated to the first order because the loaded cable can adopt a new profile that does not necessarily depend on changes in cable length for its existence. These distinctions will become clearer and will be further amplified later in this section, but it is between these limits that most cables of structural importance are found.

Properties of the Cubic for h

Before going further, let us pause and consider some important properties of the roots of the general cubic

$$\mathbf{h}^3 + a\mathbf{h}^2 + b\mathbf{h} - c,$$

where the coefficients a, b, and c are real constants. A famous result from the theory of polynomials with real coefficients, known as Descartes' rule of signs, states that the number of positive real roots of a polynomial is equal to the number of changes in the signs of its coefficients, or that number minus two [18]. In the present case a and b are positive, and c is positive if $\mathbf{P}$ is positive, as it obviously usually is. In this situation there is just one positive real root which is of course the required value of $\mathbf{h}$. Because the deflection under the load must be positive, it is necessary that $\mathbf{h} < 2\mathbf{P}$. Further investigation of (2.22) shows indeed that $0 < \mathbf{h} < 2\mathbf{P}$, although this upper bound is somewhat conservative. A sharper upper bound, obtained by considering the case when λ^2 is large, exists when the load is placed at midspan and is $\mathbf{h} < \{1 + 3\mathbf{P}(1 + \mathbf{P})\}^{1/2} - 1$. When $\mathbf{P}$ and λ^2 are given, $\mathbf{h}$ has its peak value if the load is placed at midspan.

In dealing with the less common situations when $\mathbf{P}$ is negative, we separate $\mathbf{P}$ into two ranges and simply note the following results: when $-1 < \mathbf{P} < 0$, there is just one negative real root of (2.22), and then $2\mathbf{P} < \mathbf{h} < 0$ if $-1/2 \leq \mathbf{P} < 0$, while $-1 < \mathbf{h} < 0$ if $-1 < \mathbf{P} \leq -1/2$. When $\mathbf{x}_1$ and λ^2 are given, $\mathbf{h}$ has an overall minimum at $\mathbf{P} = -1/2$. However, when $\mathbf{P} < -1$, there is again just one positive real root for the cubic, because c is again positive. A final point of interest is that when $\mathbf{P} = -1$, $\mathbf{h} = 0$, regardless of where the load is placed. The physical connotation is apparent, since application of this upward point load divides the cable into two portions from which, by joining one to the other, the original profile is recovered.

The importance of checking the solutions in this way cannot be overemphasized. We have shown (2.13), (2.14), and (2.22) to be mutually consistent and may therefore use them with confidence.

Linearized Solutions

The point load problem is linearized by neglecting all second-order terms that appear in the differential equations of equilibrium and in the cable equation. This necessitates the removal of the term $h\,dw/dx$ from (2.11) and (2.12) and $\frac{1}{2}\int_0^l (dw/dx)^2\,dx$ from (2.20). As a consequence we have [8]

$$\mathbf{w} = \left\{(1 - \mathbf{x}_1)\mathbf{x} - \frac{\mathbf{h}}{2\mathbf{P}}\mathbf{x}(1 - \mathbf{x})\right\}, \tag{2.23}$$

for $0 \leq \mathbf{x} \leq \mathbf{x}_1$, and

$$\mathbf{w} = \left\{\mathbf{x}_1(1 - \mathbf{x}) - \frac{\mathbf{h}}{2\mathbf{P}}\mathbf{x}(1 - \mathbf{x})\right\}, \tag{2.24}$$

for $\mathbf{x}_1 \leq \mathbf{x} \leq 1$. Substituting these into the reduced cable equation yields

$$\mathbf{h} = \frac{6\mathbf{P}}{(1 + 12/\lambda^2)}\mathbf{x}_1(1 - \mathbf{x}_1). \tag{2.25}$$

For taut flat cables $\lambda^2 \ll 1$ and $\mathbf{h} \to 0$. Hence the classical linear theory of the taut string is recovered. At the other end of the spectrum, when $\lambda^2 \gg 1$, we have $\mathbf{h} \to 6\mathbf{P}\,\mathbf{x}_1(1 - \mathbf{x}_1)$, which is a result that would hold typically for a suspension bridge cable. This result was produced in the mid-nineteenth century for just that purpose [13].

The point at which the response of a suspended cable becomes like that of a taut string cannot be precisely determined. But some guidance is available from the linear theory. The overall maximum additional deflection will occur at the point of load application and will involve maximizing

$$\mathbf{w}(\mathbf{x}_1) = \mathbf{x}_1(1 - \mathbf{x}_1)\left\{1 - \frac{3}{1 + 12/\lambda^2}\,\mathbf{x}_1(1 - \mathbf{x}_1)\right\}. \tag{2.26}$$

Possible turning points of this expression exist at

$$\mathbf{x}_1 = \frac{1}{2}\left[1 \mp \left\{1 - \frac{2}{3}\left(1 + \frac{12}{\lambda^2}\right)\right\}^{1/2}\right], \frac{1}{2}.$$

This leads to the following observations:

1. If $\lambda^2 \geq 24$, the additional deflection under the load has an overall maximum of

$$\mathbf{w}_{\max} = \frac{1}{12}\left(1 + \frac{12}{\lambda^2}\right), \tag{2.27}$$

when the load is placed at

$$\mathbf{x}_1 = \frac{1}{2}\left[1 \mp \left\{1 - \frac{2}{3}\left(1 + \frac{12}{\lambda^2}\right)\right\}^{1/2}\right]. \tag{2.28}$$

When $\lambda^2 \gg 24$, $\mathbf{w}_{\max} \to 1/12$ and $\mathbf{x}_1 \to (1 \mp 1/\sqrt{3})/2 = 0.211,\ 0.789$.

2. If $\lambda^2 \leq 24$, the overall maximum is

$$\mathbf{w}_{\max} = \frac{1}{4}\left\{1 - \frac{3}{4(1 + 12/\lambda^2)}\right\}, \tag{2.29}$$

when the load is placed at

$$\mathbf{x}_1 = \frac{1}{2}. \tag{2.30}$$

Again, when $\lambda^2 \ll 24$, $\mathbf{w}_{\max} \to 1/4$, in accordance with classical theory.

Therefore, we may conclude that $\lambda^2 = 24$ is a cutoff, above which response has more in common with a heavy cable and below which it is akin to that expected of a taut string. A similar cutoff exists in dynamic response.

The range over which linearized solutions are valid is something of a moot point, since the answer will depend on λ^2 as well as **P**. In a suspension bridge cable a value of $\mathbf{P} = 0.1$ may well represent an approximate upper limit beyond which changes in additional tension are judged too important for nonlinear stiffening which characterizes large displacement response to be ignored. On the other hand, a taut string may support loads in which $\mathbf{P} = 100$ and still not exhibit any appreciable nonlinearity. This in fact is the case with the experimental results shown later in figure 2.6, where central deflections of 1 cm over a span of a meter are caused by loads in which $\mathbf{P} \sim 100$, and there is no noticeable departure from the linear theory.

Example 2.2 Linearized Response of an Inclined Cable to a Point Load

In example 2.1 the profile of an inclined cable under self-weight alone was determined approximately. This provides the basis for obtaining the profile when a point load is hung from the cable. Only a linearized solution is sought. Extending it to include nonlinearity is not difficult, but we may circumvent this, as shown at the end of this example. As in example 2.1 the results here are accurate to the first power in the geometric parameter ε, so moderate departures from the chord are acceptable.

We note that on either side of the point load (see figure 2.5)

$$H\frac{dw}{dz} = V - h\frac{d}{dx}(x \tan\theta + z)$$

and

$$H\frac{dw}{dz} = V - P - h\frac{d}{dx}(x \tan\theta + z).$$

The only new variable is V, which is the increment in the vertical reaction at the lefthand support. Solutions that satisfy the requirement of zero displacement at each support are

$$w = \frac{V}{H}x - \frac{h}{H}(x \tan\theta + z),$$

for $0 \le x \le x_1$, and

$$w = \frac{(V - P)}{H}(x - l) - \frac{h}{H}\left\{(x - l)\tan\theta + z\right\},$$

for $x_1 \le x \le l$. Matching these solutions at the point of load application indicates that $V = P(1 - x_1/l) + h \tan\theta$, which checks with statics. The equations of equilibrium then assume the simple form

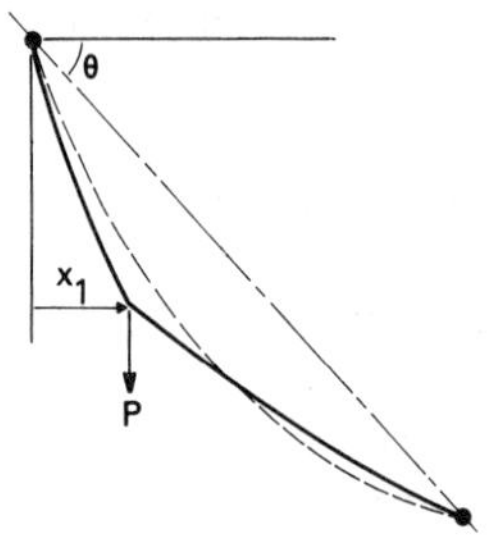

2.5 Definition diagram for a point load on an inclined cable

$$\mathbf{w} = (1 - \mathbf{x}_1)\mathbf{x} - \frac{\mathbf{h}}{\mathbf{P}}\mathbf{z},$$

for $0 \leq \mathbf{x} \leq \mathbf{x}_1$, and

$$\mathbf{w} = \mathbf{x}_1(1 - \mathbf{x}) - \frac{\mathbf{h}}{\mathbf{P}}\mathbf{z},$$

for $\mathbf{x}_1 \leq \mathbf{x} \leq 1$, where $\mathbf{w} = w/(Pl/H)$, $\mathbf{x} = x/l$, $\mathbf{h} = h/H$, $\mathbf{P} = P/(mg \sec \theta l)$, and $\mathbf{z} = z/(mg \sec \theta l^2/H)$, as given in example 2.1. The cable equation may be rearranged to the dimensionless form

$$\mathbf{h} = -\lambda^2 \mathbf{P} \int_0^1 \frac{d^2\mathbf{z}}{d\mathbf{x}^2} \mathbf{w}\, d\mathbf{x},$$

where $\lambda^2 = (mg \sec \theta l/H)^2\, l/(HL_e/EA)$, in which $L_e = l \sec^3 \theta \{1 + (mgl/H)^2/8\}$. Substitution and integration then give

$$\mathbf{h} = \frac{6\mathbf{P}\, \mathbf{x}_1(1 - \mathbf{x}_1)}{(1 + 12/\lambda^2)} \left\{1 + \frac{\varepsilon}{6}(1 - 2\mathbf{x}_1)\right\},$$

where $\varepsilon = mgl \sin \theta / H$. The form of the solution illustrates the similarity with the self-weight profile of the inclined cable and the linearized response to a point load of the horizontal cable.

In the point load case the inclination adds another parameter, ε, to the list of independent parameters. However, if the cable does not lie too far below its chord, this parameter is not of particular significance and can be ignored (the same is true in computing the dynamic response). Our general nonlinear results will apply to the inclined cable, provided we redefine $\mathbf{P}$ and λ^2: this is an important and useful conclusion. With slight adaptations these results can then be used to examine, for example, the live load profile characteristics of multispan chair lifts.

Taut Flat Cable

When a cable is initially flat (or really nearly so), we may set $z(x)$ to zero. The equations for additional displacement are therefore [8]

$$\mathbf{w} = \frac{1}{(1 + \mathbf{h})}(1 - \mathbf{x}_1)\mathbf{x}, \qquad (2.31)$$

for $0 \leq \mathbf{x} \leq \mathbf{x}_1$, and

$$\mathbf{w} = \frac{1}{(1 + \mathbf{h})}\mathbf{x}_1(1 - \mathbf{x}), \qquad (2.32)$$

for $\mathbf{x}_1 \leq \mathbf{x} \leq 1$. The cable equation becomes

$$\frac{hl}{EA} = \frac{1}{2}\int_0^l \left(\frac{dw}{dx}\right)^2 dx, \qquad (2.33)$$

which after integration by parts reduces to

$$\frac{hl}{EA} = \frac{1}{2} w \frac{dw}{dx}\bigg|_{x_1^+}^{x_1^-},$$

giving

$$\mathbf{h}(1 + \mathbf{h})^2 = \frac{\lambda^2 \mathbf{x}_1(1 - \mathbf{x}_1)\mathbf{P}^2}{2} \tag{2.34}$$

as the cubic from which **h** is determined. As might be expected, this result may be extracted from the general cubic when λ^2 is small. The required root has the straightforward form

$$\mathbf{h} = \frac{1}{(A)^{1/3}}\left\{(A)^{1/3} - \frac{1}{3}\right\}^2, \tag{2.35}$$

where $A = -q/2 + (q^2/4 + p^3/27)^{1/2}$, in which $p = -1/3$ and $q = -2/27 - \lambda^2 \mathbf{x}_1(1 - \mathbf{x}_1)\mathbf{P}^2/2$. It can be shown that $(A)^{1/3} > 1/3$ since $(A)^{1/3} = 1/3$ only in trivial cases.

Equations (2.31) and (2.34) may be rearranged to a standard classical form [7]. If the displacement under the load is denoted by $\mathbf{w}_1$, the results are

$$\frac{\lambda^2 \mathbf{P}^2}{2\mathbf{x}_1(1 - \mathbf{x}_1)} \mathbf{w}_1^3 + \mathbf{w}_1 = \mathbf{x}_1(1 - \mathbf{x}_1)\mathbf{P},$$

$$\mathbf{h} = \frac{\lambda^2 \mathbf{P}^2}{2\mathbf{x}_1(1 - \mathbf{x}_1)} \mathbf{w}_1^2. \tag{2.36}$$

The classical formulation proceeds by considering equilibrium at the load application point and using a series expansion of the Pythagorean theorem to estimate the relationship between deflection and stretching that the two inclined portions undergo. This approach is straightforward only when a point load is applied to a flat cable. This lack of generality is an obvious shortcoming.

Figure 2.6 shows the extent to which geometric nonlinearity can affect

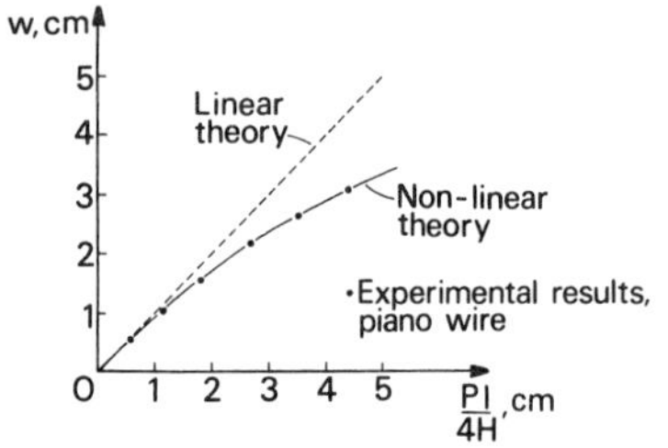

2.6 Comparison of theory and experiment for midspan deflection

response even when the loaded profile is still relatively shallow. In this instance agreement between theory and experiment is hardly surprising, for the experiments involved nothing more than loading a taut meter-long piano wire at midspan and measuring the deflection.

Table for h

Because the peak additional tension occurs when the load is applied at midspan, we consider only this case in constructing a table for **h** for a representative range of values of **P** and λ^2. In commenting on the results presented in table 2.1, we need mention only three aspects, all of which have been touched on before. When **P** is small, the linearized solution works well for all λ^2; when λ^2 is small, the results for a taut flat cable adequately encompass response for all **P**; and, when λ^2 is large, the upper bound of $\mathbf{h} < \{1 + 3\mathbf{P}(1 + \mathbf{P})\}^{1/2} - 1$ provides a reasonable estimate for all **P**. (For the present problem there is no point in solving our general cubic using one of Cardan's formulas [18]; the standard Newton-Raphson technique is simple, direct, and in this instance failsafe.) The deflection under the load is $\mathbf{w} = \{1 - (\mathbf{h}/2\mathbf{P})\}/\{4(1 + \mathbf{h})\}$.

These trends bring to the fore another feature worth emphasizing that relates to dynamic similitude, which is what occurs between a prototype flat-sag cable and its flat-sag model when λ^2 and **P** are the same in each case. Even this may be relaxed in certain instances, for if λ^2 is small in the prototype, one need check only that it is small in the model and match model and prototype values of the product parameter $\mathbf{P}^2\lambda^2$. At the other extreme, when λ^2 is large, similitude will be assured if just **P** is matched provided that the model exhibits a suitably large value of λ^2. Similar arguments apply in the case of distributed loads. This type of reduction provides welcomed latitude in experiments on a suspension bridge model, for example, where deck and tower stiffnesses may produce two more parameters that need to be matched.

Example 2.3 Behavior of a Flying Fox

A flying fox is used to transport materials across a ravine. In its free-hanging position the cable spans 100 m with a ratio of sag to span of 1:50. Anchorage compliance may be ignored. The problem is to find the additional horizontal component of cable tension and the associated deflection when a load of 20 kN is carried at midspan. Properties of the cable are $mg = 40$ N/m, $E = 150{,}000$ MPa, $A = 5 \times 10^{-4}$ m^2; thus $\mathbf{x}_1 = 0.5$, $\mathbf{P} = 5.0$, and $\lambda^2 = 75$.

From (2.22) the cubic to be solved becomes

$$\mathbf{h}^3 + 5.1\mathbf{h}^2 + 7.3\mathbf{h} - 281 = 0,$$

from which $\mathbf{h} = 5.0$ (see table 2.1 as well). The linearized theory yields $\mathbf{h} = 6.5$, while the nonlinear theory of the taut string yields $\mathbf{h} = 5.5$ (all figures are rounded). The general theory is indispensable in this case, although the other results are at least conservative.

Table 2.1 Values of **h**/**P** with load at midspan

λ^2 \ P	0.1	0.2	0.4	0.6	0.8	1.0	2.0	4.0
0.1	0.014	0.015	0.017	0.019	0.022	0.024	0.033	0.045
0.2	0.027	0.029	0.034	0.038	0.041	0.045	0.059	0.074
0.4	0.053	0.057	0.064	0.071	0.077	0.083	0.101	0.115
0.6	0.077	0.083	0.093	0.102	0.109	0.116	0.135	0.146
0.8	0.101	0.108	0.120	0.130	0.138	0.145	0.163	0.171
1.0	0.124	0.132	0.145	0.156	0.164	0.171	0.188	0.192
2.0	0.226	0.237	0.252	0.263	0.270	0.275	0.282	0.271
4.0	0.388	0.397	0.408	0.414	0.416	0.416	0.403	0.371
6.0	0.510	0.516	0.522	0.521	0.518	0.514	0.486	0.439
8.0	0.607	0.610	0.610	0.605	0.598	0.590	0.551	0.493
10.0	0.686	0.687	0.682	0.673	0.663	0.652	0.604	0.537
20.0	0.935	0.929	0.913	0.894	0.875	0.857	0.784	0.690
40.0	1.153	1.148	1.130	1.109	1.087	1.066	0.978	0.864
60.0	1.254	1.252	1.239	1.221	1.201	1.181	1.092	0.972
80.0	1.312	1.313	1.305	1.291	1.273	1.255	1.170	1.049
100.0	1.350	1.354	1.351	1.339	1.324	1.308	1.229	1.109
200.0	1.435	1.446	1.457	1.457	1.451	1.443	1.388	1.285
400.0	1.482	1.499	1.520	1.529	1.533	1.532	1.506	1.434
600.0	1.498	1.518	1.543	1.557	1.564	1.567	1.555	1.503
800.0	1.507	1.528	1.555	1.571	1.580	1.585	1.583	1.544
1,000.0	1.512	1.533	1.562	1.580	1.590	1.596	1.600	1.570

λ^2 \ P	6.0	8.0	10.0	20.0	40.0	60.0	80.0	100.0
0.1	0.051	0.054	0.056	0.057	0.053	0.049	0.046	0.044
0.2	0.079	0.082	0.082	0.079	0.070	0.064	0.060	0.057
0.4	0.118	0.118	0.116	0.106	0.092	0.084	0.078	0.073
0.6	0.146	0.144	0.141	0.126	0.108	0.097	0.090	0.085
0.8	0.169	0.164	0.160	0.142	0.120	0.108	0.100	0.094
1.0	0.188	0.182	0.176	0.155	0.131	0.117	0.108	0.101
2.0	0.258	0.246	0.236	0.202	0.169	0.150	0.138	0.129
4.0	0.346	0.326	0.311	0.262	0.216	0.192	0.176	0.164
6.0	0.406	0.382	0.362	0.303	0.249	0.221	0.202	0.189
8.0	0.454	0.425	0.403	0.336	0.275	0.244	0.223	0.208
10.0	0.493	0.461	0.437	0.363	0.298	0.263	0.241	0.225
20.0	0.631	0.589	0.557	0.461	0.377	0.333	0.304	0.284
40.0	0.791	0.739	0.699	0.580	0.474	0.419	0.383	0.357
60.0	0.893	0.836	0.791	0.659	0.540	0.478	0.438	0.408
80.0	0.968	0.908	0.861	0.720	0.591	0.524	0.480	0.448
100.0	1.026	0.965	0.917	0.770	0.634	0.562	0.515	0.481
200.0	1.207	1.146	1.096	0.936	0.780	0.696	0.640	0.599
400.0	1.370	1.316	1.270	1.112	0.945	0.850	0.785	0.737
600.0	1.451	1.404	1.364	1.214	1.047	0.948	0.880	0.828
800.0	1.500	1.460	1.423	1.284	1.120	1.020	0.950	0.896
1,000.0	1.533	1.498	1.465	1.336	1.177	1.077	1.006	0.951

The additional deflection at midspan is 1.7 m. Even though the sag of the cable has about doubled under this loading, and the tension has risen dramatically, the results are still reliable because the slopes remain small.

2.4 Response to a Uniformly Distributed Load

Consider now a uniformly distributed load of intensity p per unit span applied from $x = x_2$ to $x = x_3$ (see figure 2.7). By again exploring the analogy that exists with the simply supported beam, expressions may be written for the vertical equilibrium in the three different regions of the span. After integration and adjustment for the requisite boundary conditions, the following dimensionless equations are obtained for the additional vertical cable deflection:

$$\mathbf{w} = \frac{1}{(1+\mathbf{h})}\left[\left\{(\mathbf{x}_3 - \mathbf{x}_2) - \frac{1}{2}(\mathbf{x}_3^2 - \mathbf{x}_2^2)\right\}\mathbf{x} - \frac{\mathbf{h}}{2\mathbf{p}}\mathbf{x}(1-\mathbf{x})\right], \tag{2.37}$$

for $0 \leq \mathbf{x} \leq \mathbf{x}_2$,

$$\mathbf{w} = \frac{1}{(1+\mathbf{h})}\left[\left\{-\frac{1}{2}\mathbf{x}_2^2 + \mathbf{x}_3\mathbf{x} - \frac{1}{2}\mathbf{x}^2 - \frac{1}{2}(\mathbf{x}_3^2 - \mathbf{x}_2^2)\mathbf{x}\right\} - \frac{\mathbf{h}}{2\mathbf{p}}\mathbf{x}(1-\mathbf{x})\right], \tag{2.38}$$

for $\mathbf{x}_2 \leq \mathbf{x} \leq \mathbf{x}_3$, and

$$\mathbf{w} = \frac{1}{(1+\mathbf{h})}\left[\frac{1}{2}(\mathbf{x}_3^2 - \mathbf{x}_2^2)(1-\mathbf{x}) - \frac{\mathbf{h}}{2\mathbf{p}}\mathbf{x}(1-\mathbf{x})\right], \tag{2.39}$$

for $\mathbf{x}_3 \leq \mathbf{x} \leq 1$, where $\mathbf{w} = w/(pl^2/H)$, $\mathbf{h} = h/H$, $\mathbf{x} = x/l$, and $\mathbf{p} = p/mg$.

The increment in the horizontal component of cable tension is found from a cable equation which, because dw/dz is now continuous along the span, may be written as

$$\frac{hL_e}{EA} = -\int_0^l \left(\frac{d^2z}{dx^2} + \frac{1}{2}\frac{d^2w}{dx^2}\right) w\,dx. \tag{2.40}$$

After substituting (2.37), (2.38), and (2.39) into (2.40), integrating and rearranging the equations, the following dimensionless cubic in $\mathbf{h}$ is

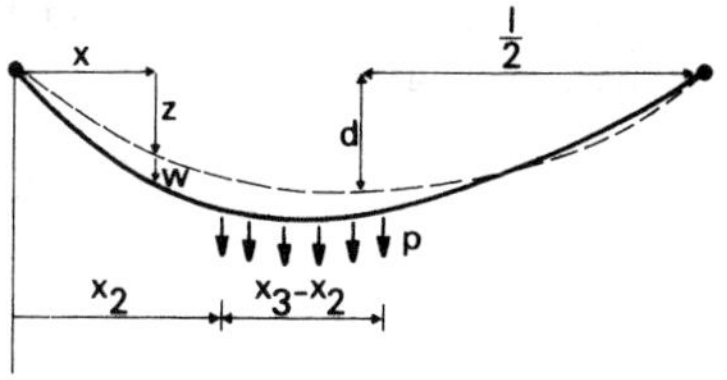

2.7 Definition diagram for a uniformly distributed load on a cable

obtained

$$\begin{aligned}
&\mathbf{h}^3 + \left(2 + \frac{\lambda^2}{24}\right)\mathbf{h}^2 + \left(1 + \frac{\lambda^2}{12}\right)\mathbf{h} \\
&\quad - \frac{\lambda^2}{2}\left\{\frac{1}{2}(\mathbf{x}_3^2 - \mathbf{x}_2^2) - \frac{1}{3}(\mathbf{x}_3^3 - \mathbf{x}_2^3)\right\}\mathbf{p} \\
&\quad - \frac{\lambda^2}{2}\left\{\frac{1}{3}(\mathbf{x}_3^3 - \mathbf{x}_2^3) - \mathbf{x}_2^2(\mathbf{x}_3 - \mathbf{x}_2) - \frac{1}{4}(\mathbf{x}_3^2 - \mathbf{x}_2^2)^2\right\}\mathbf{p}^2 \\
&\quad = 0.
\end{aligned} \tag{2.41}$$

When $\mathbf{p}$ is positive, there is just one positive real root, which is what is required. As before the general solution for $\mathbf{h}$ can be obtained using the requisite form of Cardan's equations.

There is a symmetry in the coefficient involving $\mathbf{x}_2$ and $\mathbf{x}_3$. As is to be expected, the solution is the same if $\mathbf{x}_3 = 1$, $\mathbf{x}_2 = 0.9$, as if $\mathbf{x}_3 = 0.1$, $\mathbf{x}_2 = 0$, and so on. If the loaded length $(\mathbf{x}_3 - \mathbf{x}_2)$ is allowed to become very small while $\mathbf{p}(\mathbf{x}_3 - \mathbf{x}_2)$ remains finite, equation (2.41) reduces to (2.22), the result previously obtained for a point load on the cable. It can also be shown that when $\mathbf{p}$ and λ^2 are given, $\mathbf{h}$ is largest when the loading is placed symmetrically about midspan. (The proof rests on a variation of the theme of influence lines which is well known in linear structural analysis. The method cannot of course be applied but the concept is not inappropriate.)

There are a multitude of load placements contained within these results. It suffices to note just one which is the simplest because the load is taken to extend from one support to the other. In that case

$$\begin{aligned}
&\mathbf{w} = \frac{1}{2(1 + \mathbf{h})}\left(1 - \frac{\mathbf{h}}{\mathbf{p}}\right)\mathbf{x}(1 - \mathbf{x}), \\
&\mathbf{h}^3 + \left(2 + \frac{\lambda^2}{24}\right)\mathbf{h}^2 + \left(1 + \frac{\lambda^2}{12}\right)\mathbf{h} - \frac{\lambda^2}{12}\mathbf{p}\left(1 + \frac{\mathbf{p}}{2}\right) = 0.
\end{aligned} \tag{2.42}$$

It is necessary that $\mathbf{h} < \mathbf{p}$, and this is guaranteed by the cubic. (Other explorations may be made: if $\mathbf{p} = -1$, the new profile is flat and the cable remains taut only if $\lambda^2 < 24$, for when $\lambda^2 = 24$, $\mathbf{h} = -1$; if $\mathbf{p} = -2$, the new profile is the mirror image of the original, and so on.) In fact no sharper upper bound exists because, when λ^2 is large, the solution of the resulting quadratic is $\mathbf{h} = \mathbf{p}$. In contrast to the corresponding case when a point load is applied at midspan, the cable with distributed loads cannot stretch. The additional deflection is zero, and the additional tension is linear in the applied load. With the point load at midspan the relation between additional tension and applied load was largely linear (see, for example, the results in table 2.1 for $\lambda^2 = 1{,}000$ and $\mathbf{P}$ ranging from 0.1 to 10). However, the additional deflection under the point load was certainly not linear. It started out that way but tended to a limiting value as the profile became triangular, represented by the ratio $2/(3)^{1/2}:1$, or 1.15:1, to the original sag, at which point $\mathbf{h} \rightarrow (3)^{1/2}\mathbf{P}$.

The response to distributed loads could be simplified by linearizing and restricting it to initially taut and flat profiles. But since the distributed loads solutions are analogous in every way to those pertaining to the point load, they will not be covered here.

Likewise, in tabulating the solutions of (2.41), some selection is inevitable. A wide variety of different load blocks can be envisaged. Distributed loads that grow from one end, or simultaneously from both, are not tabulated even though they may be important in practice. In fact the latter case requires a separate solution that can be produced from first principles, or an approximate solution may be constructed by utilizing the linearized theory and the principle of superposition. Tables 2.2 to 2.7 are therefore confined to loads spread evenly about midspan.

Example 2.4 Profile Adjustment in Laying the Deck of a Suspension Bridge

A single span suspension bridge is required to have a final dead load profile as shown in figure 2.8. The sag is to be 90 m over a main span of 1,000 m, and the backstays are to be inclined at 30° to the horizontal and span 200 m. After the cables are spun, laying of the deck proceeds from midspan by lifting up and fixing in place successive sections of it. As the loads increase in the backstays, they lengthen and tend to carry the flexible towers with them because the cables cannot slip over the saddles. Consequently, as the cables are hung, the towers are often pulled back a certain amount to compensate for the later placing of the deck. It is necessary to know how much the tower tops should be set back and what sag is to be specified when the cables are free hanging. In this example temperature effects, axial shortening of the tower legs, and the bending stiffness of the deck are ignored.

Each cable has a diameter of 0.5 m, E = 170,000 MPa, cable self-weight is 15 kN/m. (The cables would consist of several strands, each containing many wires. In the Forth Road Bridge, a three-span bridge of approximately 1,000 m main span, opened in 1967, 37 strands of 314, 5 mm wires each were used [1].) When the full dead is placed, each cable must support another 45 kN/m.

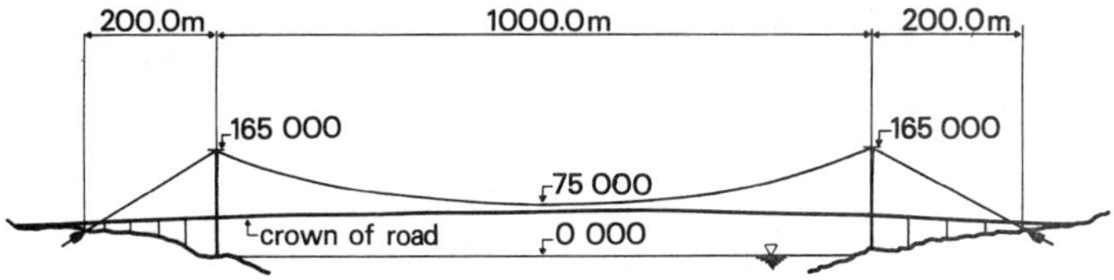

2.8 Layout of a suspension bridge cable

Table 2.2 Values of **h/p** with load uniformly distributed over the central one-tenth of the span

λ^2 \ p	0.1	0.2	0.4	0.6	0.8	1.0	2.0	4.0
0.1	0.001	0.001	0.001	0.001	0.001	0.001	0.001	0.002
0.2	0.002	0.002	0.003	0.003	0.003	0.003	0.003	0.003
0.4	0.005	0.005	0.005	0.005	0.005	0.005	0.006	0.006
0.6	0.007	0.007	0.007	0.007	0.008	0.008	0.008	0.009
0.8	0.009	0.009	0.010	0.010	0.010	0.010	0.011	0.012
1.0	0.012	0.012	0.012	0.012	0.012	0.012	0.013	0.014
2.0	0.021	0.022	0.022	0.022	0.022	0.022	0.023	0.025
4.0	0.037	0.038	0.038	0.038	0.038	0.038	0.039	0.040
6.0	0.050	0.050	0.050	0.050	0.050	0.051	0.051	0.051
8.0	0.060	0.060	0.060	0.060	0.060	0.060	0.060	0.060
10.0	0.068	0.068	0.068	0.068	0.068	0.068	0.068	0.067
20.0	0.093	0.093	0.093	0.093	0.093	0.093	0.092	0.090
40.0	0.115	0.115	0.115	0.115	0.115	0.114	0.113	0.111
60.0	0.125	0.125	0.125	0.124	0.124	0.124	0.124	0.122
80.0	0.130	0.130	0.130	0.130	0.130	0.130	0.130	0.128
100.0	0.134	0.134	0.134	0.134	0.134	0.134	0.134	0.133
200.0	0.141	0.141	0.142	0.142	0.142	0.142	0.143	0.143
400.0	0.145	0.146	0.146	0.146	0.147	0.147	0.148	0.149
600.0	0.147	0.147	0.147	0.148	0.148	0.149	0.150	0.152
800.0	0.148	0.148	0.148	0.149	0.149	0.149	0.151	0.153
1,000.0	0.148	0.148	0.149	0.149	0.150	0.150	0.151	0.154

λ^2 \ p	6.0	8.0	10.0	20.0	40.0	60.0	80.0	100.0
0.1	0.002	0.002	0.002	0.003	0.004	0.005	0.005	0.005
0.2	0.004	0.004	0.004	0.006	0.007	0.008	0.008	0.008
0.4	0.007	0.008	0.008	0.010	0.011	0.011	0.011	0.011
0.6	0.010	0.011	0.011	0.013	0.014	0.014	0.014	0.014
0.8	0.013	0.013	0.014	0.016	0.017	0.016	0.016	0.016
1.0	0.015	0.016	0.017	0.018	0.019	0.018	0.018	0.017
2.0	0.026	0.026	0.027	0.027	0.026	0.025	0.024	0.023
4.0	0.041	0.041	0.041	0.039	0.036	0.034	0.032	0.030
6.0	0.051	0.051	0.050	0.047	0.043	0.040	0.037	0.035
8.0	0.059	0.059	0.058	0.054	0.048	0.044	0.041	0.039
10.0	0.066	0.065	0.064	0.059	0.052	0.048	0.045	0.043
20.0	0.088	0.086	0.084	0.077	0.067	0.062	0.057	0.054
40.0	0.109	0.107	0.104	0.096	0.084	0.077	0.072	0.068
60.0	0.120	0.118	0.116	0.107	0.095	0.087	0.082	0.077
80.0	0.127	0.125	0.123	0.114	0.102	0.094	0.089	0.084
100.0	0.131	0.130	0.128	0.120	0.108	0.100	0.094	0.089
200.0	0.143	0.142	0.141	0.135	0.125	0.118	0.112	0.107
400.0	0.150	0.150	0.150	0.147	0.139	0.133	0.128	0.124
600.0	0.153	0.153	0.153	0.151	0.146	0.141	0.136	0.133
800.0	0.154	0.154	0.155	0.154	0.150	0.146	0.142	0.138
1,000.0	0.155	0.155	0.156	0.156	0.152	0.149	0.145	0.142

Table 2.3 Values of **h/p** with load uniformly distributed over the central two-tenths of the span

λ^2 \ **p**	0.1	0.2	0.4	0.6	0.8	1.0	2.0	4.0
0.1	0.002	0.003	0.003	0.003	0.003	0.003	0.003	0.004
0.2	0.005	0.005	0.005	0.005	0.005	0.006	0.006	0.008
0.4	0.010	0.010	0.010	0.010	0.011	0.011	0.012	0.015
0.6	0.014	0.015	0.015	0.015	0.016	0.016	0.018	0.021
0.8	0.019	0.019	0.020	0.020	0.020	0.021	0.023	0.026
1.0	0.023	0.023	0.024	0.024	0.025	0.026	0.028	0.031
2.0	0.043	0.043	0.044	0.045	0.045	0.046	0.048	0.051
4.0	0.074	0.075	0.075	0.076	0.077	0.077	0.079	0.079
6.0	0.099	0.099	0.100	0.100	0.100	0.100	0.100	0.099
8.0	0.119	0.119	0.119	0.119	0.119	0.119	0.117	0.114
10.0	0.135	0.134	0.134	0.134	0.134	0.133	0.131	0.127
20.0	0.185	0.184	0.183	0.182	0.182	0.181	0.176	0.168
40.0	0.227	0.227	0.226	0.225	0.224	0.223	0.218	0.208
60.0	0.246	0.246	0.246	0.245	0.244	0.243	0.239	0.230
80.0	0.257	0.257	0.257	0.256	0.256	0.255	0.252	0.243
100.0	0.264	0.264	0.264	0.264	0.263	0.263	0.260	0.253
200.0	0.280	0.280	0.280	0.281	0.281	0.281	0.280	0.277
400.0	0.288	0.288	0.289	0.290	0.291	0.291	0.292	0.292
600.0	0.291	0.291	0.292	0.293	0.294	0.295	0.297	0.298
800.0	0.292	0.293	0.294	0.295	0.296	0.297	0.299	0.301
1,000.0	0.293	0.294	0.295	0.296	0.297	0.298	0.300	0.303

λ^2 \ **p**	6.0	8.0	10.0	20.0	40.0	60.0	80.0	100.0
0.1	0.005	0.005	0.006	0.008	0.010	0.011	0.011	0.011
0.2	0.009	0.010	0.011	0.014	0.015	0.015	0.015	0.015
0.4	0.016	0.018	0.019	0.021	0.022	0.021	0.021	0.020
0.6	0.023	0.024	0.025	0.027	0.027	0.026	0.025	0.024
0.8	0.028	0.030	0.031	0.032	0.031	0.029	0.028	0.027
1.0	0.033	0.034	0.035	0.036	0.034	0.032	0.031	0.029
2.0	0.053	0.053	0.053	0.051	0.047	0.043	0.040	0.038
4.0	0.079	0.078	0.076	0.070	0.062	0.056	0.053	0.050
6.0	0.097	0.094	0.092	0.083	0.072	0.066	0.061	0.058
8.0	0.111	0.108	0.105	0.093	0.081	0.073	0.068	0.064
10.0	0.122	0.118	0.115	0.102	0.088	0.079	0.073	0.069
20.0	0.160	0.154	0.149	0.131	0.112	0.101	0.093	0.088
40.0	0.200	0.192	0.186	0.164	0.140	0.127	0.117	0.110
60.0	0.221	0.214	0.208	0.184	0.159	0.144	0.133	0.125
80.0	0.236	0.229	0.222	0.199	0.172	0.156	0.145	0.137
100.0	0.246	0.239	0.233	0.210	0.183	0.167	0.155	0.146
200.0	0.272	0.268	0.263	0.243	0.217	0.200	0.187	0.177
400.0	0.290	0.287	0.285	0.271	0.249	0.233	0.220	0.210
600.0	0.297	0.296	0.294	0.283	0.265	0.250	0.239	0.230
800.0	0.301	0.300	0.299	0.290	0.275	0.262	0.251	0.242
1,000.0	0.303	0.303	0.302	0.295	0.282	0.270	0.260	0.252

Table 2.4 Values of **h**/**p** with load uniformly distributed over the central four-tenths of the span

λ^2 \ p	0.1	0.2	0.4	0.6	0.8	1.0	2.0	4.0
0.1	0.005	0.005	0.005	0.006	0.006	0.006	0.007	0.010
0.2	0.010	0.010	0.010	0.011	0.011	0.012	0.014	0.018
0.4	0.019	0.019	0.020	0.021	0.022	0.023	0.027	0.032
0.6	0.028	0.028	0.030	0.031	0.032	0.033	0.038	0.044
0.8	0.036	0.037	0.039	0.040	0.042	0.043	0.048	0.054
1.0	0.045	0.046	0.047	0.049	0.051	0.052	0.058	0.064
2.0	0.082	0.084	0.086	0.088	0.089	0.091	0.096	0.099
4.0	0.143	0.144	0.145	0.147	0.147	0.148	0.149	0.145
6.0	0.190	0.190	0.190	0.190	0.190	0.190	0.186	0.177
8.0	0.227	0.227	0.226	0.225	0.223	0.222	0.215	0.202
10.0	0.257	0.256	0.255	0.253	0.251	0.248	0.239	0.222
20.0	0.353	0.350	0.346	0.341	0.337	0.333	0.316	0.290
40.0	0.434	0.432	0.427	0.422	0.417	0.412	0.392	0.362
60.0	0.471	0.469	0.465	0.460	0.456	0.452	0.432	0.402
80.0	0.492	0.490	0.487	0.483	0.479	0.476	0.458	0.429
100.0	0.506	0.504	0.501	0.498	0.495	0.492	0.476	0.449
200.0	0.536	0.535	0.534	0.533	0.531	0.529	0.520	0.500
400.0	0.552	0.552	0.553	0.553	0.552	0.552	0.548	0.537
600.0	0.558	0.558	0.559	0.560	0.560	0.560	0.558	0.551
800.0	0.561	0.561	0.562	0.563	0.564	0.564	0.564	0.559
1,000.0	0.562	0.563	0.564	0.565	0.546	0.567	0.567	0.564

λ^2 \ p	6.0	8.0	10.0	20.0	40.0	60.0	80.0	100.0
0.1	0.012	0.013	0.015	0.018	0.020	0.020	0.019	0.019
0.2	0.021	0.023	0.025	0.028	0.028	0.027	0.026	0.025
0.4	0.036	0.038	0.039	0.041	0.039	0.036	0.034	0.033
0.6	0.048	0.049	0.050	0.050	0.046	0.043	0.040	0.038
0.8	0.057	0.059	0.059	0.058	0.052	0.048	0.045	0.043
1.0	0.066	0.067	0.067	0.064	0.057	0.053	0.049	0.047
2.0	0.099	0.097	0.095	0.087	0.076	0.069	0.064	0.060
4.0	0.140	0.136	0.131	0.116	0.099	0.089	0.082	0.077
6.0	0.169	0.162	0.156	0.136	0.115	0.103	0.095	0.089
8.0	0.191	0.183	0.176	0.152	0.128	0.114	0.105	0.099
10.0	0.210	0.200	0.192	0.165	0.138	0.124	0.114	0.107
20.0	0.272	0.258	0.247	0.211	0.176	0.157	0.144	0.135
40.0	0.339	0.322	0.308	0.264	0.221	0.197	0.182	0.170
60.0	0.379	0.361	0.346	0.298	0.251	0.224	0.207	0.194
80.0	0.407	0.389	0.374	0.324	0.273	0.245	0.226	0.212
100.0	0.427	0.409	0.394	0.344	0.292	0.262	0.242	0.227
200.0	0.483	0.468	0.455	0.406	0.352	0.319	0.296	0.279
400.0	0.525	0.514	0.504	0.464	0.413	0.379	0.355	0.337
600.0	0.543	0.535	0.527	0.493	0.446	0.415	0.391	0.372
800.0	0.553	0.546	0.540	0.511	0.469	0.439	0.416	0.397
1,000.0	0.559	0.554	0.548	0.523	0.485	0.457	0.434	0.416

Table 2.5 Values of **h/p** with load uniformly distributed over the central six-tenths of the span

λ^2 \ p	0.1	0.2	0.4	0.6	0.8	1.0	2.0	4.0
0.1	0.007	0.007	0.008	0.008	0.009	0.009	0.011	0.015
0.2	0.013	0.014	0.015	0.016	0.017	0.018	0.022	0.028
0.4	0.026	0.027	0.029	0.031	0.032	0.034	0.040	0.048
0.6	0.039	0.040	0.042	0.045	0.047	0.048	0.056	0.064
0.8	0.051	0.052	0.055	0.058	0.060	0.062	0.070	0.078
1.0	0.063	0.064	0.067	0.070	0.073	0.075	0.083	0.090
2.0	0.115	0.117	0.121	0.124	0.126	0.128	0.134	0.135
4.0	0.200	0.201	0.203	0.204	0.204	0.205	0.202	0.193
6.0	0.264	0.264	0.264	0.262	0.261	0.259	0.250	0.233
8.0	0.316	0.314	0.311	0.308	0.305	0.302	0.288	0.265
10.0	0.358	0.355	0.350	0.346	0.341	0.337	0.318	0.290
20.0	0.489	0.484	0.474	0.465	0.457	0.449	0.418	0.377
40.0	0.603	0.597	0.585	0.575	0.565	0.557	0.520	0.471
60.0	0.654	0.649	0.639	0.629	0.620	0.611	0.576	0.525
80.0	0.684	0.679	0.670	0.662	0.653	0.645	0.612	0.563
100.0	0.703	0.699	0.691	0.683	0.676	0.669	0.638	0.590
200.0	0.745	0.743	0.738	0.734	0.729	0.724	0.703	0.666
400.0	0.768	0.767	0.765	0.763	0.761	0.758	0.745	0.721
600.0	0.776	0.776	0.775	0.774	0.772	0.771	0.762	0.744
800.0	0.780	0.780	0.780	0.779	0.778	0.777	0.771	0.757
1,000.0	0.783	0.783	0.783	0.782	0.782	0.781	0.776	0.765

λ^2 \ p	6.0	8.0	10.0	20.0	40.0	60.0	80.0	100.0
0.1	0.018	0.021	0.022	0.026	0.027	0.026	0.025	0.024
0.2	0.032	0.034	0.036	0.038	0.037	0.035	0.033	0.032
0.4	0.052	0.054	0.055	0.055	0.050	0.047	0.044	0.042
0.6	0.067	0.069	0.069	0.066	0.060	0.055	0.051	0.048
0.8	0.080	0.081	0.080	0.076	0.067	0.061	0.057	0.054
1.0	0.091	0.091	0.090	0.083	0.073	0.067	0.062	0.059
2.0	0.132	0.129	0.125	0.112	0.096	0.087	0.080	0.075
4.0	0.184	0.177	0.170	0.148	0.124	0.111	0.103	0.096
6.0	0.220	0.209	0.201	0.172	0.144	0.129	0.119	0.111
8.0	0.248	0.235	0.225	0.192	0.160	0.142	0.131	0.123
10.0	0.271	0.256	0.245	0.208	0.173	0.154	0.141	0.132
20.0	0.350	0.330	0.314	0.265	0.219	0.195	0.179	0.167
40.0	0.437	0.413	0.393	0.332	0.276	0.245	0.225	0.211
60.0	0.490	0.464	0.443	0.376	0.313	0.279	0.257	0.240
80.0	0.528	0.501	0.479	0.409	0.342	0.306	0.281	0.263
100.0	0.556	0.529	0.507	0.436	0.366	0.327	0.301	0.282
200.0	0.636	0.612	0.591	0.520	0.444	0.400	0.370	0.348
400.0	0.700	0.681	0.664	0.601	0.526	0.480	0.448	0.423
600.0	0.728	0.713	0.699	0.644	0.574	0.529	0.496	0.470
800.0	0.744	0.731	0.719	0.671	0.606	0.562	0.530	0.504
1,000.0	0.754	0.743	0.733	0.690	0.630	0.588	0.556	0.530

Table 2.6 Values of **h/p** with load uniformly distributed over the central eight-tenths of the span

λ^2 \ p	0.1	9.2	0.4	0.6	0.8	1.0	2.0	4.0
0.1	0.008	0.009	0.009	0.010	0.011	0.011	0.014	0.019
0.2	0.016	0.017	0.018	0.019	0.021	0.022	0.027	0.035
0.4	0.032	0.033	0.035	0.037	0.040	0.041	0.049	0.058
0.6	0.047	0.048	0.051	0.054	0.057	0.059	0.068	0.077
0.8	0.061	0.063	0.067	0.070	0.073	0.076	0.085	0.093
1.0	0.075	0.077	0.081	0.085	0.088	0.091	0.100	0.107
2.0	0.138	0.141	0.145	0.149	0.151	0.154	0.159	0.158
4.0	0.238	0.239	0.241	0.242	0.242	0.242	0.237	0.224
6.0	0.315	0.314	0.313	0.310	0.308	0.305	0.292	0.269
8.0	0.375	0.373	0.368	0.364	0.359	0.354	0.334	0.304
10.0	0.425	0.421	0.414	0.407	0.400	0.394	0.369	0.333
20.0	0.581	0.573	0.559	0.546	0.535	0.524	0.484	0.432
40.0	0.716	0.707	0.690	0.676	0.662	0.650	0.602	0.540
60.0	0.778	0.769	0.754	0.740	0.727	0.715	0.667	0.603
80.0	0.813	0.806	0.792	0.779	0.767	0.756	0.711	0.647
100.0	0.836	0.829	0.817	0.805	0.795	0.784	0.742	0.680
200.0	0.886	0.882	0.875	0.867	0.860	0.853	0.821	0.772
400.0	0.914	0.912	0.908	0.903	0.899	0.895	0.875	0.841
600.0	0.924	0.923	0.920	0.917	0.914	0.911	0.896	0.871
800.0	0.929	0.928	0.926	0.923	0.921	0.919	0.908	0.887
1,000.0	0.932	0.931	0.929	0.928	0.926	0.924	0.915	0.898

λ^2 \ p	6.0	8.0	10.0	20.0	40.0	60.0	80.0	100.0
0.1	0.023	0.026	0.028	0.031	0.031	0.030	0.028	0.027
0.2	0.039	0.042	0.043	0.045	0.043	0.040	0.038	0.036
0.4	0.063	0.064	0.065	0.063	0.057	0.053	0.050	0.047
0.6	0.080	0.081	0.081	0.076	0.068	0.062	0.058	0.055
0.8	0.095	0.095	0.094	0.087	0.076	0.069	0.064	0.061
1.0	0.107	0.106	0.105	0.096	0.083	0.075	0.070	0.066
2.0	0.154	0.149	0.144	0.127	0.108	0.097	0.090	0.085
4.0	0.212	0.202	0.194	0.167	0.140	0.125	0.115	0.108
6.0	0.252	0.239	0.229	0.195	0.162	0.145	0.133	0.124
8.0	0.284	0.268	0.255	0.216	0.180	0.160	0.147	0.137
10.0	0.310	0.292	0.278	0.235	0.194	0.173	0.159	0.148
20.0	0.399	0.375	0.356	0.299	0.246	0.219	0.201	0.187
40.0	0.499	0.469	0.446	0.375	0.310	0.275	0.252	0.236
60.0	0.560	0.528	0.503	0.425	0.352	0.313	0.288	0.269
80.0	0.604	0.571	0.545	0.463	0.385	0.343	0.315	0.295
100.0	0.637	0.604	0.577	0.493	0.412	0.367	0.338	0.316
200.0	0.734	0.703	0.677	0.591	0.501	0.451	0.416	0.391
400.0	0.812	0.788	0.766	0.688	0.598	0.544	0.506	0.477
600.0	0.848	0.828	0.809	0.740	0.654	0.600	0.561	0.531
800.0	0.868	0.851	0.836	0.774	0.693	0.640	0.601	0.571
1,000.0	0.882	0.867	0.853	0.797	0.722	0.670	0.632	0.602

Table 2.7 Values of **h/p** with load uniformly distributed over the whole span

λ^2 \ **p**	0.1	0.2	0.4	0.6	0.8	1.0	2.0	4.0
0.1	0.009	0.009	0.010	0.011	0.011	0.012	0.016	0.021
0.2	0.017	0.018	0.019	0.021	0.022	0.023	0.029	0.037
0.4	0.034	0.035	0.038	0.040	0.042	0.044	0.053	0.063
0.6	0.050	0.051	0.055	0.058	0.061	0.063	0.073	0.082
0.8	0.065	0.067	0.075	0.075	0.078	0.081	0.091	0.099
1.0	0.080	0.082	0.087	0.090	0.094	0.097	0.107	0.113
2.0	0.146	0.149	0.154	0.158	0.161	0.163	0.168	0.167
4.0	0.252	0.254	0.256	0.256	0.256	0.256	0.250	0.235
6.0	0.333	0.333	0.331	0.328	0.325	0.322	0.307	0.282
8.0	0.398	0.395	0.389	0.384	0.379	0.373	0.351	0.319
10.0	0.450	0.446	0.437	0.429	0.422	0.415	0.387	0.349
20.0	0.615	0.606	0.590	0.576	0.563	0.551	0.507	0.452
40.0	0.758	0.748	0.729	0.712	0.697	0.684	0.631	0.565
60.0	0.823	0.814	0.796	0.781	0.766	0.753	0.701	0.632
80.0	0.861	0.852	0.837	0.822	0.809	0.797	0.747	0.678
100.0	0.885	0.877	0.863	0.850	0.838	0.827	0.780	0.713
200.0	0.939	0.934	0.925	0.916	0.908	0.900	0.865	0.811
400.0	0.968	0.966	0.960	0.955	0.950	0.946	0.923	0.885
600.0	0.979	0.977	0.973	0.970	0.966	0.963	0.946	0.917
800.0	0.984	0.982	0.980	0.977	0.974	0.971	0.959	0.935
1,000.0	0.987	0.986	0.984	0.981	0.979	0.977	0.966	0.947

λ^2 \ **p**	6.0	8.0	10.0	20.0	40.0	60.0	80.0	100.0
0.1	0.025	0.028	0.030	0.033	0.033	0.031	0.030	0.029
0.2	0.042	0.045	0.046	0.048	0.045	0.042	0.039	0.038
0.4	0.067	0.068	0.069	0.067	0.060	0.055	0.052	0.049
0.6	0.085	0.086	0.085	0.080	0.071	0.065	0.060	0.057
0.8	0.100	0.100	0.099	0.091	0.079	0.072	0.067	0.063
1.0	0.113	0.112	0.110	0.100	0.087	0.079	0.073	0.069
2.0	0.162	0.156	0.151	0.133	0.113	0.101	0.094	0.088
4.0	0.222	0.212	0.203	0.174	0.146	0.130	0.120	0.112
6.0	0.264	0.250	0.239	0.203	0.169	0.150	0.138	0.129
8.0	0.297	0.280	0.267	0.225	0.187	0.166	0.153	0.143
10.0	0.324	0.305	0.290	0.244	0.202	0.180	0.165	0.154
20.0	0.417	0.391	0.371	0.311	0.256	0.227	0.208	0.195
40.0	0.521	0.490	0.465	0.390	0.322	0.286	0.262	0.245
60.0	0.586	0.552	0.525	0.443	0.366	0.326	0.299	0.279
80.0	0.631	0.596	0.569	0.482	0.400	0.357	0.328	0.306
100.0	0.667	0.631	0.603	0.514	0.428	0.382	0.351	0.329
200.0	0.769	0.736	0.709	0.617	0.522	0.469	0.433	0.407
400.0	0.854	0.827	0.804	0.719	0.624	0.566	0.527	0.496
600.0	0.892	0.870	0.850	0.775	0.683	0.626	0.585	0.553
800.0	0.914	0.896	0.879	0.811	0.725	0.668	0.627	0.595
1,000.0	0.929	0.913	0.898	0.837	0.755	0.701	0.660	0.628

The solution cannot proceed directly from (2.41) because the initial profile is unknown. However, the final dead load profile may be used for a preliminary assessment and corrections made subsequently. An initial estimate for the horizontal component of tension in one free-hanging cable is then 2.1×10^4 kN (see equation 2.3). Thus $\lambda^2 = 500$ (calculation of L_e includes an important contribution from the backstays), and $\mathbf{p} = 3$. The cubic to be solved is

$$\mathbf{h}^3 + 23\mathbf{h}^2 + 43\mathbf{h} - 313 = 0,$$

from which $\mathbf{h} = 2.75$. An improved estimate of the initial cable tension is $H = 2.25 \times 10^4$ kN, so that an estimate of the initial sag is 84 m. Going through the calculations again gives $\lambda^2 = 410$, and so $\mathbf{h} = 2.7$ (see table 2.7). Therefore the sag of the cables in their free-hanging position should be about 83 m. Put another way, the cables will have dropped about 7 m at midspan after the full dead load has been placed.

In placing the full dead load, the tension in the backstays increases by some 7×10^4 kN, causing them to lengthen by 0.24 m. As a result, the distance each tower top should be set back toward each anchorage is about 0.28 m.

A final item of interest is to keep track of the sag at midspan as the deck is placed. Because of catenary action the sag will exceed the final value during most of the hanging process, as figure 2.9 indicates. This example will be taken up again when we consider the postelastic response.

2.5 Energy Relations

The linearized theory provides a convenient framework within which to explore the way energy is stored by a suspended cable when it responds to applied load. When the loads are applied statically, the work done by the load is stored in two forms: gravitational potential energy and elastic strain energy of extension.

Suppose that a distributed patch of loading $p(x)$ is gradually applied

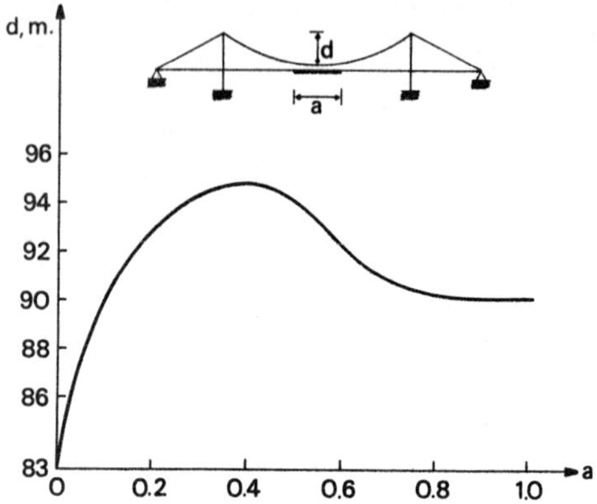

2.9 Sag registered as laying of the deck proceeds

to part of the span. The work done by it is

$$W = \frac{1}{2}\int_0^l p(x)\, w dx, \tag{2.43}$$

and it is recoverable. The extensional strain energy is given by (see the triangular area in figure 2.10)

$$V_e = \frac{1}{2}\int_0^l \tau\left(\frac{ds'}{ds} - 1\right) ds = \frac{1}{2} h \int_0^l \left(\frac{ds}{dx}\right)^2 \left(\frac{ds'}{ds} - 1\right) dx,$$

or

$$V_e = \frac{1}{2} h \int_0^l \frac{dz}{dx}\frac{dw}{dx} dx, \tag{2.44}$$

which is accurate to second order. The gravitational potential energy arises as a second-order difference in two larger terms:

$$V_g = \int_0^l H\frac{ds}{dx}\left(\frac{ds'}{ds} - 1\right) ds - \int_0^l mgw\, ds, \tag{2.45}$$

where the first term corresponds to the rectangular area in figure 2.10, and the second accounts for the lowering of the cable by the applied load. Retaining second-order quantities in the strain-displacement relation allows us to rewrite (2.45) as

$$V_g = \frac{1}{2} H \int_0^l \left(\frac{dw}{dx}\right)^2 dx + \int_0^l \left(H\frac{dz}{dx}\frac{dw}{dx} - mgw\frac{ds}{dx}\right) dx, \tag{2.46}$$

in which the first-order terms are conveniently grouped on the right. Since z and dz/dx are continuous along the cable, and w is continuous, integration by parts yields

$$V_g = \frac{1}{2} H \int_0^l \left(\frac{dw}{dx}\right)^2 dx + H\frac{dz}{dx} w\Big|_0^l - \int_0^l \left(H\frac{d^2z}{dx^2} + mg\frac{ds}{dx}\right) w dx. \tag{2.47}$$

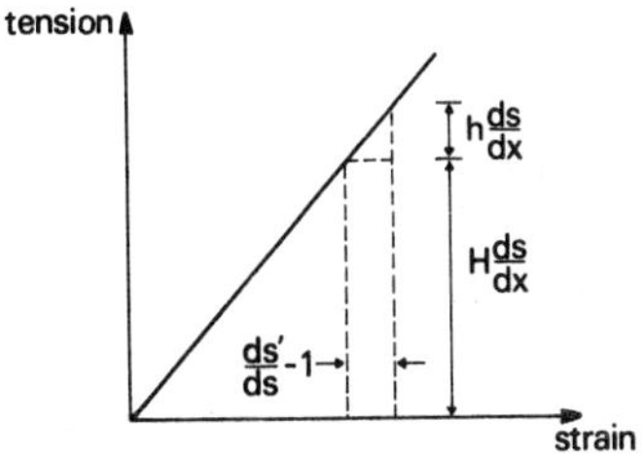

2.10 Tension-strain details for an element of the cable

The second term contributes nothing because the additional deflection is zero at each end, and neither does the third term because the integrand is identically zero on account of the equilibrium requirements under self-weight alone. Thus the expression for the gravitational potential energy assumes the form

$$V_g = \frac{1}{2} H \int_0^l \left(\frac{dw}{dx}\right)^2 dx, \tag{2.48}$$

while the strain energy of extension is similarly reduced to

$$V_e = \frac{1}{2} mg \frac{h}{H} \int_0^l w dx. \tag{2.49}$$

Conservation of energy requires that

$$W = V_e + V_g. \tag{2.50}$$

Notice that the work done and the potential energy stored involve second-order terms, as expected.

This argument differs somewhat in form, but little in substance, from that advanced by Pugsley [14], who was one of the first to consider the energy relations of a suspended cable. Pugsley's concern was with relatively inextensible suspension bridge cables, so only gravitational potential energy was considered. Additional discussion of this, together with other relevant work, may be found in Pugsley [13] and Huedey et al. [6].

Using the linear theory of section 2.3, we can examine a whole range of cable profiles under various applied loads [10], although the investigation here is confined to the action of a point load.

In dimensionless form the work done by a point load is

$$\mathbf{W} = \frac{1}{2} \mathbf{P} \mathbf{w}_1, \tag{2.51}$$

while the potential energy stored consists of

$$\begin{aligned} \mathbf{V}_g &= \frac{1}{2} \int_0^1 \left(\frac{d\mathbf{w}}{d\mathbf{x}}\right)^2 d\mathbf{x}, \\ \mathbf{V}_e &= \frac{1}{2} \mathbf{h} \int_0^1 \mathbf{w} d\mathbf{x}, \end{aligned} \tag{2.52}$$

where the new nondimensionalizations are $\mathbf{W} = W/\{(mgl^2/H)\, mgl\}$ and $\mathbf{V} = V/\{(mgl^2/H)\, mgl\}$. Therefore

$$\mathbf{W} = \frac{1}{2} \mathbf{P}^2 \mathbf{x}_1(1 - \mathbf{x}_1) \left\{1 - \frac{3\mathbf{x}_1(1 - \mathbf{x}_1)}{1 + 12/\lambda^2}\right\}, \tag{2.53}$$

which has its peak value when $\mathbf{w}_1$ has its peak value. The extensional strain energy is

$$\mathbf{V}_e = \frac{1}{2}\frac{\mathbf{h}^2}{\lambda^2} = \frac{18\lambda^2}{(12 + \lambda^2)^2}\mathbf{P}^2\mathbf{x}_1^2(1 - \mathbf{x}_1)^2, \tag{2.54}$$

and $\mathbf{V}_e \to 0$ as $\lambda^2 \to 0$ and as $\lambda^2 \to \infty$: both results are as expected.

Extensional strain energy storage is greatest when the load is placed at midspan. In addition, with a load placed at midspan, the overall peak occurs when $\lambda^2 = 12$. A final point is obviously that the greatest values of $\mathbf{V}_e/\mathbf{W}$ occur when the load is placed at midspan, and $\lambda^2 = 24$ is the value of the characteristic cable parameter to which its overall maximum corresponds, resulting in $(\mathbf{V}_e/\mathbf{W})_{max} = 1/3$. There is an analogous result in the free vibrational response.

2.6 Postelastic Response

After the elastic limit of the cable is reached, the problem of the response becomes more difficult since it is nonlinear both with respect to geometry and material properties. The loading that causes a cable segment to reach the elastic limit may be found with the previous theory, but, to determine the largest load, beyond which any increment of it would lead to failure, consideration must be given to the postelastic response. Although ultimate strength calculations are frequently useful, a cautionary note must be sounded, concerning what is to be inferred from such calculations (see example 2.6).

The load causing failure is always greater, sometimes substantially so, than the load pertaining to the elastic limit. A suspended cable resists applied load by changes in tension and geometry. Beyond the elastic limit changes in tension and geometry can still occur to accommodate the increased load. Tension changes occur because of the pronounced strain-hardening characteristic of the high strength steels frequently used in cables, for which there is no clearly defined yield plateau. Failure will occur when the ultimate strain is reached in some portion of the cable. (A "weakest link in the chain" argument obviously explains the actual failure and must be considered in deciding what cumulative inelastic extension is reasonable for the whole cable.) For the steel typically used in cables, the ratio of strain at ultimate to strain at elastic limit is small, being of the order of 10. This may be contrasted with mild steels (which have a clearly defined yield plateau) where typically this ratio may be of the order of 100 or more. A characteristic of relatively flat suspended cables is that small changes in cable length give rise to substantial changes in cable geometry. Therefore, even though the strain ratio at ultimate may be small (and the nonrecoverable strain itself be small), this behavior together with the strain-hardening effect makes it possible for the load that causes failure to be often substantially in excess of that which just exceeds the elastic limit.

Quite a lot of numerical work has been done on this problem and has usually involved an incremental load approach based on tangent modulus concepts. But, since it is difficult to be precise about the conditions at failure, the results of hand methods are probably just as meaningful. By reversing the role of dependent and independent variables, analytical solutions may be derived in which direct use is made of experimental data on stress-strain behavior. Such a theory provides a natural complement to the nonlinear elastic theory [9].

Stress-Strain Properties of Steel Cables

Sizable quantities of structural cable were not manufactured under reasonable quality control until the middle of the nineteenth century. But cables were used for structural purposes long before that, as may be seen from this quotation from Scalzi and McGrath [15]:

> A copper cable found in the ruins of Ninevah near Babylon indicates that cables have been used as a structural component since 685 B.C. History also tells us that the Romans used bronze wire rope as evidenced by the pieces found in the excavations of Pompeii, which was buried by the eruption of Mount Vesuvius in A.D. 79. Samples of these early ropes and strands are on display in an English Museum and at the Museo Barbonico at Naples, Italy.
>
> Records indicate that machine-drawn wire made its first appearance in Europe during the fourteen century. Credit is given to A. Albert of Germany for producing the first wire rope in 1834 which closely resembles the product of today. Others claim that an Englishman, named Wilson, produced the first rope in 1832. The first machine-made American steel-wire rope was produced by John A. Roebling and placed in service shortly before 1850.

Three different types of steel cable are common in structural applications: structural strand with individual wires wound around a core, structural rope with several strands of wires somewhat finer in diameter than in the strand wound around a core, and parallel wire strands with the wires bundled, not twisted. Due to the differing methods of manufacture the effective modulus of elasticity varies. The more pronounced influence of the helices gives the rope the smallest modulus, with an approximate value of $E = 140{,}000$ MPa. This may be contrasted with 170,000 MPa for a structural strand and 190,000 to 210,000 MPa for a parallel wire strand [11, 15].

Because the elastic limit is not clearly defined (see figure 2.11), the modulus of elasticity is usually calculated from the slope of the straight line that connects the 10 percent breaking load with the 90 percent prestretching load of the cable specimen. A prestretching load of about

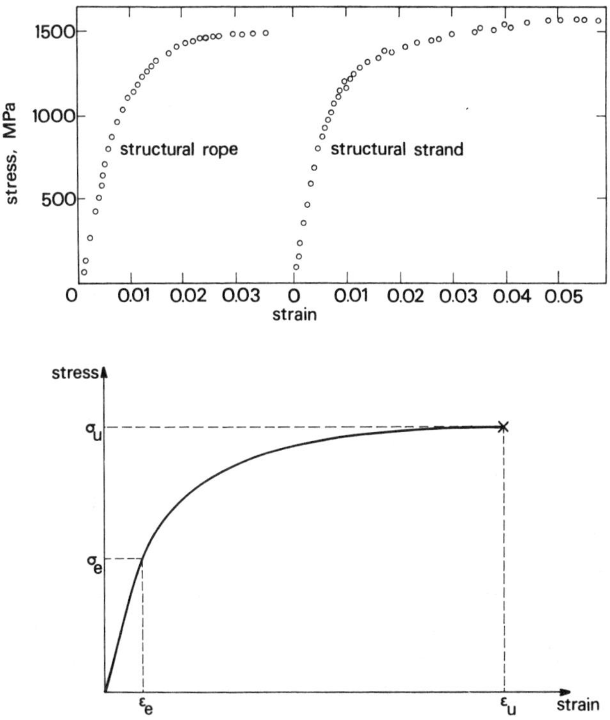

2.11 Typical stress-strain properties of structural rope and strand (from Jonatowski and Birnstiel [11], courtesy ASCE)

55 percent of the breaking load is usually applied to remove constructional looseness in the cables—a very important prerequisite. Typically the elastic limit σ_e is reached at about 50 percent of the ultimate tensile strength σ_u. Ultimate tensile strengths of 1,500 MPa are regularly achieved, with ultimate strains around 3 percent for rope and 6 percent for strand; tests show that strand is stronger than the more flexible rope of the same size. Strength and stiffness, based as they are on the nominal cross-sectional areas, are affected by the class of zinc coating. Factors of safety vary, but a working stress of approximately 500 MPa (giving a factor of 3 on the ultimate tensile strength) is routinely used for a rope or a strand. A factor of safety of 2.5 (corresponding to a working stress of around 600 MPa) may be specified in applications calling for preassembled parallel wire strands, such as cable-stayed bridges. This same figure is typical for the cables of long-span suspension bridges where the individual wires are spun and grouped into strands, and these strands are bundled

together before compaction and wrapping produces the final circular cross section.

Further details may be found in Scalzi and McGrath and in the trade literature of the steel industry. In the United States the relevant specifications are ASTM A586-68 (*Standard Specification for Zinc-Coated Steel Structural Strand*) and ASTM A603-70 (*Standard Specification for Zinc-Coated Steel Structural Wire Rope*).

Elastic Capacity

In an elastic analysis **h** and **w** are generally the dependent variables, while **P** or **p** and λ^2 are the independent variables. However, in postelastic analyses the situation is best reversed. The elastic limit is known, so the independent variables are λ^2 and the cable tension; calculation of the additional load carried at the elastic limit $\mathbf{P}_e$ or $\mathbf{p}_e$, and the additional deflection is straightforward. A similar situation exists in the postelastic range. The cable stress and accompanying strain are read off a plot of the stress-strain properties of the particular cable, and the load and associated deflection may be uniquely determined.

The horizontal component of cable tension may be used in lieu of the true cable tension. For the geometries under consideration the cable tension is, at most, about 10 percent higher than its horizontal component: this is a small variation. Consequently the horizontal component of cable tension at the elastic limit H_e will be assumed to be

$$H_e = \sigma_e A,$$

where σ_e is the stress at the elastic limit and A is the nominal area of the cable. Therefore at this elastic limit, which is the limit of applicability of (2.13), (2.14), and (2.22), we have for a point load applied at midspan (which gives the smallest $\mathbf{P}_e$)

$$\mathbf{P}_e^2 + \mathbf{P}_e - \alpha = 0, \tag{2.55}$$

where $\alpha = 8\{(\mathbf{H}_e - 1)^3 + (2 + \lambda^2/24)(\mathbf{H}_e - 1)^2 + (1 + \lambda^2/12)(\mathbf{H}_e - 1)\}/\lambda^2$, in which $\mathbf{H}_e = H_e/H$ and $\mathbf{P}_e = P_e/mgl$. Thus

$$\mathbf{P}_e = \frac{1}{2}\{(1 + 4\alpha)^{1/2} - 1\}, \tag{2.56}$$

and the additional deflection under this load is

$$\mathbf{w}_e = \frac{1}{4\mathbf{H}_e}\left\{1 - \frac{(\mathbf{H}_e - 1)}{2\mathbf{P}_e}\right\}, \tag{2.57}$$

where $\mathbf{w}_e = w_e/(P_e l/H)$.

In the case of a distributed load, the smallest $\mathbf{p}_e$ occurs when the load covers the whole span. The results are

$$\mathbf{p}_e = (1 + 3\alpha)^{1/2} - 1,$$
$$\mathbf{w}_e = \frac{1}{2\mathbf{H}_e}\left\{1 - \frac{(\mathbf{H}_e - 1)}{\mathbf{p}_e}\right\}\mathbf{x}(1 - \mathbf{x}), \tag{2.58}$$

where $\mathbf{p}_e = p_e/mg$ and $\mathbf{w}_e = w_e/(p_e l^2/H)$. Comparison of (2.56) and the first equation of (2.58) indicates that regardless of α

$$\frac{3}{2} < \frac{\mathbf{p}_e}{\mathbf{P}_e} < (3)^{1/2}. \tag{2.59}$$

In other words, the ratio of total loads lies between 1.5 and 1.732.

Ultimate Capacity

As the loading is increased past the elastic limit, nonrecoverable strains arise. Equilibrium must still be maintained, so the equations for additional deflection are of the same form as previously. The cable equation is altered, but not substantially. In the case of a point load at ultimate, it reads

$$\frac{\phi_u - 1}{\lambda^2} = \mathbf{P}_u \int_0^1 \frac{d\mathbf{z}}{d\mathbf{x}}\frac{d\mathbf{w}_u}{d\mathbf{x}}\, d\mathbf{x} + \frac{1}{2}\mathbf{P}_u^2 \int_0^1 \left(\frac{d\mathbf{w}_u}{d\mathbf{x}}\right)^2 d\mathbf{x}, \tag{2.60}$$

where $\mathbf{P}_u = P_u/mgl$, $\mathbf{w}_u = w_u/(P_u l/H)$ and $\phi_u = \varepsilon_u/\varepsilon$, in which ε_u is the strain assigned to the whole cable as representative of ultimate conditions and ε is the strain in the initial self-weight profile. This equation is not restricted to ultimate conditions but with suitable adjustment may be used for any part of the postelastic range. The quadratic for $\mathbf{P}_u$ is

$$\mathbf{P}_u^2 + \mathbf{P}_u - \beta = 0, \tag{2.61}$$

where $\beta = 8\mathbf{H}_u^2(\phi_u - 1)/\lambda^2 + (\mathbf{H}_u^2 - 1)/3$, and $\mathbf{H}_u = H_u/H$. Thus

$$\mathbf{P}_u = \frac{1}{2}\{(1 + 4\beta)^{1/2} - 1\}. \tag{2.62}$$

Inspection of (2.61) shows that $\mathbf{P}_u > (\mathbf{H}_u - 1)/2$, which is a lower bound for the ultimate load-carrying capacity and analogous to the expression $\mathbf{h} < 2\mathbf{P}$. The additional deflection under the load is

$$\mathbf{w}_u = \frac{1}{4\mathbf{H}_u}\left\{1 - \frac{(\mathbf{H}_u - 1)}{2\mathbf{P}_u}\right\}. \tag{2.63}$$

With the distributed load the results are

$$\mathbf{p}_u = (1 + 3\beta)^{1/2} - 1,$$
$$\mathbf{w}_u = \frac{1}{2\mathbf{H}_u}\left\{1 - \frac{(\mathbf{H}_u - 1)}{\mathbf{p}_u}\right\}\mathbf{x}(1 - \mathbf{x}). \tag{2.64}$$

A lower bound of $\mathbf{p}_u > (\mathbf{H}_u - 1)$ exists, and again

$$\frac{3}{2} < \frac{\mathbf{p}_u}{\mathbf{P}_u} < (3)^{1/2}, \tag{2.65}$$

which is a tidy result.

It is obviously advisable to be conservative in calculating the ultimate load. It seems reasonable to reduce σ_u by about 10 percent, so that the actual cable tension rather than its horizontal component approaches the ultimate. Also the ultimate strain as measured in tests should not be used. There is no consensus on a suitable value, but a value one-half as large as the test results is probably acceptable. Just what modified value of ε_u is adopted is to a certain extent arbitrary, but the principle is important because allowance may then be made for material imperfections, kinking at points of load application, local necking, and so on.

Unloading after a Postelastic Excursion

If the live loading that was sufficient to cause postelastic behavior is removed, there exists a nonrecoverable strain that causes a lengthening of the cable. The new equilibrium position under dead load is altered because the deflection is increased by Δz and the horizontal component of cable tension is reduced by ΔH. In dimensionless terms this additional deflection is

$$\Delta\mathbf{z} = \frac{\Delta\mathbf{H}}{2(1 - \Delta\mathbf{H})}\mathbf{x}(1 - \mathbf{x}), \tag{2.66}$$

where $\Delta\mathbf{z} = \Delta z/(mgl^2/H)$ and $\Delta\mathbf{H} = \Delta H/H$.

The nonrecoverable strain is $(\Delta\varepsilon - \Delta H/EA)$, where $\Delta\varepsilon = \varepsilon_* - \sigma_e/E$ (and $\sigma_e/E < \varepsilon_* < \varepsilon_u$). The cable equation is

$$\frac{\Delta\phi - \Delta\mathbf{H}}{\lambda^2} = \int_0^1 \Delta\mathbf{z}\, d\mathbf{x} + \frac{1}{2}\int_0^1 \left(\frac{d\Delta\mathbf{z}}{d\mathbf{x}}\right)^2 d\mathbf{x}, \tag{2.67}$$

where $\Delta\phi = \Delta\varepsilon/\varepsilon$. Substituting (2.66) into (2.67), integrating, and rearranging yields the cubic

$$\Delta\mathbf{H}^3 - \left(2 + \Delta\phi + \frac{\lambda^2}{24}\right)\Delta\mathbf{H}^2 + \left(1 + 2\Delta\phi + \frac{\lambda^2}{12}\right)\Delta\mathbf{H} - \Delta\phi = 0. \tag{2.68}$$

From Descartes' rule of signs there are either three or one positive real roots of (2.68). However, it is clear from (2.66) that, because $\Delta\mathbf{z}$ must be positive, $\Delta\mathbf{H}$ must be less than unity, and this is guaranteed by (2.68). But if $\Delta\phi < 1$, (2.67) requires in addition that $\Delta\mathbf{H} < \Delta\phi$. By writing (2.68) in the alternative form

$$\left(\Delta\phi + \frac{\lambda^2}{24}\right) - \Delta\mathbf{H} = \frac{\lambda^2}{24(1 - \Delta\mathbf{H})^2}, \tag{2.69}$$

and graphing the two portions, it is seen that their intersection always gives one positive real root that meets these interlacing requirements. This then is the required value of $\Delta\mathbf{H}$, and the problem is therefore self-consistent.

Example 2.5 Ultimate Load Capacity of a Flying Fox

Returning to example 2.3, we note that under the point live load of 20 kN the stress in the cable is about 300 MPa, which corresponds to a factor on the ultimate stress of about 5. The calculation of the ultimate point load carried is straightforward. The cable is a structural rope. We assign an ultimate strain of $\varepsilon_u =$ 1.5 percent to it; hence $\phi_u = 45$. Similarly $\mathbf{H}_u = 25$. Equation (2.61) becomes

$$\mathbf{P}_u^2 + \mathbf{P}_u - 3140 = 0,$$

from which $\mathbf{P}_u = 55$, so the point load carried at ultimate is about 220 kN. The cable then dips by 7.3 m at midspan. Working conditions correspond to a factor on the ultimate load of about 11.

Example 2.6 Ultimate Load Capacity of a Suspension Bridge Cable

Let us take as the initial profile the suspension bridge cable of example 2.4 under full dead load. In this case $\lambda^2 = 125$. The backstays experience the highest tensions. We suppose that the stresses are highest in the backstays too, although often cable areas are increased there. Interpreting this for the whole cable implies $\varepsilon_u = 1.2$ percent; thus $\phi_u = 4.8$. Calculations indicate that $\mathbf{H}_u = 3.0$ is reasonable. From the first equation of (2.64) the ultimate uniformly distributed live load that may be applied is found from

$$\mathbf{p}_u^2 + 2\mathbf{p}_u - 14.6 = 0,$$

from which $\mathbf{p}_u = 3.0$, and from the second equation of (2.64) the sag has increased to an alarming 120 m. The load and stress factors are of similar magnitude here. This is a manifestation of the linear relation between applied load and cable tension when large loads are applied to cables with relatively deep profiles.

Several comments must be made about these calculations. Most important, there is no way that the bridge could ever receive such a live load, even though theoretically the load might be higher because the flexural capacity of the deck has been ignored. The calculations are in any event misleading: the tower tops would be required to move toward each other by several meters, which is not possible. Therefore an ultimate load calculation for the cable is quite meaningless. Actual live loads are far lower, corresponding to an overall stress factor whose minimum value will be about 2.5. The cable will never enter the postelastic region, nor should it.

2.7 Stress Factors, Load Factors and Factors of Safety

Horne and Merchant [5] make some interesting comments on the distinction between stress factors and load factors for nonlinear systems. Their comments relate to a slender column under axial and transverse loads, a system that is of the nonlinear softening variety when the axial load is plotted against the maximum stress induced. With reference to figure 2.12, point A could correspond to working stress conditions, whereas point B might correspond to conditions close to collapse of the member. The stress σ_2 grows rapidly, as the axial load P_2 approaches the Euler load, because of accelerating importance of the bending stresses that result from the increased bowing of the member. Clearly the load factor P_2/P_1 will be substantially less than the stress factor σ_2/σ_1. Therefore, to make sure that an adequate factor of safety exists, it must be based on specification of a load factor for design of such members.

This situation is the reverse of that obtained with a suspended cable. There, as figure 2.13 shows, applied load and cable tension (or nominal stress) exhibit a stiffening relation of one to the other. Conditions at

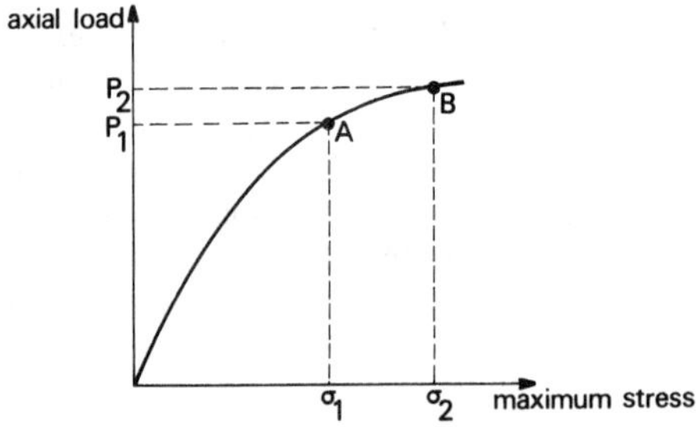

2.12 Behavior of a column under axial and transverse loads (from Horne and Merchant [5], courtesy Pergamon Press)

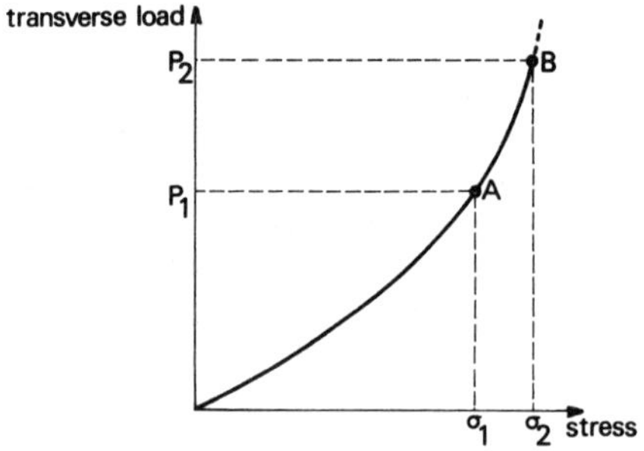

2.13 Behavior of a cable under transverse loads

point B could be considered to correspond to fracture. The difference between the ultimate stress σ_2 and the working stress σ_1 is significantly less than the difference between the corresponding applied loads. Consequently the factor of safety must be based on the stress factor.

To amplify this, consider example 2.5 where a stress factor of 5 corresponded to a load factor of 11. A factor of safety on stress of 5 is certainly safe, but a factor of safety on load of 5 might not be. The inevitable conclusion seems to be that the design of cable structures must be based on a working stress approach.

Exercises

2.1 Although the overall behavior of a chain composed of many links is cablelike, bending stresses are important in a local sense because each link is an elongated ring, and the tensile force is transmitted from one link to the next along each link's major axis. In figure 2.14 it is supposed that the diameter of the rod used for the chain link is small compared to the two other dimensions. This being the case, show that, if the nominal direct stress is $\sigma_a(= T/2\pi r^2)$, the worst tensile stresses occur at either end of the major axis, and are

$$\sigma_{\max} = \frac{8(1 + b/a)(a/r)}{(\pi + 2b/a)}\sigma_a,$$

and that when allowance is made for both bending and axial effects the equivalent modulus of elasticity for a chain is

$$\frac{E(1 + b/a - r/a)}{[2(a/r)^2\{3\pi/2 - 4 - (\pi - 2)^2/(\pi + 2b/a)\} + \pi/4 + b/a]}.$$

The relevant cross-sectional area is then that based on both legs of the link.

Because of its symmetry the problem is statically indeterminate only to the first degree, and moment area provides a quick means of establishing the bending moment diagram around the ring. Calculation of the strain energy due to bending and axial effects then allows the diametral

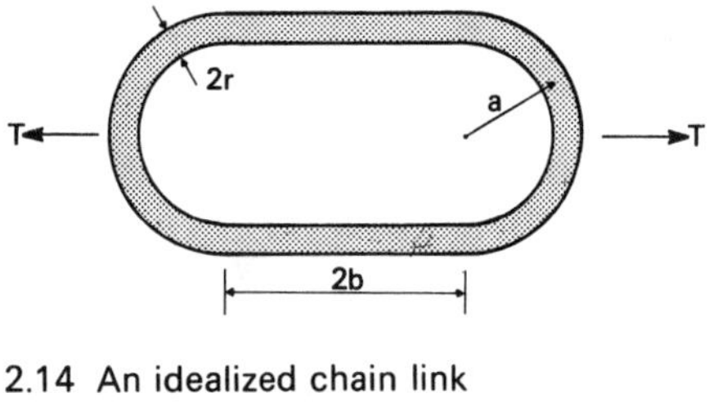

2.14 An idealized chain link

lengthening to be found, and from that the reduced modulus follows. When, for example, $b = a$ and $a/r = 4$, $\sigma_{max} \simeq 13\sigma_a$ and $E' \simeq E/10$, which is striking evidence of the importance of flexural effects in this context.

2.2 Consider a uniform chain attached by one end to a vertical axis and rotating steadily about it. With reference to figure 2.15, the statements of horizontal and vertical equilibrium of an element are

$$\frac{d}{ds}\left(T\frac{dx}{ds}\right) + m\Omega^2 x = 0,$$

$$\frac{d}{ds}\left(T\frac{dz}{ds}\right) + mg = 0,$$

together with the geometrical constraint that $(dx/ds)^2 + (dz/ds)^2 = 1$. These equations are difficult to solve as they stand, although approximate solutions are possible.

If $\Omega^2 l/g$ is large, the length of the chain l is little different from its horizontal projection. Show then that

$$z = \frac{2g}{\Omega^2} \ln (1 + x/l).$$

When $\Omega^2 l/g$ is small, the length of the cable is little different from its vertical projection, and the equations of equilibrium reduce to a classical linear eigenproblem, namely, the whirling of a vertical chain or, as originally presented by Daniel Bernoulli, the closely related problem of its small oscillations. The lowest eigenvalue is [19]

$$\Omega_1 = 1.20\left(\frac{g}{l}\right)^{1/2}.$$

Thus below this angular velocity the chain will remain vertical. A linear analysis cannot correctly predict what happens when $\Omega > \Omega_1$. Whirling will probably occur, but the analysis is extremely difficult, because it

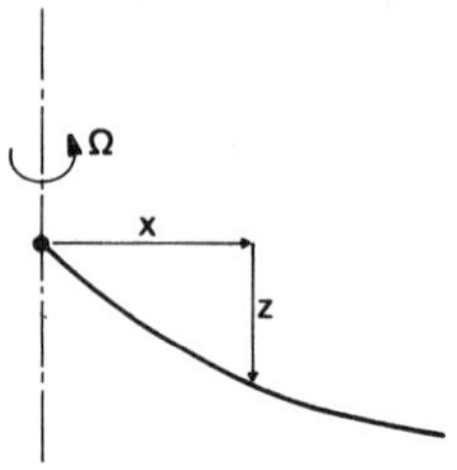

2.15 Steady whirling of a cable about a fixed axis

involves the solution of a nonlinear eigenproblem. Caughey has produced exact solutions in certain special cases, which lend support to this conclusion [2, 3]. When $\Omega \gg \Omega_1$, the logarithmic profile presented here would appear to be the correct limiting form.

2.3 Figure 2.16 shows a moment applied to a cable. Setting aside for the time being how such a loading can arise, show that the linearized solution is

$$\mathbf{w} = -\mathbf{x}\left\{1 + \frac{\mathbf{h}}{2\mathbf{M}}(1 - \mathbf{x})\right\},$$

for $0 \leq \mathbf{x} < \mathbf{x}_1$,

$$\mathbf{w} = (1 - \mathbf{x})\left\{1 - \frac{\mathbf{h}}{2\mathbf{M}}\mathbf{x}\right\},$$

for $\mathbf{x}_1 < \mathbf{x} \leq 1$, and

$$\mathbf{h} = \frac{6\mathbf{M}}{(1 + 12/\lambda^2)}(1 - 2\mathbf{x}_1),$$

where $\mathbf{w} = w/(M/H)$, $\mathbf{M} = M/mgl^2$, and so on.

Note that there is a discontinuity in the profile at $\mathbf{x}_1$. Mathematically speaking, the cable can accommodate this, although in practice the finite flexural rigidity of a real cable would smooth it. In the case of a taut string (when, $\lambda^2 \to 0$), the profile for additional deflection is analogous to the influence line for the shear force at $\mathbf{x}_1$ in a simply supported beam. In the same way the result for a point load is analogous to the influence line for bending moment at $\mathbf{x}_1$.

When the moment is applied at midspan, no additional tension is generated, as the cable rises and falls to either side by the same amount. The largest additional tension occurs when the moment is applied at an end, and, when two equal and opposite moments are applied to the ends, as in figure 2.17, the details are

$$\mathbf{w} = -1 + \frac{6\mathbf{x}(1 - \mathbf{x})}{(1 + 12/\lambda^2)},$$

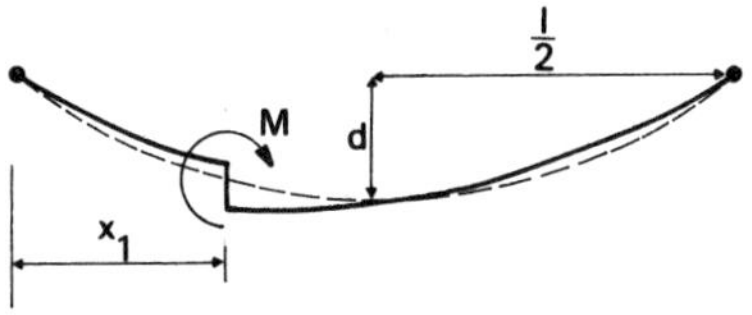

2.16 Definition diagram for a moment applied to a cable

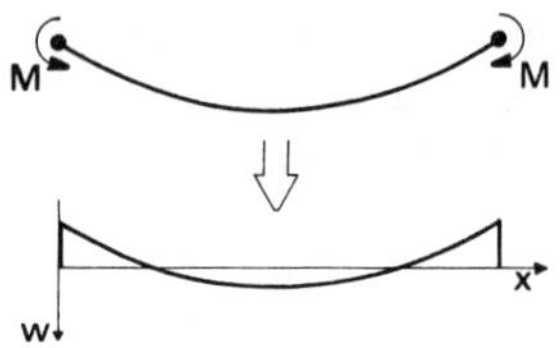

2.17 Profile of additional deflection due to equal moments applied to the ends of a cable

and

$$\mathbf{h} = \frac{-12\mathbf{M}}{(1 + 12/\lambda^2)}.$$

This application is as a limiting case of a suspension bridge's response to a summer midafternoon temperature gradient through its deck. The cable's response to a pair of end moments gives an upper limit to the deflection response. In a long-span bridge midspan deflections of 2 to 3 m have been recorded on hot days.

2.4 Derive the following solution for the response of a suspended cable to an antisymmetric distributed load (see figure 2.18):

$$\mathbf{w} = \frac{\mathbf{x}}{2(1 + \mathbf{h})}\left\{\frac{1}{2}(1 - 2\mathbf{x}) - \frac{\mathbf{h}}{\mathbf{p}}(1 - \mathbf{x})\right\},$$

for $0 \leq \mathbf{x} \leq 1/2$, and

$$\mathbf{w} = \frac{(1 - \mathbf{x})}{2(1 + \mathbf{h})}\left\{\frac{1}{2}(1 - 2\mathbf{x}) - \frac{\mathbf{h}}{\mathbf{p}}\mathbf{x}\right\},$$

for $1/2 \leq \mathbf{x} \leq 1$, where $\mathbf{h}^3 + (2 + \lambda^2/24)\mathbf{h}^2 + (1 + \lambda^2/12)\mathbf{h} - \mathbf{p}^2\lambda^2/96 = 0$. Show that a general upper bound is $h < (1 + \mathbf{p}^2/4)^{1/2} - 1$. The midspan always rises under this loading, although this may be ignored when **p** is sufficiently small for the linear theory to hold. Then, on account of the antisymmetric pattern of additional deflection, no additional tension is generated.

2.5 Changes in the ambient temperature cause changes in the cable profile and may therefore be thought of as a loading on the cable.

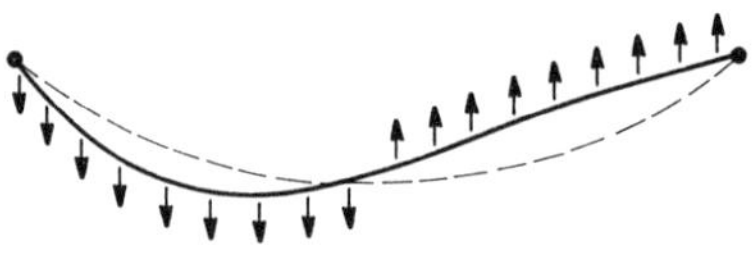

2.18 Antisymmetric distributed load on a cable

(a) Uniform temperature rise. With reference to figure 2.19, show that the additional deflection is

$$\mathbf{w} = \frac{\mathbf{h}}{2(1 - \mathbf{h})}\mathbf{x}(1 - \mathbf{x})$$

and that the fractional amount by which the original horizontal component of tension falls is found from the solution of

$$\mathbf{h}^3 - \left(2 + \theta + \frac{\lambda^2}{24}\right)\mathbf{h}^2 + \left(1 + 2\theta + \frac{\lambda^2}{12}\right)\mathbf{h} - \theta = 0,$$

where $\theta = \alpha|\Delta T|L_t/(HL_e/EA)$, in which α is the coefficient of expansion per degree, ΔT is the temperature increment, and $L_t = l(1 + (mgl/H)^2/12)$. In practice θ is almost invariably found to be less than unity. The problem is therefore analogous to that of unloading after a postelastic excursion: consequently there are no difficulties with the roots of this cubic.

(b) Uniform temperature fall. With reference to figure 2.20, show that the additional deflection is

$$\mathbf{w} = \frac{\mathbf{h}}{2(1 + \mathbf{h})}\mathbf{x}(1 - \mathbf{x}),$$

where the fractional amount by which the original horizontal component of cable tension rises is found from the solution of

$$\mathbf{h}^3 + \left(2 - \theta + \frac{\lambda^2}{24}\right)\mathbf{h}^2 + \left(1 - 2\theta + \frac{\lambda^2}{12}\right)\mathbf{h} - \theta = 0.$$

Show that there is always just one positive real root and that an upper bound to it is $\mathbf{h} < \theta$.

2.6 Show that the additional deflection due to stretch, when an initially

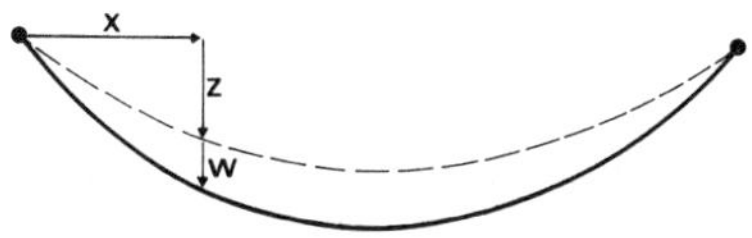

2.19 Definition diagram for a uniform temperature rise

2.20 Definition diagram for a uniform temperature fall

unstressed length of cable is hung between two supports, is

$$\mathbf{w} = \frac{\Delta \mathbf{H}}{2(1 - \Delta \mathbf{H})}\mathbf{x}(1 - \mathbf{x}),$$

where $\mathbf{w} = w/(mgl^2/H)$ and $\Delta\mathbf{H} = \Delta H/H$, in which the reduction in cable tension is found from the cubic

$$(1 - \Delta \mathbf{H})^3 = \frac{\lambda^2}{24}(2\Delta \mathbf{H} - \Delta \mathbf{H}^2).$$

Written in this form, it is clear that only one positive real root exists and that the upper bound is $\Delta\mathbf{H} < 1$, an intuitively obvious result. Show further that a general lower bound is $\Delta\mathbf{H} > 1/(3 + \lambda^2/12)$. This latter result tends to be most accurate when λ^2 is large, while the upper bound is best when λ^2 is small.

In the case of the suspension bridge cable of example 2.4 the increase in sag during the spinning process is about 2 m. Adjustments are made to the initial lengths of the wires to accommodate this.

References

1. Anderson, K. A., et al. 1965. *Proc. Instn. Civ. Engrs.*, 32:321–512.

2. Caughey, T. K. 1969. *Int. J. Non-Linear Mech.*, 4:61–75.

3. Caughey, T. K. 1970. *SIAM J. Appl. Math.*, 18:210–237.

4. Cole, J. D. 1968. *Perturbation methods in applied mathematics.* Waltham, Massachusetts: Blaisdell, ch. 2.

5. Horne, M. R., and Merchant, W. 1965. *The Stability of Frames.* Oxford: Pergamon, ch. 1.

6. Huedey, T. M., et al. 1973. *Nat. Res. Council, Canada*, Rep. No. LTR-St-656.

7. Inglis, Sir C. 1963. *Applied Mechanics for Engineers.* New York: Dover, ch. 3.

8. Irvine, H. M. 1975. *J. Eng. Mech. Div., Proc. ASCE*, 101:187–205.

9. Irvine, H. M. 1975. *J. Eng. Mech. Div., Proc. ASCE*, 101:725–737.

10. Irvine, H. M. 1980. *Q. J. Mech. Appl. Math.*, 33:227–234.

11. Jonatowski, J. J., and Birnstiel, C. 1970. *J. Struct. Div., Proc. ASCE*, 96:1143–1166.

12. Porter-Goff, R. F. D. 1974. *Proc. Instn. Civ. Engrs.*, 56:303–321.

13. Pugsley, A. G. 1952. *Q. J. Mech. Appl. Math*, 5:385–394.

14. Pugsley, A. G. 1968. *The Theory of Suspension Bridges.* 2nd ed. London: Edward Arnold.

15. Scalzi, J. B., and McGrath, W. K. 1971. *J. Struct. Div., Proc. ASCE*, 97:2837–2844 (see also discussion by Birdsall, B. 1972. 98:1883–1884.

16. Timoshenko, S. P. 1943. *J. Franklin Inst.*, 253:327–349.

17. Timoshenko, S. P., and Young, D. H. 1965. *Theory of Structures*. 2nd ed. New York: McGraw-Hill, ch. 11.

18. Uspensky, J. V. 1948. *Theory of Equations*. New York: McGraw-Hill, chs. 5–6.

19. Watson, G. N. 1966. *Theory of Bessel Functions*. 6th ed. Cambridge: Cambridge University Press, ch. 1.

3

Dynamics of a Suspended Cable

3.1 Introduction

Cable vibration problems are of considerable antiquity. Witness the Aeolian harp of the Greeks which, due to the strumming effect of the shed vortices, gave out musical sounds on exposure to the wind. Pythagoras and his disciples were keenly interested in string vibrations and understood at least qualitatively the relations between pitch of the note and the tension, length, and mass of the string that produced it.

The basic laws governing taut string vibrations were found experimentally by Mersenne and stated by him in 1636. (This subject was also of interest to Galileo who around the same time came up with an experimentally determined formula for the period of the pendulum; Huygens came along with the necessary mathematics a little later.) The idea that a given string has many modes of vibration and the associated concept of internal nodes were established by Noble and Pigott in 1676 [42]. Only later, starting with Taylor in 1713 and continuing with the work of D'Alembert, Euler, and Daniel Bernoulli during the first half of the eighteenth century, did its mathematical theory become firmly established. The crowning achievement was Bernoulli's controversial demonstration in 1755 that a general compound vibration could be broken down into independent constituent modes [24]. It is on this foundation that dynamics is based.

An important contribution to the theory of cable vibrations came in 1820 when Poisson published a paper that gave the general cartesian partial differential equations of motion of a cable element under the action of a general force system. (These provide a natural complement to the string equation first given by D'Alembert in 1750 [24].) Poisson used these equations to improve the solutions previously obtained for the vertical freehanging cable and the taut string [35].

Thus by 1820 correct solutions had been given for the linear vibrations of uniform cables, the geometries of which were the limiting forms of the catenary. But no results had been given for cables where the ratio of sag to span was not either zero or infinite.

In 1851 Rohrs, in a rambling paper in which some of the crucial results seem to have been derived by Stokes, obtained an approximate solution for the symmetric vertical vibrations of a uniform suspended cable where the ratio of sag to span, although appreciable, was small. He arrived at his solution using a form of Poisson's general equations, correct to first order, and in addition used another equation that he termed the equation of continuity of the chain. He assumed the chain to be inextensible, so this continuity equation related only to geometric compatibility [34].

In 1868 Routh gave an exact solution for the symmetric vertical vibrations (and associated longitudinal motion) of a heterogeneous cable that hung in a cycloid. Like Rohrs he assumed the cable was inextensible. He showed that his results reduced to Rohr's solution when the ratio of sag to span was small. Routh also obtained an exact solution for the antisymmetric modes [35].

At this point the subject seems to have been laid to rest until 1941 when Rannie and von Kármán independently derived results for both the symmetric and antisymmetric in-plane modes of an inextensible three-span cable [32]. In work done in 1945 Vincent extended this analysis to include the effects of cable elasticity in the symmetric modes [43]. However, he did not explore the nature of the solutions so obtained. These works, and several others besides, were prompted by the aerodynamic failure of the Tacoma Narrows suspension bridge in 1940. (Indeed Bleich et al. wrote a book on the mathematical theory of vibration in suspension bridges, the impetus for which was provided by the demise of this bridge [5].) In them the lack of emphasis on cable stretch is hardly surprising since, as has been amply demonstrated in the previous chapter, cables with relatively deep profiles will not be much affected by stretching, at least in the lower modes.

Pugsley's semiempirical theory for the natural frequencies of the first three in-plane modes was put forward in 1949. He demonstrated the applicability of the results by conducting experiments on cables in which the ratio of sag to span ranged from 1:10 up to about 1:4 [30].

By assuming again that the cable was inextensible, Saxon and Cahn made a major contribution to the theory of in-plane vibrations in 1953. They obtained solutions that effectively reduced to the previously known results for inextensible cables of small sag to span and for which asymptotic solutions gave extremely good results for large ratios of sag to span [36]. Goodey has also investigated deep-sag cables, but by different methods [15], and Smith and Thompson have shown how the analysis may be extended to include inclined cables [39].

But discrepancies are apparent. For example, for small ratios of sag to span the theories based on the imposition of inextensibility show that the first symmetric in-plane mode, primarily involving vertical motion, occurs at a frequency contained in the first nonzero root of

$$\tan\frac{\omega}{2} = \frac{\omega}{2},$$

namely,

$$\omega_1 = 2.86\pi.$$

Nevertheless it has long been known that the frequency of the first symmetric mode of the transverse vibration of a taut string is contained in the first root of

$$\cos\frac{\omega}{2} = 0,$$

or

$$\omega_1 = \pi.$$

This discrepancy, which amounts to almost 300 percent, cannot be resolved with analyses restricted by the assumption of inextensibility.

In recent years there have been several analytical studies [28, 38, 40, 18], all performed independently and with varying objectives in mind, that are capable of demonstrating that the inclusion of cable stretch resolves this dilemma. A review article by Luongo and Rega, [25] indicates that these approaches generally exhibit good agreement with more sophisticated numerical methods.

At the other end of the geometric spectrum, when the profile is very deep, Saxon and Cahn's results and Goodey's results are in conformity with experimental results except that, as will be discussed, they cannot pass to the limit of the vertical cable. However, by retaining more terms in analysis, it should be possible to incorporate the case of the vertical cable in their work. In fact, by including Hooke's law as well, it may be possible to span the whole range of cable profiles, although that has not yet been done.

This chapter will describe in some detail the linear theory of free vibrations of structurally important flat-sag cables and extend this to cover cables that dip substantially (which on that account are of somewhat less structural importance). A section is also devoted to the linearized dynamic response of flat-sag cables. The chapter closes with some consideration of geometrically nonlinear vibrations and of elastic-plastic stress waves in wires. The contrast between the earlier and later sections, in terms of a body of definitive results, arises because of the relative paucity of knowledge concerning the nonlinear theory of vibration in suspended cables where much analytical work remains to be done.

3.2 The Linear Theory of Free Vibrations of a Suspended Cable

If a uniform flat-sag suspended cable anchored on supports at the same level is given a slight disturbance, the subsequent equilibrium of an element requires that (see figure 3.1)

$$\frac{\partial}{\partial s}\left\{(T+\tau)\left(\frac{dx}{ds}+\frac{\partial u}{\partial s}\right)\right\} = m\frac{\partial^2 u}{\partial t^2},$$

$$\frac{\partial}{\partial s}\left\{(T+\tau)\left(\frac{dz}{ds}+\frac{\partial w}{\partial s}\right)\right\} = m\frac{\partial^2 w}{\partial t^2} - mg, \tag{3.1}$$

$$\frac{\partial}{\partial s}\left\{(T+\tau)\frac{\partial v}{\partial s}\right\} = m\frac{\partial^2 v}{\partial t^2},$$

where u and w are the longitudinal and vertical components of the in-plane motion, respectively, v is the out-of-plane, or swinging component, and τ is the additional tension generated. All are functions of both position and time.

These equations may be simplified for the problem at hand. Each equation is expanded, substitutions are made for the equations of static equilibrium, and terms of the second order are neglected. Most important, because the profile is shallow, the longitudinal component of the equations of motion is considered unimportant and is dropped. We shall return to this point from time to time in subsequent pages. Consequently the equations of motion are reduced to

$$H\frac{\partial^2 w}{\partial x^2} + h\frac{d^2 z}{dx^2} = m\frac{\partial^2 w}{\partial t^2}, \tag{3.2}$$

and

$$H\frac{\partial^2 v}{\partial x^2} = m\frac{\partial^2 v}{\partial t^2}, \tag{3.3}$$

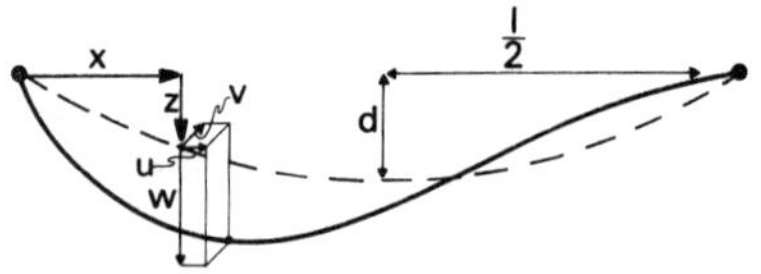

3.1 Definition diagram showing components of displacement in disturbed profile where in conformity with the definition for static response, h is the additional horizontal component of tension and is a function of time alone. To first order the cable equation is

$$\frac{h(ds/dx)^3}{EA} = \frac{\partial u}{\partial x} + \frac{dz}{dx}\frac{\partial w}{\partial x}, \tag{3.4}$$

which we integrate to the form

$$\frac{hL_e}{EA} = \frac{mg}{H}\int_0^l w\,dx. \tag{3.5}$$

Equations (3.3) and (3.5) constitute a linear homogeneous system in w. With (3.2), (3.3), and (3.5) the fundamental features of the linear theory of free vibrations may be explored.

For a start it will be noticed that the swinging motion has uncoupled from the in-plane motion because, to first order, no additional tension is generated by it (viewed in plan the cable is a taut string). This uncoupling is consistent, for instance, with a chain hanging across a driveway, in which case the only easily excited mode of vibration is its first pendulum mode. Therefore to first order, a disturbance that has no in-plane components will induce only out-of-plane motion, and vice-versa.

Under the restrictions placed here on profile geometry, the vertical component of motion is most apparent when the cable vibrates in an in-plane mode. The amplitude of the corresponding longitudinal modal component is always substantially less than the amplitude of the vertical motion. (For this reason (3.4) is by itself adequate to describe the longitudinal motion.) Consequently a symmetric in-plane mode is defined as one in which the vertical component of the mode is symmetric, and vice-versa.

The Out-of-Plane Motion

The swinging modes will be considered first because they are the easiest to analyze. By writing $v(x, t) = \tilde{v}(x)e^{i\omega t}$, where ω is the natural circular frequency of vibration (3.3) is reduced to

$$H\frac{d^2\tilde{v}}{dx^2} + m\omega^2\tilde{v} = 0. \tag{3.6}$$

The boundary conditions are $\tilde{v}(0) = \tilde{v}(l) = 0$, from which it is found that the natural frequencies and associated modes are

$$\begin{aligned} \omega_n &= \frac{n\pi}{l}\left(\frac{H}{m}\right)^{1/2}, \\ \tilde{v}_n &= A_n \sin\frac{n\pi x}{l}, \end{aligned} \tag{3.7}$$

where $n = 1, 2, 3, \ldots$ signify the first, second, third modes, respectively, and so on. The frequency of the first out-of-plane mode is the lowest of any given flat-sag suspended cable.

The In-Plane Motion

From equation (3.5) it is seen that $h = 0$ when $\int_0^l w\,dx = 0$. Modes of vibration that meet this condition and induce no overall additional tension may be called antisymmetric. (One has to be careful here because, when λ^2 is very large, $\int_0^l w\,dx \to 0$ for the symmetric modes. The difference is that then overall additional tension is generated.) All other modes induce overall additional tension and are symmetric. As defined previously, antisymmetric in-plane modes consist of antisymmetric vertical components and symmetric longitudinal components, while symmetric in-plane modes consist of symmetric vertical components and antisymmetric longitudinal components.

In the case of the *antisymmetric in-plane modes* (3.2) becomes

$$H\frac{d^2\tilde{w}}{dx^2} + m\omega^2\tilde{w} = 0, \tag{3.8}$$

where the substitution $w(x, t) = \tilde{w}(x)e^{i\omega t}$ has been made. The cable equation is reduced to a statement of geometric compatibility

$$\frac{d\tilde{u}}{dx} + \frac{dz}{dx}\frac{d\tilde{w}}{dx} = 0, \tag{3.9}$$

where the substitution $u(x, t) = \tilde{u}(x)e^{i\omega t}$ has also been made. Together with the boundary conditions $\tilde{w}(0) = \tilde{w}(l/2) = 0$, (3.8) and (3.9) are sufficient to obtain the natural frequencies and modal components of the antisymmetric in-plane modes.

It is easily shown then that

$$\omega_n = \frac{2n\pi}{l}\left(\frac{H}{m}\right)^{1/2}, \tag{3.10}$$

where $n = 1, 2, 3, \ldots$ signify natural frequencies of the first, second, third antisymmetric modes, and so on. The vertical modal components are given by

$$\tilde{w}_n = A_n \sin\left(\frac{2n\pi x}{l}\right), \quad n = 1, 2, 3, \ldots. \tag{3.11}$$

The longitudinal components are conveniently found from (3.9), which was why we expressed it in the form shown. These components are symmetric because dz/dx is zero at midspan and the product dz/dx

$d\tilde{w}/dx$ changes sign at midspan. After substitution, integration, and rearrangement

$$\tilde{u}_n = -\frac{1}{2}\left(\frac{mgl}{H}\right)A_n\left\{\left(1 - \frac{2x}{l}\right)\sin\left(\frac{2n\pi x}{l}\right) + \frac{1 - \cos(2n\pi x/l)}{n\pi}\right\}, \quad (3.12)$$

where as before A_n is the amplitude of the vertical component of the nth antisymmetric in-plane mode.

It is clear that the amplitudes of the longitudinal components become very small as the cable becomes flatter (as $mgl/H \to 0$). These components have some peculiar properties (see figure 3.2) because the maximum displacement occurs at the quarter-span points and not at midspan in the case of the first modal component (the displacement is a local minimum at midspan). Also both the slope and displacement are zero at midspan for the second component. This pattern repeats itself for the higher modes. In fact $\tilde{u}_{n,\max}/A_n = (mgl/H)\{1 - (1/2 - 1/\pi)/n\}/2$, so that the peak longitudinal displacement is no more than half the vertical component as the limit applicability is reached in the flat-sag assumption. For flatter profiles it is a good deal less.

Luongo and Rega's comparative numerical study of reference [18] indicates that the essential features of the antisymmetric in-plane modes have been preserved with one exception. Even though the overall, or mean, additional tension is zero, at the scale of the cable element it is nonzero. Luongo and Rega indicate that an antisymmetric distribution

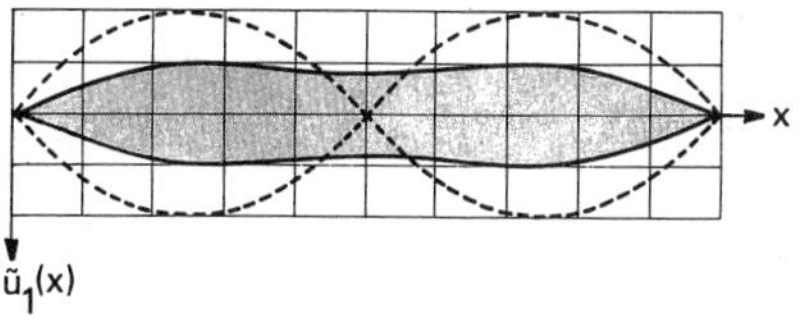

(a) First symmetric longitudinal modal component

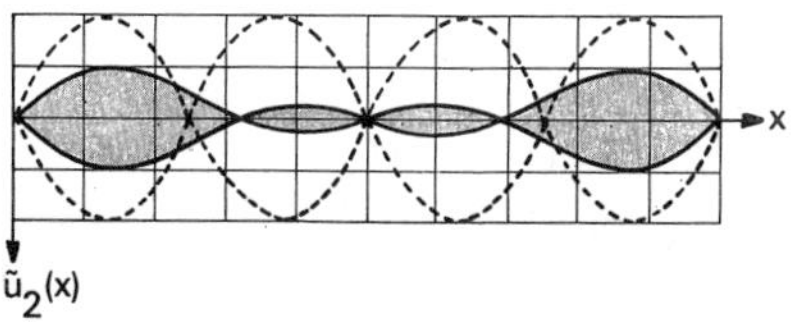

(b) Second symmetric longitudinal modal component

3.2 Longitudinal components and associated vertical components (shown dotted) of the first two antisymmetric in-plane modes. The vertical scale is arbitrary.

of additional tension $\tilde{h}(x)$ is possible, so that $\int_0^l \tilde{h}(x)\,dx = 0$. In essence the task of demonstrating this involves use of the longitudinal equation of motion, which reads

$$\frac{\partial h}{\partial x} = -H\frac{\partial^2 u}{\partial x^2} + m\frac{\partial^2 u}{\partial t^2},$$

or

$$\frac{d\tilde{h}}{dx} = -\left(H\frac{d^2\tilde{u}}{dx^2} + m\omega^2\tilde{u}\right), \tag{3.13}$$

when time is removed from it. Substitution, integration, and the meeting of the requirement of antisymmetry in $\tilde{h}(x)$ yields

$$\frac{\tilde{h}_n(x)}{H} = -\frac{mgl}{H}\frac{A_n}{l}\left\{n\pi\left(1 - \frac{2x}{l}\right) + \sin\left(\frac{2n\pi x}{l}\right)\right\}. \tag{3.14}$$

The sine function supplies a modulation on a basic sawtooth profile.

The effect is less pronounced for flatter profiles but seems to be more marked for higher modes. However, this latter case is misleading, for if we use results for the symmetric in-plane modes and compare peak additional tensions on the basis of the same amplitude of vibration in both the antisymmetric and symmetric in-plane modes, the result is a ratio of magnitude $(mgl/H)^2 n\pi\{1 - \sec(\omega/2)\}/\omega^2$. Thus, on comparing the first antisymmetric mode and first symmetric mode (see equations 3.16 and 3.17) for a cable of deep profile, the ratio of additional tensions is about 0.22, which indicates that this source of additional tension is not of profound importance.

It is interesting though how a little juggling of the equations of motion can provide useful leads on such items. There is no doubt that a more refined analysis of the antisymmetric in-plane modes could be undertaken, but most features of interest seem to have been highlighted with this rudimentary approach.

In the case of the *symmetric in-plane modes*, additional tension is induced (and is substantially constant with span [25]), so that (3.2) becomes

$$H\frac{d^2\tilde{w}}{dx^2} + m\omega^2\tilde{w} = \frac{mg}{H}\tilde{h}. \tag{3.15}$$

A solution that satisfies zero boundary conditions is (here we revert to suitable dimensionless forms)

$$\tilde{\mathbf{w}} = \frac{\tilde{\mathbf{h}}}{\omega^2}\left(1 - \tan\frac{\omega}{2}\sin\omega\mathbf{x} - \cos\omega\mathbf{x}\right), \tag{3.16}$$

where $\tilde{\mathbf{w}} = w/(mgl^2/H)$, $\mathbf{x} = x/l$, $\tilde{\mathbf{h}} = \tilde{h}/H$, and $\boldsymbol{\omega} = \omega l/(H/m)^{1/2}$, and its value specifies the particular (symmetric) vertical modal component. Use is now made of (3.5) to eliminate $\tilde{\mathbf{h}}$ and obtain the following transcendental equation from which the natural frequencies of the symmetric in-plane modes may be found:

$$\tan\frac{\boldsymbol{\omega}}{2} = \frac{\boldsymbol{\omega}}{2} - \frac{4}{\lambda^2}\left(\frac{\boldsymbol{\omega}}{2}\right)^3, \tag{3.17}$$

where as before $\lambda^2 = (mgl/H)^2 l/(HL_e/EA)$. (Support flexibility, often a feature to be contended with in practice, may be allowed for in precisely the same way as in section 2.3.) This equation is of fundamental importance in the theory of cable vibrations. In table 3.1 the frequencies of the first eight symmetric in-plane modes are tabulated for a wide range of values of the parameter involving cable elasticity and geometry.

It is perhaps not redundant to note that just this one independent parameter is present. Many purely numerical studies have faltered on this point and isolated two and sometimes three parameters upon which parameter studies have been undertaken. An all-out numerical effort would then be required to discover that these quasi-independent parameters may be replaced by just one independent parameter. In this instance

Table 3.1 Natural frequencies of the first eight symmetric in-plane modes as a function of λ^2 [20]

λ^2	ω_1/π	ω_2/π	ω_3/π	ω_4/π	ω_5/π	ω_6/π	ω_7/π	ω_8/π
∞	2.86	4.92	6.94	8.95	10.96	12.97	14.97	16.98
$256\pi^2$	2.86	4.91	6.93	8.93	10.93	12.91	14.81	16.0$\dot{0}$
$196\pi^2$	2.85	4.91	6.92	8.92	10.91	12.81	14.0$\dot{0}$	15.15
$144\pi^2$	2.85	4.90	6.91	8.90	10.81	12.0$\dot{0}$	13.15	15.05
$100\pi^2$	2.85	4.89	6.89	8.80	10.0$\dot{0}$	11.15	13.04	15.02
$64\pi^2$	2.84	4.87	6.79	8.0$\dot{0}$	9.14	11.04	13.02	15.01
$36\pi^2$	2.82	4.78	6.0$\dot{0}$	7.14	9.04	11.02	13.01	15.01
$16\pi^2$	2.74	4.0$\dot{0}$	5.12	7.03	9.01	11.01	13.00	15.00
100	2.60	3.48	5.05	7.01	9.01	—	—	—
80	2.48	3.31	5.04	7.01	9.01	—	—	—
60	2.29	3.18	5.03	7.01	—	—	—	—
$4\pi^2$	2.0$\dot{0}$	3.09	5.02	7.01	—	—	—	—
20	1.61	3.04	5.01	7.00	—	—	—	—
10	1.35	3.02	5.00	—	—	—	—	—
8	1.28	3.01	—	—	—	—	—	—
6	1.22	—	—	—	—	—	—	—
4	1.15	—	—	—	—	—	—	—
2	1.08	—	—	—	—	—	—	—
1	1.04	—	—	—	—	—	—	—
0	1.0$\dot{0}$	3.0$\dot{0}$	5.0$\dot{0}$	7.0$\dot{0}$	9.0$\dot{0}$	11.0$\dot{0}$	13.0$\dot{0}$	15.0$\dot{0}$

the analytical approach is of considerable help in determining physical characteristics.

To illustrate the following discussion, reference should be made to figures 3.3 and 3.4. When λ^2 is very large the cable may be assumed

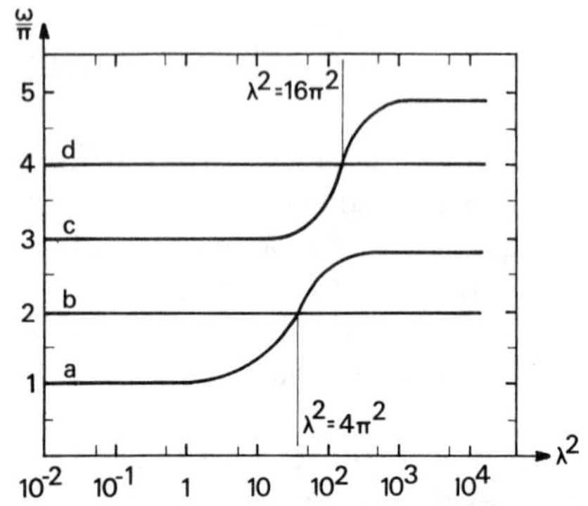

3.3 General dimensionless curves for the first four natural frequencies of a flat-sag suspended cable: (a) first symmetric in-plane mode, (b) first antisymmetric in-plane mode, (c) second symmetric in-plane mode, (d) second antisymmetric in-plane mode.

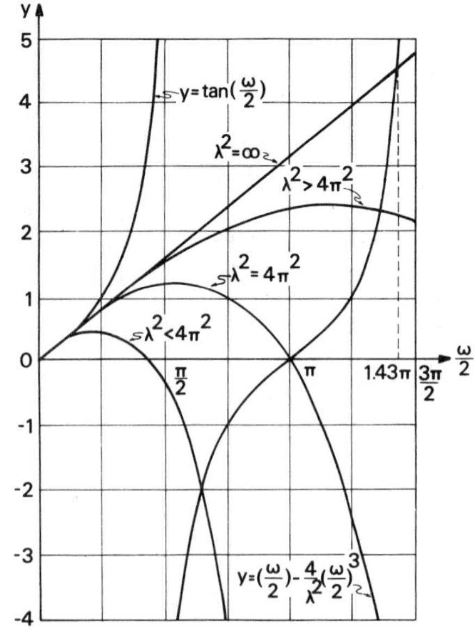

3.4 Graphical solution for first nonzero root of equation (3.17)

inextensible, and (3.17) is reduced to

$$\tan \frac{\omega}{2} = \frac{\omega}{2}. \tag{3.18}$$

This is the transcendental equation first given by Rohrs in 1851. The equation occurs in other problems in structural mechanics, including the flexural and torsional buckling of struts under certain boundary conditions. The roots are given quite accurately by $\omega_n = (2n + 1)\pi\{1 - 4/((2n + 1)^2\pi^2)\}$, $n = 1, 2, 3, \ldots$.

The other limiting value of λ^2 occurs as the cable profile approaches that of a taut string. The roots of the transcendental equation correspond to the frequencies of the symmetric modes of the taut string, namely, $\omega_n = (2n - 1)\pi$, $n = 1, 2, 3, \ldots$. Thus the condition of inextensibility causes a shift of almost 2π in the roots obtained from the frequency equation governing the symmetric modes of the taut string, since the first nonzero root of (3.17) lies between $\pi/2$ and 1.43π, the second root lies between $3\pi/2$ and 2.46π, and so on: the actual values depend on λ^2.

In the case of the first modes three important ranges may be established:

1. If $\lambda^2 < 4\pi^2$, the frequency of the first symmetric mode is less than the frequency of the first antisymmetric mode. The vertical component of the first symmetric mode has no internal nodes (see figure 3.5).

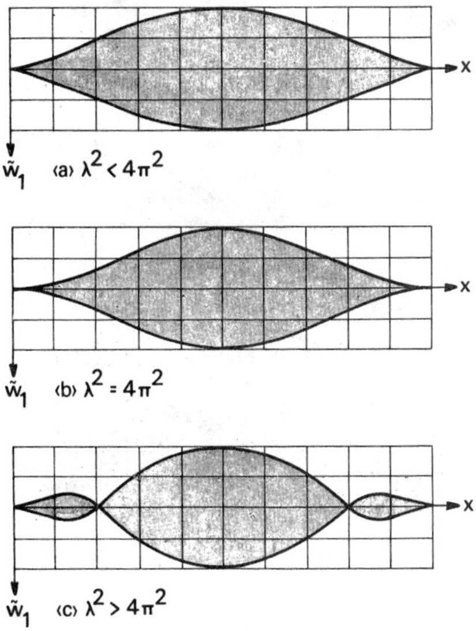

3.5 Possible forms for the vertical component of the first symmetric in-plane mode

2. If $\lambda^2 = 4\pi^2$, the frequencies of these modes are equal. The vertical modal component is tangential to the profile at the supports. This value of λ^2 gives the first modal crossover. Behavior changes from that akin to a taut string to that in keeping with catenary action. In an analogous physical system—that of the buckling of a pin-ended strut with an elastic restraint at midheight—the mode of buckling changes from a C-curve to an S-curve when the spring stiffness bears a certain relation to the flexural stiffness.

3. If $\lambda^2 > 4\pi^2$, the frequency of the symmetric mode is now greater, and two internal nodes appear.

The first modal crossover was isolated in a series of qualitative experiments run by exciting a fine copper wire that spanned some 2 m with an alternating electromagnetic force. The long exposure photographs in figure 3.6 are the result. When λ^2 was close to $4\pi^2$ the cable would vibrate in the symmetric mode and then jump to the antisymmetric mode [18]. Quantitative results available from Ramberg and Griffin and based on data supplied by Richardson are depicted in figure 3.7. [31]

It may also be noted that, if $4\pi^2 < \lambda^2 < 16\pi^2$, both the first and second symmetric modes have two internal nodes. When $\lambda^2 = 16\pi^2$, the frequency of the second symmetric mode is equal to the frequency of the

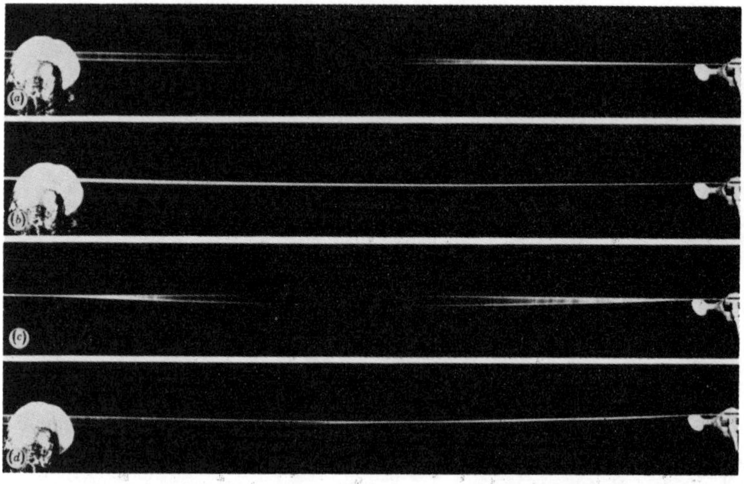

3.6 Long exposure photographs of the first in-plane modes of vibration of a flat-sag cable: (a) $\lambda^2 \ll 4\pi^2$, first symmetric mode of a taut string; (b) $\lambda^2 \simeq 4\pi^2$, first symmetric in-plane mode at the first crossover point; (c) $\lambda^2 \simeq 4\pi^2$, first antisymmetric in-plane mode at the first crossover point; (d) $\lambda^2 > 4\pi^2$, first symmetric in-plane mode. Midspan is at the left-hand edge.

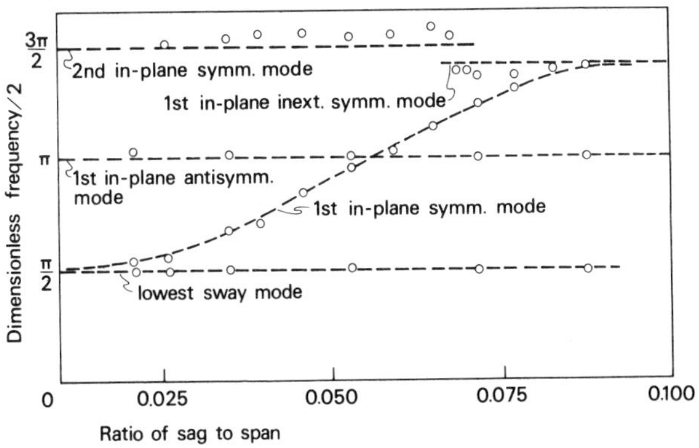

3.7 Ramberg and Griffin's comparison of the present theory with the experimental results of Richardson for a particular model cable (from Ramberg and Griffin [31], courtesy ASCE)

second antisymmetric mode. This value of λ^2 gives the second crossover, and so on.

The associated longitudinal modal components may be found with little extra difficulty. They read

$$\tilde{\mathbf{u}} = \frac{\tilde{\mathbf{h}}}{\omega^2}\left[\frac{\omega^2}{\lambda^2}\frac{L_x}{L_e} - \frac{1}{2}(1 - 2\mathbf{x})\left\{1 - \tan\frac{\omega}{2}\sin\omega\mathbf{x} - \cos\omega\mathbf{x}\right\}\right.$$
$$\left. -\frac{1}{\omega}\left\{\omega\mathbf{x} - \tan\frac{\omega}{2}(1 - \cos\omega\mathbf{x}) - \sin\omega\mathbf{x}\right\}\right], \tag{3.19}$$

where $\tilde{\mathbf{u}} = \tilde{u}/((mgl/H)(mgl^2/H))$ and $L_x = l[\mathbf{x} + \frac{3}{8}(mgl/H)^2\{\mathbf{x} - 2\mathbf{x}^2 + 4\mathbf{x}^3/3\}]$. These longitudinal components are antisymmetric, since (3.19) and (3.4) show that the longitudinal displacement and slope are always zero and nonzero at midspan, respectively.

Example 3.1 Free Vibrations of an Inclined Cable

The purpose of this example is to show that we may readily extend the present investigation to cover the case of an inclined cable hanging under self-weight. In example 2.1 we obtained an approximate solution for the static profile of a guywire which lay close to its chord. However, the equation describing that profile is not a convenient form from which to start the dynamic analysis because, with axes directed horizontally and vertically, horizontal inertia assumes increasing importance as the chord inclination steepens. The difficulty may be overcome by transforming axes, so that x_* measures distance along the chord from A (see figure 3.8) and z_* measures distance to the profile from the chord and perpendicular to it. A similar approach was used by Dean in his investigation of guywire vibrations [13]. Now $x_* = x\sec\theta + z\sin\theta$, $z_* = z\cos\theta$, $l_* = l\sec\theta$, and $H_* = H\sec\theta$, so that

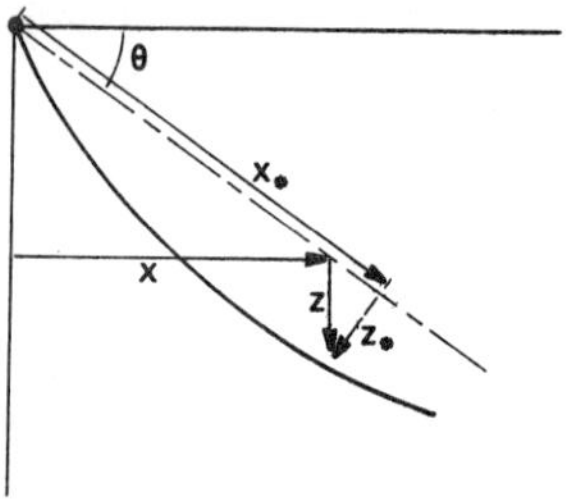

3.8 Transformation of axes for an inclined cable

$$\mathbf{z}_* = \frac{1}{2}\mathbf{x}_*(1 - \mathbf{x}_*)\left\{1 - \frac{\varepsilon_*}{3}(1 - 2\mathbf{x}_*)\right\},$$

where $\mathbf{z}_* = z_*/(mgl_*^2 \cos\theta/H_*)$, $\mathbf{x}_* = x_*/l_*$ and $\varepsilon_* = mgl_* \sin\theta/H_*(=\varepsilon)$. We shall suppose that ε_* is sufficiently small for a parabola to portray the profile accurately. This is often the case with guyed masts where the tension does not vary much over the length of a guy.

With axes chosen in this way, the linearized three-dimensional free vibrations are relatively easy to analyze. To first order, the out-of-plane motion is uncoupled from the in-plane motion. The chordwise components of the in-plane modes are judged of secondary importance to those in the perpendicular direction.

Therefore with minor changes the results of the previous pages may be applied directly to this more general problem [21]. In the case of the natural frequencies, those of the out-of-plane modes are

$$\omega_{*n} = n\pi, \quad n = 1, 2, 3, \ldots,$$

and those of the antisymmetric in-plane modes are

$$\omega_{*n} = 2n\pi, \quad n = 1, 2, 3, \ldots,$$

while those of the symmetric in-plane modes are contained in the nonzero roots of the transcendental equation

$$\tan\frac{\omega_*}{2} = \frac{\omega_*}{2} - \frac{4}{\lambda_*^2}\left(\frac{\omega_*}{2}\right)^3,$$

where $\omega_* = \omega_* l_*/(H_*/m)^{1/2}$ and $\lambda_*^2 = (mgl_* \cos\theta/H_*)^2 l_*/(HL_{e*}/EA)$, in which $L_{e*} = l_*\{1 + (mgl_* \cos\theta/H_*)^2/8\}$.

These equations are of rather general applicability and contain as a special case results of the earlier work in which the supports were taken to be level ($\theta = 0$). At the other extreme, when $\theta = 90°$, $\lambda_*^2 = 0$, and results for a vertical taut string are obtained. In this situation it is clearly necessary for the cable tension to be much greater than the cable weight; otherwise the cable would tend to be supported at one end only and give rise to the other classical eigenvalue problem in cable theory. Thus, as the inclination increases, ever tighter restrictions have to be placed on ε_*, but angles of inclination greater than 60° are not common.

Perhaps the chief advantage is that no extra independent parameters are required. A parameter identical to λ_*^2 was used by Hartmann and Davenport in their studies of the dynamic response of guywires [16]. The whole subject of guyed masts will be considered at more length in chapter 4.

Example 3.2 Cable Vibrations via Difference Equations

Lagrange, as an example in *Mécanique analytique*, considered the vibration of a taut string from the point of view of its analogy with a string of beads—a problem that was also discussed at length by Rayleigh [33]. It is of interest to consider the extension of this to the free vibrations of a string of heavy beads. The example focuses on some aspects of the solution for one discretization, which is due to Segedin [37]. Other discretizations yield essentially similar results.

The profile shown in figure 3.9 consists of N equal links and the mass of each link is concentrated at the $(N - 1)$ internal nodes. Because there are $(N - 1)$ moving masses, there are $(N - 1)$ in-plane modes of vibration and $(N - 1)$ out-of-plane modes.

Static equilibrium is expressed by

$$\mathbf{z}_{n+1} - 2\mathbf{z}_n + \mathbf{z}_{n-1} = -\frac{1}{N^2},$$

where $\mathbf{z} = z/(mgl^2/H)$, so that each concentrated weight is mgl/N. The boundary conditions are $\mathbf{z}_0 = \mathbf{z}_N = 0$, and the solution is

$$\mathbf{z}_n = \frac{n(N-n)}{2N^2}, \quad n = 0, 1, 2, \ldots, N.$$

For small amplitude vibrations the out-of-plane motion may again be considered separately from the in-plane motion. The out-of-plane motion, however, need not be considered explicitly, since it is of precisely the same form as that for the taut string, and therefore the associated discrete problem is the classical one. For given N the frequency of the first out-of-plane mode is the lowest for that discrete problem.

The difference equations governing the in-plane vibrations are

$$\mathbf{w}_{n+1} - 2\left(1 - \frac{\boldsymbol{\omega}^2}{2N^2}\right)\mathbf{w}_n - \mathbf{w}_{n-1} = \frac{\mathbf{h}}{N^2},$$

where $\mathbf{w} = w/(mgl^2/H)$, $\boldsymbol{\omega} = \omega l/(H/m)^{1/2}$, and $\mathbf{h} = h/H$. The boundary conditions are $\mathbf{w}_0 = \mathbf{w}_N = 0$, and the relationship between the additional tension and the additional deflection takes the form (several steps are involved here)

$$\mathbf{h} = \frac{\lambda^2}{N}\sum_{n=1}^{N} \mathbf{w}_n,$$

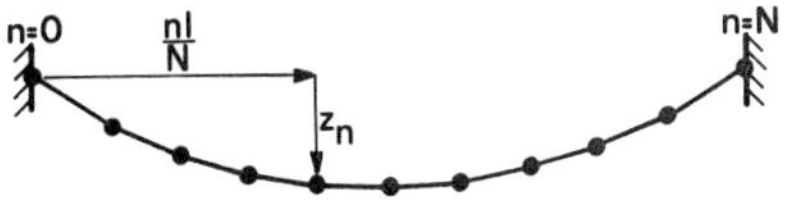

3.9 A discrete representation of a flat-sag suspended cable

where as before $\lambda^2 = (mgl/H)^2 l/(HL_e/EA)$ and $L_e = l\{1 + (1 - 1/N^2)(mgl/H)^2/8\}$. This closure condition for the symmetric modes is based on the trapezoidal rule and is clearly analogous to the result for the continuous formulation.

For the antisymmetric in-plane modes no additional tension is generated, and the requirement of antisymmetry is expressed by $w_n = -w_{N-n}$, $n = 0, 1, 2, \ldots, N$. We make the substitution $\cos\theta = 1 - \omega^2/2N^2$, so that the solution of the difference equation leads to antisymmetric in-plane modes of the form $\mathbf{w}_n = A \sin n\theta$, provided that $\sin N\theta = 0$. This last equation is the frequency equation that has roots that allow values of $\cos\theta$, and consequently the natural frequencies, to be found. The requisite roots are $\theta_i = 2i\pi/N$, and the ith antisymmetric in-plane mode is then

$$\mathbf{w}_{n,i} = A_i \sin\left(\frac{2in\pi}{N}\right),$$

with associated natural frequency

$$\omega_i = 2N \sin\left(\frac{i\pi}{N}\right),$$

where i ranges from 1 to $(N - 1)/2$, when N is odd, and from 1 to $(N - 2)/2$, when N is even.

For the symmetric in-plane modes the vertical modal components are given by

$$\mathbf{w}_n = \frac{\mathbf{h}}{4N^2 \sin^2(\theta/2)}\left[1 - \frac{\cos\{(N - 2n)\theta/2\}}{\cos(N\theta/2)}\right],$$

and the frequency equation is

$$\left\{1 - \frac{\tan(N\theta/2)}{N\tan(\theta/2)}\right\} = \frac{4N^2 \sin^2(\theta/2)}{\lambda^2}.$$

Solving this equation for θ, and then obtaining the natural frequencies through the cosine substitution, works well for nearly all the modes. It turns out that the highest mode may not be captured if λ^2 is below a certain value. However, if N is moderate to large, this restriction does not arise, and in any event the higher modes are bound to be suspect on purely physical grounds, quite aside from any mathematical considerations.

When λ^2 is small, the frequencies are

$$\omega_i = 2N \sin\left\{\frac{(2i - 1)\pi}{2N}\right\},$$

as expected, while, when λ^2 is large, the transcendental equation reduces to

$$\tan\left(\frac{N\theta}{2}\right) = N \tan\left(\frac{\theta}{2}\right).$$

Crossovers occur when

$$\lambda^2 = 4N^2 \sin^2\left(\frac{i\pi}{N}\right).$$

As we pass to the limit where N is large, the results of the continuous theory are reproduced. Table 3.2 gives a few results for the frequency of the first symmetric in-plane mode for differing N and for a range of λ^2. As might be expected, discrepancies between the discrete theory and the continuous theory are rather slight, except for cables of appreciable sag modeled by four elements where obviously agreement is poor.

Example 3.3 Energy Partition in the Vibration System

This example arises as a natural extension of the work in section 2.5 [22]. Consider a free vibration of the form

$$\mathbf{w}(\mathbf{x}, \mathbf{t}) = \frac{\mathbf{h}}{\omega^2}\left(1 - \tan\frac{\omega}{2}\sin\omega\mathbf{x} - \cos\omega\mathbf{x}\right)e^{i(\omega\mathbf{t} + \phi)},$$

$$\mathbf{h}(\mathbf{t}) = \mathbf{h}e^{i(\omega\mathbf{t} + \phi)},$$

where ϕ is a phase angle. The potential energy is

$$\mathbf{V} = \frac{1}{2}\int_0^1 \left(\frac{\partial\mathbf{w}}{\partial\mathbf{x}}\right)^2 d\mathbf{x} + \frac{1}{2}\mathbf{h}\int_0^1 \mathbf{w}\, d\mathbf{x},$$

while the kinetic energy is

$$\mathbf{T} = \frac{1}{2}\int_0^1 \left(\frac{\partial\mathbf{w}}{\partial\mathbf{t}}\right)^2 d\mathbf{x}.$$

Making the necessary substitutions prior to integration leads to temporal peak values of the form

$$\mathbf{V}_e = \frac{\mathbf{h}^2}{16}\frac{8}{\lambda^2}$$

for the strain energy of extension,

$$\mathbf{V}_g = \frac{\mathbf{h}^2}{16}\left\{\frac{4}{\lambda^2} + \left(\frac{\tan\omega/2}{\omega/2}\right)^2\right\}$$

for the gravitational potential energy, and

Table 3.2 Natural frequency of the first symmetric in-plane mode of a discrete system [37]

λ^2 \ N	4	10	20	∞
0	3.06	3.13	3.14	π
0.1	3.07	3.14	3.15	3.15
1.0	3.18	3.25	3.26	3.27
10	4.08	4.20	4.22	4.23
100	7.19	8.05	8.13	8.16
1,000	7.30	8.68	8.88	8.95
∞	7.30	8.72	8.92	8.99

$$\mathbf{T} = (= \mathbf{V}_e + \mathbf{V}_g) = \frac{\mathbf{h}^2}{16}\left\{\frac{12}{\lambda^2} + \left(\frac{\tan \omega/2}{\omega/2}\right)^2\right\}$$

for the kinetic energy. The ratio

$$\frac{\mathbf{V}_e}{\mathbf{V}} = \frac{2/3}{\{1 + \lambda^2(\tan(\omega/2)/(\omega/2))^2/12\}},$$

has its largest value of 2/3 when $\lambda = 2j\pi, j = 1, 2, 3$, when the modal crossovers occur. Thus in general, of the total potential energy stored by a vibrating suspended cable, two-thirds at most can go into strain energy of extension. Numerical work on the closely related problem of suspension bridge vibration bears out this conclusion [1].

This energy ratio is of interest in another context because it is identical to the expression for the modal participation factor for additional tension.

Extension to Cables of Deep Profile

By an ingenious approximate method of solution Saxon and Cahn were able to analyze the in-plane modes of vibration of deep-sag cables. Since their analysis is somewhat involved, we shall be content to list the asymptotic solutions for the natural frequencies and compare them with known limits and existing experimental results. Details may be pursued in the original article [36].

First for a catenary of length L hanging between two level supports the angle of inclination to the horizontal at those supports is given by

$$\tan \psi_0 = \frac{4d/L}{\{1 - (2d/L)^2\}}, \tag{3.20}$$

where d is the sag. The span is

$$l = \frac{L \sinh^{-1} (\tan \psi_0)}{\tan \psi_0}. \tag{3.21}$$

Defining a new dimensionless frequency according to the equation

$$\omega_*^2 = \frac{\omega^2 L}{(2g \tan \psi_0)} \tag{3.22}$$

allows (after much working) the following frequency equations to be written:

1. for the antisymmetric in-plane modes, where by definition the modal component perpendicular to the cable is odd about midspan and the tangential component is even,

$$\tan\left(\omega_* A + \frac{B}{\omega_*}\right) = -\frac{C}{\omega_*}, \tag{3.23}$$

2. for the symmetric in-plane modes, in which the perpendicular modal component is even and the tangential component is odd,

$$\tan\left(\omega_* A + \frac{B}{\omega_*}\right) = \frac{\omega_*}{D}, \tag{3.24}$$

where $A(\psi_0) = \int_0^{\psi_0} \sec^{3/2} \psi d\psi$, $B(\psi_0) = 11/8 \int_0^{\psi_0} (1 + (1/44) \tan^2 \psi)$ $\cos^{3/2} \psi d\psi$, $C(\psi_0) = \psi_0 \cos^{5/2} \psi_0/(\cos \psi_0 + \psi_0 \sin \psi_0)$, and $D(\psi_0) = \cos^{5/2} \psi_0/\sin \psi_0$. (The equations in this form emphasize the similarity with our previous results. When ψ_0 is small, (3.23) and (3.24) become $\tan \omega/2 \simeq 0$ and $\tan \omega/2 \simeq \omega/2$ because $\omega_* \psi_0 \to \omega/2$. Goodey, using a different method of analysis based on equations originally presented by Routh, confirms these results. He also shows that A and B may be expressed in terms of trigonometric functions and elliptic integrals of the first and second kinds [15].)

Both solutions (3.23) and (3.24) depend on just one geometric parameter, a point that Pugsley had noted from his work [30]. There is no parameter involving elasticity because the cable was taken to be inextensible at the outset.

Each of these equations is transcendental in ω_*, and, if ω_* is sufficiently large for terms of order $1/\omega_*^2$ to be ignored, there is obtained

$$\omega_{*n} \simeq \frac{n\pi}{A}\left\{1 - \frac{A(B + C)}{(n\pi)^2}\right\}, \quad n = 1, 2, 3, \ldots, \tag{3.25}$$

for the frequencies of the antisymmetric modes, and

$$\omega_{*n} \simeq \frac{(n + 1/2)\pi}{A}\left\{1 - \frac{A(B + D)}{(n + 1/2)^2\pi^2}\right\}, \quad n = 1, 2, 3, \ldots, \tag{3.26}$$

for the symmetric modes. As far as this approximation goes, these frequencies are all distinct, and the lowest is always that of the first antisymmetric in-plane mode.

Table 3.3 contains results for the first three antisymmetric modes and the first two symmetric modes. Values in parentheses in the upper rows refer to solutions obtained from the flat-sag theory (see equations 3.10 and 3.18, with suitable amendments to notation). Agreement is excellent when ψ_0 is small, which is not surprising as equations (3.25) and (3.26) reduce to $\omega_{*n}\psi_0 \simeq n\pi$ and $\omega_{*n}\psi_0 \simeq (n + 1/2)\pi\{1 - 1/((n + 1/2)^2\pi^2)\}$, in keeping with that earlier theory. Indeed up to a sag to span of 1:4 agreement is still good. However, notwithstanding this, the upper two rows for the symmetric modes are suspect on account of the neglect of elasticity.

The experimental results and the theoretical curves of figure 3.10 bear a close resemblance one to the other and offer irrefutable evidence of the

Table 3.3 Frequencies of the first five in-plane modes of a deep-sag cable [36]

ψ_0	d/L	Antisymmetric			Symmetric	
		ω_{*1}/π	ω_{*2}/π	ω_{*3}/π	ω_{*1}/π	ω_{*2}/π
5°	0.022	11.42 (11.36)	22.86 (22.73)	34.31 (34.09)	16.38 (16.25)	28.13 (28.0)
10°	0.044	5.64 (5.68)	11.35 (11.36)	17.04 (17.04)	8.14 (8.12)	13.98 (13.97)
20°	0.088	2.70 (2.76)	5.52 (5.54)	8.30 (8.31)	3.96 (4.00)	6.82 (6.85)
30°	0.13	1.68 (1.78)	3.50 (3.56)	5.29 (5.34)	2.52 (2.55)	4.36 (4.39)
40°	0.18	1.13 (1.26)	2.44 (2.52)	3.71 (3.79)	1.76 (1.81)	3.06 (3.11)
50°	0.23	0.79 (0.95)	1.76 (1.88)	2.70 (2.82)	1.27 (1.35)	2.23 (2.32)
60°	0.29	0.54	1.27	1.95	0.91	1.61
70°	0.35	0.34	0.87	1.35	0.61	1.11
75°	0.38	0.25	0.68	1.07	0.47	0.88
80°	0.42	0.15	0.50	0.80	0.34	0.65
82°	0.44	0.12	0.42	0.68	0.28	0.55
84°	0.45	0.075	0.34	0.56	0.22	0.45
86°	0.47	0.031	0.25	0.43	0.15	0.34
88°	0.48	0.021	0.15	0.27	0.074	0.21

Note: Values in parentheses were obtained with the flat-sag theory.

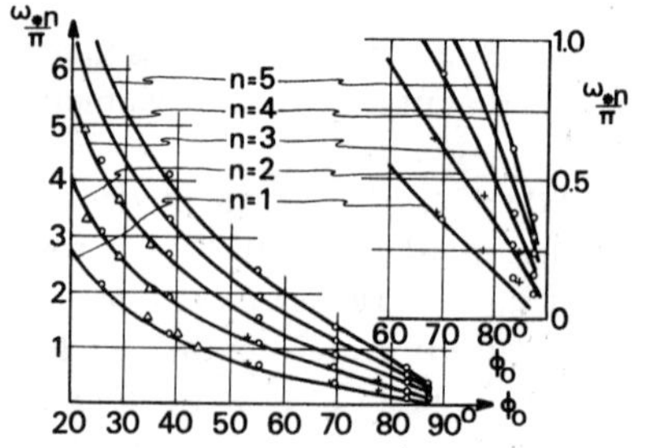

3.10 Agreement between Saxon and Cahn's theory and the experimental results of Rudnick and Leonard (0), Pugsley (Δ) and Saxon and Cahn (+) for the first five in-plane modes of a deep-sag suspended cable [36]. Note that n even signifies the symmetric modes while n odd signifies the antisymmetric modes.

theory's success. The experimental results were gleaned from several sources [36].

As the sag becomes particularly pronounced, the theoretical results are likely to be in error, because we are at this point severely stressing the assumption of large ω_* made in constructing the solution. When, for example, $\psi_0 = 88°$, all the frequencies are shown to be quite distinct, but this cannot be the case when we pass to the limit of the vertical cable. There the first two frequencies provided by classical theory are $\omega_{1,2} = 1.20$ and $2.76(2g/L)^{1/2}$ [44] and should correspond to the first two symmetric modes as defined by Saxon and Cahn's work (profiles for the modal component perpendicular to the cable are even about midspan). The fact that the bottom entries of the first two symmetric modes turn out to be 1.23 and $3.57(2g/L)^{1/2}$ might add weight to this argument, but this is illusory.

Goodey makes the interesting point that, when the cable is vertical, there can be no distinction between pairs of frequencies of the symmetric and antisymmetric modes. Thus when a vertical cable has its lower end constrained to lie vertically below the point of support (as would occur with an antisymmetric mode according to Saxon and Cahn's definition), the natural frequencies are still contained in the nonzero roots of the Bessel function of first kind and zero order—they are the same as for the cable with the lower end free. The modes differ only in a small region near that lower end. The physical reason for this is related to the absence of cable tension there. This trend is not in evidence from Saxon and Cahn's work nor should we expect it, given the variables used. To demonstrate this conclusively, it may be best to consider their problem as a perturbation on a vertical cable.

The particular point that this latter discussion brings out is that modal crossovers occur when the cable is vertical in somewhat the same way as happens when it is nearly horizontal. This adds in an unexpected way a degree of completeness to the theory.

3.3 The Linearized Dynamic Response of a Flat-Sag Suspended Cable

We take as the equation of motion

$$\frac{\partial^2 \mathbf{w}}{\partial \mathbf{x}^2} - \frac{\partial^2 \mathbf{w}}{\partial \mathbf{t}^2} = \mathbf{h}(\mathbf{t}) + \mathbf{p}(\mathbf{x}, \mathbf{t}), \tag{3.27}$$

where

$$\mathbf{h}(\mathbf{t}) = \lambda^2 \int_0^1 \mathbf{w}(\mathbf{x}, \mathbf{t})\, d\mathbf{x}, \tag{3.28}$$

and we intend to construct a general solution for **w** and **h** by superposing modal responses. The new dimensionless terms introduced are $\mathbf{t} = t(H/m)^{1/2}/l$ and $\mathbf{p} = p/mg$. Boundary conditions are that the additional displacements are zero at each end of the cable, and initial conditions may if necessary be prescribed. Rather than applying a distributed dynamic load, we could consider the forcing to be provided by displacements or accelerations imposed at the cable ends. However, few distinctive features are introduced, so we shall not give explicit consideration to such loadings at this stage. Damping is to be included later.

Our interest is with the in-plane dynamic response and in particular with the additional tension generated. Therefore we shall confine our attention to cases in which $\mathbf{p}(\mathbf{x}, \mathbf{t})$ is even about midspan.

The solution to (3.27) may be written [20]

$$\mathbf{w}(\mathbf{x}, \mathbf{t}) = \sum_n \xi_n(\mathbf{x})\phi_n(\mathbf{t}), \tag{3.29}$$

where $\xi_n(\mathbf{x})$ is the vertical component of the nth symmetric in-plane mode of vibration and $\phi_n(\mathbf{t})$ is the associated normal coordinate. For convenience we reiterate that ξ_n is the solution of

$$\frac{d^2\xi_n}{d\mathbf{x}^2} + \omega_n^2\xi_n = \mathbf{h}_n, \tag{3.30}$$

where

$$\mathbf{h}_n = \lambda^2 \int_0^1 \xi_n d\mathbf{x}, \tag{3.31}$$

with $\xi_n(0) = \xi_n(1) = 0$. The solutions are given by (3.16) and (3.17).

Substitution of (3.29), (3.30), and (3.31) into (3.27) and (3.28) and the use of orthogonality (see problem 3.3) lead in the usual way to the normal coordinate equations

$$\frac{d^2\phi_n}{d\mathbf{t}^2} + 2\omega_n\zeta_n\frac{d\phi_n}{d\mathbf{t}} + \omega_n^2\phi_n = -\frac{\displaystyle\int_0^1 \mathbf{p}(\mathbf{x}, \mathbf{t})\xi_n(\mathbf{x})d\mathbf{x}}{\displaystyle\int_0^1 \xi_n^2(\mathbf{x})d\mathbf{x}}, \tag{3.32}$$

and the additional tension is

$$\mathbf{h}(\mathbf{t}) = \sum_n \left\{\lambda^2 \int_0^1 \xi_n(\mathbf{x})d\mathbf{x}\right\}\phi_n(\mathbf{t}) = \sum_n \mathbf{h}_n\phi_n(\mathbf{t}), \tag{3.33}$$

where ζ_n is the fraction of critical damping in the nth mode.

It will be noticed that provision for viscous damping has been made at this step: this is probably as good a way to do it as any other. Damping due to structural causes is usually small in cables, although it is affected be such factors as the lay of the strands. In this situation rubbing between individual helices can supply a reasonable amount of energy absorption. For example, tests by Yu on power transmission lines suggest values of $\zeta = 2$ percent or more [45]. In this way too the hangers of suspension bridges provide what is suspected to be the bridges' principal source of energy absorption. But in terms of the complete bridge damping is low: typical values reported are rarely above $\zeta = 1$ percent to 2 percent, and often they are a good deal less [2, 3, 6].

Yet recent tests by Ramberg and Griffin on the damping characteristics of suspended cables have some interesting ramifications [31]. They found that an order of magnitude increase in structural damping could be expected (say from $\zeta \simeq 0.4$ percent to $\zeta \simeq 4$ percent) if the cable was slack rather than taut. The distinction between a taut cable and a slack one is obviously semiquantitative at best, but it is related to the ability to generate additional tension. (Perhaps a cutoff of $\lambda^2 = 4\pi^2$ is appropriate?) The explanation would seem to center in part on the tension fluctuations possible in slack cables. Work can be done on the supports which are never completely rigid, so that a portion of the energy may be radiated away. This mechanism is certainly available in some cable systems.

The other major source of damping in cables is that due to the surrounding air or water. In still air this fluid component is negligible, but in strong winds Davenport and Steels suggest the additional damping may be as much as 4 percent of critical for guywires [12]. In water Ramberg and Griffin's experiments indicate additional damping amounting to some 1 to 2 percent. Thus, even though damping levels in the cables themselves may be low, their interaction with winds, waves, or currents may substantially increase these levels. (Of course in offshore applications involving, say, surface moored structures, the drag on and wave-making ability of the body itself, may lead to an overall damping capability in which the cable contributions may be swamped.) By and large then, detailed investigations may be necessary in those situations where a precise knowledge is required of the damping characteristics of a particular cable structure.

In the special but important case when the loading is uniform along the span, the general solution to (3.32) is

$$\phi_n(\mathbf{t}) = \frac{\alpha_n \omega_n^2}{\mathbf{h}_n} D_n(\mathbf{t}), \tag{3.34}$$

where

$$D_n(\mathbf{t}) = \exp(-\zeta_n \omega_n \mathbf{t}) \Big\{ \phi_{n0} \cos\left[(1 - \zeta_n^2)^{1/2} \omega_n \mathbf{t}\right]$$

$$+ \frac{\dot{\phi}_{n0} + \zeta_n \omega_n \phi_{n0}}{(1 - \zeta_n^2)^{1/2}\omega_n} \sin [(1 - \zeta_n^2)^{1/2}\omega_n \mathbf{t}]\Bigg\}$$

$$- \int_0^t \frac{1}{(1 - \zeta_n^2)^{1/2}\omega_n} \exp\{-\zeta_n \omega_n (\mathbf{t} - \tau)\}$$

$$\sin [(1 - \zeta_n^2)^{1/2}\omega_n(\mathbf{t} - \tau)]\mathbf{p}(\tau)d\tau,$$

in which $\phi_{n0} = \phi_n(0)$ and $\dot{\phi}_{n0} = d\phi_n(0)/d\mathbf{t}$. The integral is Duhamel's integral, and usually it must be evaluated numerically. Two different exact methods have been developed for piecewire linear loading [29, 14], although usually equations like (3.32) are solved directly using step-by-step numerical integration procedures [4, 10]. By employing the continuum approach to start with, and by supposing that only a few modes are important, the size of the problem remains manageable, and modal analysis techniques are efficient.

We define the parameter α_n as the participation factor for additional tension; it is given by

$$\alpha_n = \frac{\displaystyle\int_0^{\omega_n/2} (1 - \sec \omega_n/2 \cos \mathbf{z})d\mathbf{z}}{\displaystyle\int_0^{\omega_n/2} (1 - \sec \omega_n/2 \cos \mathbf{z})^2 d\mathbf{z}}, \tag{3.35}$$

which, on carrying out the integration and using (3.17), leads to

$$\alpha_n = \frac{2/3}{[1 + \lambda^2\{\tan(\omega_n/2)/(\omega_n/2)\}^2/12]}. \tag{3.36}$$

Values of α_n are listed in table 3.4 for the first eight modes from a wide range of values of λ^2. Two interesting identities may be established:

$$\sum_n \alpha_n \omega_n^2 = \lambda^2 \quad \text{and} \quad \sum_n \alpha_n = \frac{1}{(1 + 12/\lambda^2)}, \tag{3.37}$$

proofs of which are left as an exercise (see problem 3.4).

An indication of the validity of the second equation of (3.37) may be found by adding the entries in a particular row of table 3.4. It is noted further that the participation factors reach maximum values when tan $(\omega_n/2) = 0$, which occurs when $\lambda^2 = 4\pi^2$, $16\pi^2$, $32\pi^2$, and so on. Then $\alpha_n = 2/3$, and the results are sharply peaked in the immediate vicinity of those points.

When λ^2 is very large, $\sum \alpha_n \to 1$, and the cable is behaving essentially as a rigid body. Alternatively, when λ^2 is small, $\alpha_n \to \lambda^2/(2(\omega_n/2)^4)$, and, as expected, the additional tension becomes vanishingly small. These trends are apparent in the table.

Table 3.4 Participation factors for additional tension for the first eight symmetric in-plane modes as a function of λ^2 [20]

λ^2	α_1	α_2	α_3	α_4	α_5	α_6	α_7	α_8
$256\pi^2$	0.003	0.004	0.005	0.007	0.011	0.025	0.126	0.666
$196\pi^2$	0.004	0.005	0.007	0.012	0.026	0.127	0.666	0.116
$144\pi^2$	0.006	0.008	0.012	0.027	0.118	0.666	0.115	0.017
$100\pi^2$	0.010	0.014	0.028	0.129	0.666	0.114	0.016	0.005
$64\pi^2$	0.016	0.031	0.131	0.666	0.112	0.015	0.005	0.002
$36\pi^2$	0.035	0.135	0.666	0.110	0.014	0.004	0.002	0.001
$16\pi^2$	0.141	0.666	0.104	0.011	0.003	0.001	0.000	0.000
100	0.345	0.506	0.033	0.005	0.001	0.001	—	—
80	0.480	0.362	0.021	0.004	0.001	0.001	—	—
60	0.610	0.206	0.013	0.003	0.001	0.000	—	—
$4\pi^2$	0.666	0.090	0.007	0.002	0.001	—	—	—
20	0.590	0.302	0.003	0.001	0.000	—	—	—
10	0.440	0.012	0.001	0.000	—	—	—	—
8	0.389	0.009	0.001	0.000	—	—	—	—
6	0.325	0.07	0.001	—	—	—	—	—
4	0.245	0.004	0.001	—	—	—	—	—
2	0.140	0.002	0.000	—	—	—	—	—
1	0.076	0.001	—	—	—	—	—	—
0.1	0.008	0.000	—	—	—	—	—	—

From the evidence presented, it appears that, when λ^2 is moderate to large, the higher modes are likely to participate most strongly in the generation of additional tension. This behavior may be traced to the enhanced extensional strain storage capability characteristic of these cables modes.

The expression for the additional tension generated is

$$\mathbf{h}(\mathbf{t}) = \sum_n \alpha_n \omega_n^2 D_n(\mathbf{t}), \tag{3.38}$$

while a convenient way of writing the expression for the additional deflection at midspan is

$$\mathbf{w}\left(\frac{1}{2}, \mathbf{t}\right) = \sum_n \beta_n \omega_n^2 D_n(\mathbf{t}), \tag{3.39}$$

where the participation factor is

$$\beta_n = \frac{\alpha_n}{\omega_n^2}\left\{1 - \sec\left(\frac{\omega_n}{2}\right)\right\}. \tag{3.40}$$

Values of β_n are listed in table 3.5, and the trends observed there are readily explained.

Table 3.5 Participation factors for additional vertical midspan deflection for the first eight symmetric in-plane modes as a function of λ^2 [20]

λ^2	$\beta_1 \times 10^3$	$\beta_2 \times 10^3$	$\beta_3 \times 10^3$	$\beta_4 \times 10^3$	$\beta_5 \times 10^3$	$\beta_6 \times 10^3$	$\beta_7 \times 10^3$	$\beta_8 \times 10^3$
$256\pi^2$	0.23	−0.10	0.10	−0.07	0.10	−0.09	0.26	0.0
$196\pi^2$	0.30	−0.13	0.14	−0.11	0.17	−0.19	0.69	−0.16
$144\pi^2$	0.42	−0.19	0.22	−0.18	0.48	0.0	0.36	−0.09
$100\pi^2$	0.62	−0.29	0.41	−0.38	0.14	−0.31	0.16	−0.07
$64\pi^2$	1.02	−0.51	1.19	0.0	0.75	−0.19	0.10	−0.05
$36\pi^2$	2.06	−1.15	3.75	−0.82	0.32	−0.13	0.08	−0.05
$16\pi^2$	6.68	0.0	2.52	−0.52	0.22	−0.11	0.07	−0.04
100	14.0	−1.96	1.71	−0.46	0.20	−0.11	0.06	−0.04
80	18.8	−3.67	1.52	−0.44	0.20	−0.10	0.06	−0.04
60	24.9	−5.13	1.37	−0.42	0.19	−0.10	0.06	−0.04
$4\pi^2$	33.8	−5.59	1.23	−0.40	0.19	−0.10	0.06	−0.04
20	51.3	−5.31	1.13	−0.39	0.18	−0.10	0.06	−0.04
10	72.4	−5.05	1.08	−0.38	0.18	−0.10	0.06	−0.04
8	79.2	−5.00	1.07	−0.38	0.18	−0.10	0.06	−0.04
6	87.5	−4.95	1.06	−0.38	0.18	−0.10	0.06	−0.04
4	97.9	−4.89	1.05	−0.38	0.18	−0.10	0.06	−0.04
2	111	−4.83	1.04	−0.38	0.18	−0.10	0.06	−0.04
1	119	−4.81	1.04	−0.38	0.18	−0.10	0.06	−0.04
0.1	128	−4.78	1.03	−0.38	0.18	−0.10	0.06	−0.04

Example 3.4 Additional Tension in the Cables of a Suspension Bridge Responding to Earthquake Excitation

The suspension bridge cable of example 2.4 is used again. In this case it is subjected to the vertical component of ground acceleration of the El Centro 1940 earthquake. Many assumptions are made that are, strictly speaking, inadmissible, but they do allow a preliminary estimate to be made of the likely importance of additional tension of seismic origin in a particular long-span suspension bridge. This whole matter is taken up again in section 4.4, where a simplified design formula is produced. Nevertheless it is as well to be aware of the numerous shortcomings in the present approach: for example, the stiffening influence of the deck is ignored, as is tower compliance; phase differences between ground motion inputs at various support points are also ignored, and the example is confined to just one suspended span. Our object is simply to see what is the difference between the peak additional cable tension (expressed as a fraction of the static tension due to self-weight) and the peak vertical ground acceleration (expressed as a fraction of the gravitational acceleration).

Under vertical ground displacements that are the same at each end of the cable the normal coordinate equations may be shown to assume the form

$$\frac{d^2\phi_n}{dt^2} + 2\omega_n\zeta_n\frac{d\phi_n}{dt} + \omega_n^2\phi_n = \frac{\alpha_n\omega_n^2}{\mathbf{h}_n}(2\omega_n\zeta_n\dot{\mathbf{w}}_g + \ddot{\mathbf{w}}_g),$$

where $\dot{\mathbf{w}}_g$ and $\ddot{\mathbf{w}}_g$ are the dimensionless vertical ground velocity and ground acceleration, respectively. Because the damping is small, we choose to ignore the first term in parentheses on the right. Converting to dimensional quantities suitable for calculation, we have

$$\frac{d^2\phi_n}{dt^2} + 2\omega_n\zeta_n\frac{d\phi_n}{dt} + \omega_n^2\phi_n = \alpha_n\omega_n^2\frac{\ddot{w}_g}{g},$$

where the additional tension is

$$\frac{h(t)}{H} = \sum_n \phi_n(t).$$

Under full dead load of 60 kN/m per cable the horizontal component of tension in each cable is 8.33×10^4 kN. Since $\lambda^2 = 125$, extrapolation of the results in tables 3.1 and 3.4 leads to the following values for the first four symmetric in-plane modes:

	Mode 1	Mode 2	Mode 3	Mode 4
ω	1.0 rad/sec	1.4 rad/sec	1.9 rad/sec	2.6 rad/sec
α	0.23	0.61	0.05	0.01
ζ	0.005	0.005	0.005	0.005

The normal coordinate equations were solved using a standard step-by-step numerical technique that produced results at intervals of 0.02 seconds throughout the duration of the record [10]. These results were then superposed to produce a trace of the additional tension. Figure 3.11a is the input acceleration record. Below it is the trace of additional tension for just the second modal contribution. Figure 3.11c is the result for superposition of the four modes considered.

It is clear that the second mode is the most strongly excited insofar as additional tension is concerned, as might be inferred from the size of its participation factor. (In the case of the additional deflection generated at midspan, by contrast, the first mode provides the bulk of it. But even so, the peak is miniscule, being less than 2 cm.) A peak of some 3 percent of the dead load tension indicates that earthquake-generated additional tension is not a particularly significant live load condition in this instance, and it is a small fraction of the peak vertical ground acceleration recorded of 0.21 g. However, this particular result does not justify a complacent attitude, for strong motion records taken in the vicinities of the causative faults of such recent shocks as San Fernando (1971) and Imperial Valley (1979) show that peak vertical components may be 1 g, or more. It hardly needs to be mentioned that peak additional tensions of 15 percent of the dead load tension are a significant live load condition in a long-span bridge!

Example 3.5 Resonant Effects in Additional Tension

The additional tension generated under a forcing of $\mathbf{p(t)} = \mathbf{p}_0 \sin \boldsymbol{\omega}\mathbf{t}$, when damping is ignored and the cable starts from rest, may be shown to be

$$\mathbf{p}_0 A(\omega_n, \lambda)\mathbf{t} \cos \boldsymbol{\omega}_n\mathbf{t},$$

where

$$\mathbf{A}(\omega, \lambda) = \frac{\omega}{\{3 + \lambda^2(1 - (\omega/\lambda)^2)^2/4\}}.$$

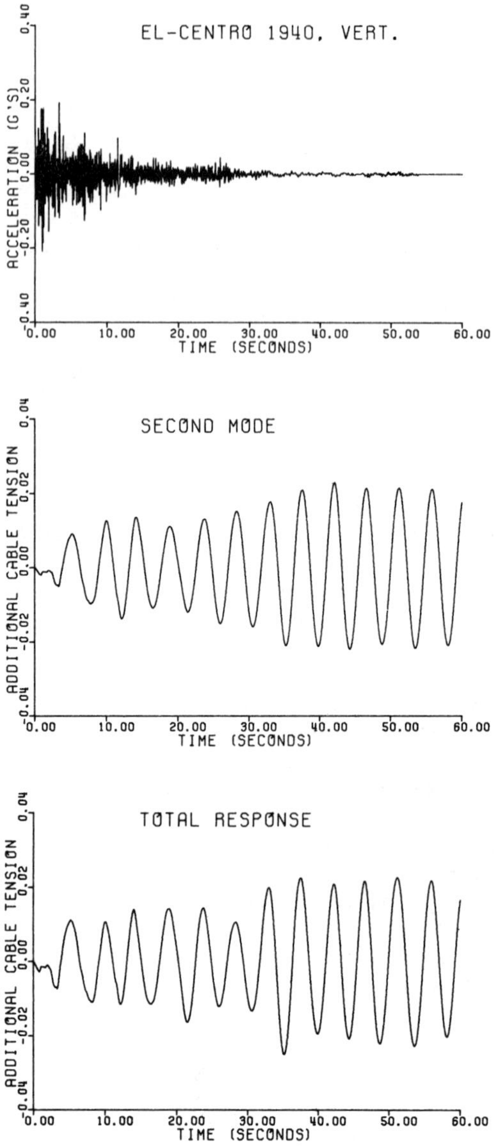

3.11 (a) Vertical component of the El Centro (1940) earthquake. (b) Second mode contribution to the additional tension. (c) Dimensionless additional tension in the cables of a long-span suspension bridge responding to earthquake excitation.

The most critical frequency of the system is that natural frequency for which **A** is a maximum.

Suppose for argument's sake that λ^2 is fixed and $\boldsymbol{\omega}$ is a continuous variable. Then **A** has a maximum when

$$\boldsymbol{\omega}^2 = \frac{\lambda^2}{6}\left\{2 + 4\left(1 + \frac{9}{\lambda^2}\right)^{1/2}\right\},$$

and it increases as λ^2 increases. Thus when λ^2 is large, the most rapidly growing additional tensions are generated if the cable is forced at a natural frequency near λ. If that frequency nearest λ is $\boldsymbol{\omega}_m$, the amplitude is

$$\mathbf{A} = \frac{\lambda}{\{3 + (\lambda - \boldsymbol{\omega}_m)^2\}}.$$

It turns out that, when λ^2 is large, there is always a frequency such that $|\lambda - \boldsymbol{\omega}_m| < 0.42\pi$, with the result that

$$0.21 < \frac{\mathbf{A}_{\max}}{\lambda} < 0.33.$$

However, if forcing occurs at one of the adjacent natural frequencies, the response is much less marked. Consequently, when λ^2 is large, and the forcing frequency is not in a range such as $\lambda \pm 2\pi$, resonant effects in additional tension will not be appreciable.

We are tempted to conclude from this that a special natural frequency exists equal to λ. For example, if (3.27) is integrated over the span, and substitutions are made for (3.28), there is derived

$$\frac{d^2\mathbf{h}}{d\mathbf{t}^2} + \lambda^2\mathbf{h} = \lambda^2 f(\mathbf{t}),$$

where

$$f(\mathbf{t}) = \int_0^1 \mathbf{p}(\mathbf{x}, \mathbf{t})d\mathbf{x} + \frac{\partial \mathbf{w}}{\partial \mathbf{x}}(1, \mathbf{t}) - \frac{\partial \mathbf{w}}{\partial \mathbf{x}}(0, \mathbf{t}).$$

(This result, and much of the discussion on which this example is based, is due to Jerry H. Griffin.) What we have here is an oscillator equation for the additional tension, and from it might be inferred that resonance in the additional tension will occur when $\boldsymbol{\omega} = \lambda$. This can be shown to be incorrect, but for very short times a wave-type solution indicates that, when $\boldsymbol{\omega} = \lambda$, and λ is large, a strong excitation is possible in which $\mathbf{A} = \lambda/2$. In other words, it is 50 percent stronger than for true resonance. This peak fades for longer times.

However, the crux of the matter is that we have assumed λ^2 to be very large, and the analysis, although mathematically correct, is not physically motivated. When λ^2 is very large, the natural frequencies of the system near λ will in all probability be spurious, since we are entering the range where dilatational effects are of importance.

It should be emphasized that this whole analysis to date is based on the assumption that the disturbance wave speed, equal to $(H/m)^{1/2}$, is much less than that of the stress waves which propagate along the cable at a velocity equal to $(EA/m)^{1/2}$.

Under this assumption changes in additional tension are considered to occur instantaneously throughout the cable. This shortcoming means that it would be a mistake to place too much faith on the limiting trends evident in this example.

3.4 Nonlinear Theories

Two topics are covered here, the first being the vibration of a taut flat cable. In this case the nonlinearity is geometric; the equivalent spring is of the hardening variety, and the additional tension generated is never negative. In the case of stress waves in a cable the linearized theory is first discussed as an aid to the discussion of elastic-plastic stress waves in which the nonlinearity is due to softening behavior in the material. An example is also presented on geometrically nonlinear wave propagation.

The Nonlinear Theory of Vibration in a Taut Flat Cable

If the amplitude of the vibration is such that its square is no longer negligible, appreciable levels of additional tension may be generated, and the equation of motion of free planar vibrations takes the form

$$(1 + \mathbf{h}(\mathbf{t}))\frac{\partial^2 \mathbf{w}}{\partial \mathbf{x}^2} - \frac{\partial^2 \mathbf{w}}{\partial \mathbf{t}^2} = 0, \tag{3.41}$$

with

$$\mathbf{h}(\mathbf{t}) = \frac{\lambda^2}{2}\int_0^1 \left(\frac{\partial \mathbf{w}}{\partial \mathbf{x}}\right)^2 d\mathbf{x}, \tag{3.42}$$

where $\mathbf{w}(0, \mathbf{t}) = \mathbf{w}(1, \mathbf{t}) = 0$ and $\mathbf{w}(\mathbf{x}, 0) = \mathbf{w}_0(\mathbf{x})$, $\partial \mathbf{w}(\mathbf{x}, 0)/\partial \mathbf{t} = \dot{\mathbf{w}}_0(\mathbf{x})$. This problem was first studied by Carrier [7, 8].

Two points may be noted. First, the frequencies of the system will be to some extent amplitude-dependent, in fact dependent on the second power of the amplitude. This amplitude dependence is usual in nonlinear vibrations. Second, the modes of vibration are identical to those of the classical linear system; therefore the governing partial differential equation may be reduced to a system of ordinary differential equations. The ease with which the variables may be separated in this instance is rather uncommon, but unfortunately it does not work when sag is introduced.

To illustrate these points, let an initial displacement $\mathbf{w}_0(\mathbf{x})$ be prescribed, and for convenience let the initial velocity $\mathbf{w}_0(\mathbf{x})$ be zero. Suppose the solution of the ensuing vibration is of the form

$$\mathbf{w}(\mathbf{x}, \mathbf{t}) = \sum_{n=1}^{\infty} \mathbf{A}_n \sin n\pi\mathbf{x}\ \xi_n(\mathbf{t}), \tag{3.43}$$

where $\mathbf{A}_n = A_n/(mgl^2/H)$.

On account of the orthogonality of the trigonometric functions, there are no cross terms in the expression for the additional tension. It is simply a sum of squares:

$$\mathbf{h}(\mathbf{t}) = \frac{\lambda^2}{4} \sum_{n=1}^{\infty} \mathbf{A}_n^2 n^2 \pi^2 \xi_n(\mathbf{t}). \tag{3.44}$$

Substituting (3.43) and (3.44) into (3.41), and following standard procedures, leads to a series of equations in time alone of the form

$$\frac{d^2\xi_n}{dt^2} + n^2\pi^2\mathbf{A}_n^2\left\{1 + \frac{\lambda^2}{4}\sum_{n=1}^{\infty}\mathbf{A}_i^2 i^2\pi^2\xi_i^2\right\}\xi_n = 0, \tag{3.45}$$

where the coefficients are $\mathbf{A}_n = 2\int_0^1 \mathbf{w}_0(\mathbf{x}) \sin n\pi\mathbf{x}\, d\mathbf{x}$, and the initial conditions are $\xi_n(0) = 1$, $\dot{\xi}_n(0) = 0$, $n = 1, 2, 3, \ldots$. Thus, as predicted, the modes of vibration of the nonlinear system are identical to those of the linear system.

These coupled equations do not possess an exact solution. Therefore perturbation techniques are required: the two variable expansion procedure of Chikwendu and Kevorkian may be adapted [9]. Expressions for the time history of displacement and additional tension may be shown to be

$$\mathbf{w}(\mathbf{x}, \mathbf{t}) \simeq \sum_{n=1}^{\infty} \mathbf{A}_n \sin n\pi\mathbf{x} \cos n\pi\left[1 + \frac{\lambda^2}{32}\left\{n^2\pi^2\mathbf{A}_n^2 + 2\sum_{n=1}^{\infty} i^2\pi^2\mathbf{A}_i^2\right\}\right]\mathbf{t}, \tag{3.46}$$

$$\mathbf{h}(\mathbf{t}) \simeq \frac{\lambda^2}{4}\sum_{n=1}^{\infty} n^2\pi^2\mathbf{A}_n^2 \cos^2 n\pi\left[1 + \frac{\lambda^2}{32}\left\{n^2\pi^2\mathbf{A}_n^2 + 2\sum_{n=1}^{\infty} i^2\pi^2\mathbf{A}_i^2\right\}\right]\mathbf{t}, \tag{3.47}$$

and the only proviso is that $\lambda^2 n^2 \pi^2 \mathbf{A}_n^2$ should not become too large. This interferes with the geometric approximations made in setting up the equation of motion and will also lead to tensions that do not lie within the confines of linear elasticity. As might be expected, the solutions are periodic. (These results and those of (3.52), (3.53), and (3.54) are due to Jerry H. Griffin.)

In the particular case when the initial displacement is a mode such as

$$\mathbf{w}_0(\mathbf{x}) = \mathbf{A}_m \sin m\pi\mathbf{x}, \tag{3.48}$$

the exact solution of (3.45) may be found in terms of the elliptic integral of the first kind. However, in view of the comments made in the previous paragraph, the solution is best extracted from (3.46) and (3.47). It reads

$$\mathbf{w}(\mathbf{x}, \mathbf{t}) \simeq \mathbf{A}_m \sin m\pi\mathbf{x} \cos m\pi\left(1 + \frac{3\lambda^2 m^2\pi^2\mathbf{A}_m^2}{32}\right)\mathbf{t}, \tag{3.49}$$

$$\mathbf{h(t)} \simeq \frac{\lambda^2 m^2 \pi^2 \mathbf{A}_m^2}{4} \cos^2 m\pi \left(1 + \frac{3\lambda^2 m^2 \pi^2 \mathbf{A}_m^2}{32}\right) \mathbf{t}. \tag{3.50}$$

Clearly the amplitude-dependent natural frequency of the mth mode is

$$\omega_m = m\pi \left(1 + \frac{3\lambda^2 m^2 \pi^2 \mathbf{A}_m^2}{32}\right). \tag{3.51}$$

In the case of forced response the problem is rather more complicated. Suppose sinusoidal forcing of the form

$$\mathbf{p} \sin (1 + \kappa)^{1/2} n\pi\mathbf{t}, \quad n = 1, 3, 5, \ldots,$$

is applied, where $\mathbf{p} = p/mg$ and κ is a small parameter, either positive or negative (which indicates how close the forcing frequency is to the nth natural frequency of the linear system). It may be shown that just one mode is excited in the steady state response. (Since this loading is uniform across the span, only symmetric modes may be excited, and consequently n is odd.) If the cable is constrained to vibrate in a single plane, that response is

$$\mathbf{w(x, t)} = -\mathbf{A}_n \sin n\pi\mathbf{x} \sin (1 + \kappa)^{1/2} n\pi\mathbf{t}, \tag{3.52}$$

with

$$\mathbf{h(t)} = \frac{\lambda^2 n^2 \pi^2 \mathbf{A}_n^2}{4} \sin^2 (1 + \kappa)^{1/2} n\pi\mathbf{t}, \tag{3.53}$$

where the level of the response is found from the cubic

$$q^3 - rq - 2 = 0, \tag{3.54}$$

in which $q = \mathbf{A}_n/\{32\mathbf{p}/(3\lambda^2 n^5 \pi^5)\}^{1/3}$ and $r = \{4\kappa^3 n^4 \pi^4/(3\lambda^2 \mathbf{p}^2)\}^{1/3}$. (Such a situation could conceivably arise in a one-way cable roof where λ^2 may be sufficiently small for its natural frequencies to be quite close to those of the taut string, see table 3.1.) Written in this form, the roots depend on just one independent parameter that may be positive or negative, depending on whether κ is positive or negative. Consequently there is always one positive real root. A point of vertical tangency occurs for $r = 3$ when the solutions $(-1, -1, 2)$ are found. If $r > 3$, there are two distinct negative real roots, while for $r < 3$ these two roots are complex. When $|r|$ is large, the roots tend to $\pm\sqrt{r}$, $-2/r$. Figure 3.12 presents a plot of $|q|$ versus r. Clearly the undamped nonlinear steady-state response of a taut flat cable constrained to vibrate in a plane reduces to the solution of Duffing's equation for an undamped mass-hardening spring system [26].

If the response is not constrained to remain in a single plane, the cable may whirl, that is, significant out-of-plane components are present. This

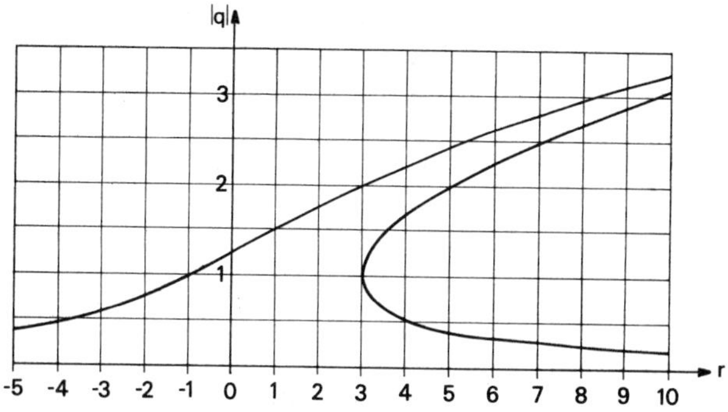

3.12 Universal plot of the roots of equation (3.54)

fact, well known from experimentation, has been demonstrated mathematically by a number of investigators. A comprehensive account is that by Miles, in whose study references to the other investigations may be found [27].

It is difficult to say what importance should be attached to nonlinear vibrations in cable structures in practice. The temptation to use linearized theories is hard to resist because, when nonlinear response is likely to be important, the nature of the loading and damping is likely to be equally important and considerably more uncertain. Nevertheless there is a definite need for more work in this area. It may well be that most progress will be made using the increasingly powerful numerical techniques currently being implemented.

Elastic and Plastic Stress Waves in Cables

The items touched on here are fundamental to the assessment of the strength and service life of hawsers and various types of lifting tackle such as the cables of elevators, winding ropes in mine shafts, and draglines. The idealized situations treated may be regarded as merely an introduction to the substantially more complicated problems that occur in practice where the prescribed initial conditions, support compliance, and the effects of multiple reflections make for a certain degree of imprecision.

One of the assumptions made in developing the theory of transverse vibration of cable systems is that temporal changes in cable tension occur in unison with fluctuations in the transverse deflection: spatial changes in tension occur instantaneously throughout the cable. But in reality these spatial changes are brought about by the rapid passage of stress waves up and down the cable. Consequently the concept of uniformization of cable tension, implicit in our earlier work, depends on the stress wave velocity being very large compared to the disturbance wave velocity. Under many types of dynamic excitation this assumption is rarely called into question, but if the excitation has important high frequency components, or if loads are suddenly applied along the axis of the cable, the stress waves induced may profoundly affect the ability of the cable to function adequately as a tensile member.

For our purposes therefore we consider the cable to be horizontal, taut, and very long. The longitudinal dynamic equilibrium of an element of this cable is expressed by

$$\frac{\partial h}{\partial x} = \rho A \frac{\partial^2 u}{\partial t^2}, \tag{3.55}$$

where ρA is its mass per unit length. If the material remains linearly elastic, Hooke's law applies and is of the form

$$h = EA\frac{\partial u}{\partial x}, \tag{3.56}$$

since the strain is $\partial u/\partial x$. The resulting equation—the wave equation—may be written as

$$E\frac{\partial^2 u}{\partial x^2} - \rho\frac{\partial^2 u}{\partial t^2} = 0. \tag{3.57}$$

The dilatational or stress wave velocity is

$$c = (E/\rho)^{1/2}, \tag{3.58}$$

and the solution of (3.57)—D'Alembert's solution—may be written as

$$u = f(x - ct) + g(x + ct), \tag{3.59}$$

where f and g are functions determined by the initial conditions and refer to waves travelling in the positive and negative x directions, respectively.

For example, suppose at $t = 0$, u has a distribution $f(x)$ along the cable. This is the wave profile. Suppose subsequently at time t the profile has moved a distance ct to the right. According to an origin now placed at this new point, and supposing the profile to have been unchanged during passage, $u = f(X)$, where $x = X + ct$. As a result the wave profile, when referred to the original axes, is $u = f(x - ct)$. Similar arguments may be advanced for $g(x + ct)$, because we need only to reverse the sign of c.

For a very long cable subject to the initial conditions $u(x, 0) = u_0(x)$, $\partial u(x, 0)/\partial t = \dot{u}_0(x)$, we have

$$f(x) + g(x) = u_0(x), \tag{3.60}$$

and, applying the rules for differentiation by parts,

$$-c\frac{df(x)}{dx} + c\frac{dg(x)}{dx} = \dot{u}_0(x). \tag{3.61}$$

Integrating this equation yields

$$-f(x) + g(x) = \frac{1}{c}\int_0^x \dot{u}_0(x)\,dx, \tag{3.62}$$

and so

$$\begin{aligned} f(x) &= \frac{1}{2}\left\{u_0(x) - \frac{1}{c}\int_0^x \dot{u}_0(x)\,dx\right\}, \\ g(x) &= \frac{1}{2}\left\{u_0(x) + \frac{1}{c}\int_0^x \dot{u}_0(x)\,dx\right\}. \end{aligned} \tag{3.63}$$

Therefore at later times we have the completely general solution [11]

$$u(x, t) = \frac{1}{2}\left\{u_0(x - ct) + u_0(x + ct) + \frac{1}{c}\int_{x-ct}^{x+ct} \dot{u}_0(x)\,dx\right\}. \tag{3.64}$$

If we look at the righthand end of the cable, and consider a wave passing back along it to the left, we have

$$u = g(x + ct), \tag{3.65}$$

from which it quickly follows that

$$\frac{\partial u}{\partial t} = c\frac{\partial u}{\partial x}. \tag{3.66}$$

From (3.56) we then establish that

$$h = \rho Ac\frac{\partial u}{\partial t}, \tag{3.67}$$

and the additional tension depends on the particle velocity at each point. It the origin of the cable is a rigid support, and the wave incident on it is

$$u_i = g(x + ct), \tag{3.68}$$

while the wave reflected from it is

$$u_r = f(x - ct), \tag{3.69}$$

the total displacement is

$$u_i + u_r = f(x - ct) + g(x + ct). \tag{3.70}$$

At this origin $u_i + u_r = 0$, so that

$$f(-ct) = -g(ct). \tag{3.71}$$

The resultant additional cable tension at the support is

$$h_i + h_r = EA\left(\frac{\partial u_i}{\partial x} + \frac{\partial u_r}{\partial x}\right) = EA\left\{-\frac{df(-ct)}{d(ct)} + \frac{dg(ct)}{d(ct)}\right\}, \tag{3.72}$$

which means that the reflection has doubled the additional tension [23].

If, for example, an initial velocity $\partial u/\partial t = \dot{u}_0$ is imposed and maintained the peak additional tensile force after the first reflection is $2\rho Ac\dot{u}_0$. This situation may be ameliorated by the insertion of a spring as in figure 3.13. If u now measures displacement in the cable itself, while $\dot{u}_0$ is imposed and maintained on the free end of the spring, we have for the spring force

$$h = k(\dot{u}_0 t - u). \tag{3.73}$$

Incorporating (3.67) yields as the equation of motion

$$\rho Ac\frac{du}{dt} + ku = k\dot{u}_0 t. \tag{3.74}$$

This is a linear equation of the first order. The solution that satisfies the initial condition of $u(0) = 0$ is

$$u = \frac{\dot{u}_0 \rho Ac}{k}\left\{\exp\left(-\frac{kt}{\rho Ac}\right) - 1 + \frac{kt}{\rho Ac}\right\}, \tag{3.75}$$

and the additional tension is

$$h = \rho Ac\dot{u}_0\left\{1 - \exp\left(-\frac{kt}{\rho Ac}\right)\right\}. \tag{3.76}$$

If $kt/\rho Ac$ remains small during the time of interest, the stress wave propagates as a triangular wedge of markedly reduced amplitude. Such spring-loaded couplings are frequently used to reduce stress wave peaks.

Example 3.6 Deceleration of a Mass by the Propagation of Stress Waves

In a series of classic experiments performed in 1872 Hopkinson (the elder) studied the tensile fracture of a vertically suspended wire caused by a sliding weight that was allowed to drop and be arrested by a collar attached to the lower end of the wire. For a given wire he found, surprisingly, that fracture was not related to the mass of the falling weight (provided that it was always substantially greater than the mass of the collar) but rather to the height from which it fell—the velocity imparted on impact [23]. The situation after the initial impulse is that

$$M\frac{d^2u}{dt^2} = -\rho Ac\frac{du}{dt},$$

which may be integrated directly to

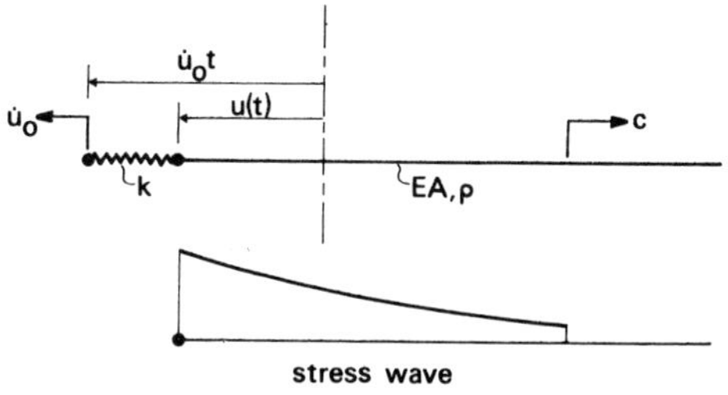

3.13 Stress wave propagation in a cable with a spring-loaded coupling

$$\frac{du}{dt} = \dot{u}_0 \exp\left(-\frac{\rho Act}{M}\right),$$

where M is the mass of the falling weight and $\dot{u}_0$ is its initial velocity, as it starts to come to rest. Hence the additional tension is

$$h = \rho Ac\dot{u}_0 \exp\left(-\frac{\rho Act}{M}\right),$$

the peak of which occurs at the front of the wave and is indeed independent of mass of the weight. This simple analysis was originally put forward by Hopkinson to explain what he had observed. Fracture, if it occurred, was generally found near the top at the support, in conformity with the addition of the tensile stress waves there.

The analysis may be extended by the insertion of a spring at the lower end of the cable. If we let u_1 describe the displacement of the mass, while u_2 describes the displacement of the cable end, the equation of motion is (see figure 3.14)

$$M\frac{d^2u_1}{dt^2} = -k(u_1 - u_2),$$

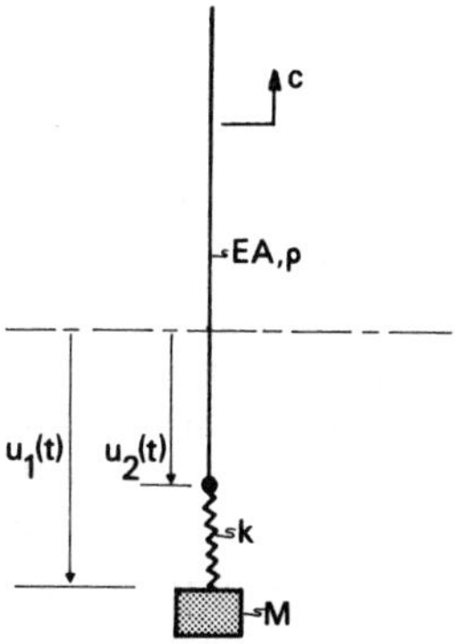

3.14 A simple model of a system exhibiting radiation damping with

$$k(u_1 - u_2) = \rho A c \frac{du_2}{dt}.$$

Inserting one in the other, integrating, and noting that $du_1/dt = \dot{u}_0$ and $u_2 = 0$ at $t = 0$ yields

$$M\frac{du_1}{dt} = -\rho A c u_2 + M\dot{u}_0.$$

The equation of motion of the mass becomes

$$M\frac{d^2 u_1}{dt^2} + \frac{kM}{\rho A c}\frac{du_1}{dt} + k u_1 = \frac{kM}{\rho A c}\dot{u}_0,$$

which is in the form of a forced, viscously damped oscillator equation, the damping being provided by the stress waves that transport energy away from the spring-mass system. (So, regardless of the stiffness of the spring, the peak displacement is at most $2M\dot{u}_0/\rho A c$.)

By definition critical damping is present when

$$\frac{kM}{\rho A c} = 2(kM)^{1/2},$$

or when

$$\frac{l/c}{(M/k)^{1/2}} = \frac{2}{(M/\rho A l)}.$$

The lefthand side is a ratio of time scales for the stress waves and for the spring-mass system. This ratio need not be large to ensure critical damping if, as may easily be envisaged, the righthand side is small. Nonetheless, reflections would then in all probability play an important role in the subsequent behavior.

This is one of the simplest systems exhibiting radiation damping. It is clear that several potentially useful investigations could be based on this oscillator equation. For example, the same equation arises in pile-driving problems if a cushion is inserted between pile-head and hammer [10].

Returning to our previous examples, we note that an initially elastic stress wave (which propagates away from an end which is suddenly grabbed and moved forward) may cause fracture at the far end if the addition of stresses there exceeds the ultimate tensile strength of the material. For typical high-strength steel cables this would require velocities of the order of 10 m/sec, which are not achievable in practice, although velocities of, say, 3 m/sec are regularly encountered with the buckets of draglines. But we should investigate the situation when the grabbed end is on the verge of being torn from the rest of the cable if only to be satisfied that this eventuality is at least as unlikely as fracture on reflection. To attempt this, we need results from the theories of plastic waves in a cable—theories developed independently during World War II by G. I. Taylor and by von Kármán [23].

If it is assumed that the stress is a known, univalued function of the strain and that this nonlinear stress-strain relation is independent of the rate of loading, the equation of motion may be written as

$$S\frac{\partial^2 u}{\partial x^2} = \rho\frac{\partial^2 u}{\partial t^2}, \tag{3.77}$$

where S is the modulus of deformation, namely, the slope of the stress-strain curve. We further assume that the strain $\partial u/\partial x$ is a function of x/t, say, $F(x/t)$. Now

$$u = \int \frac{\partial u}{\partial x} dx = t\int F\left(\frac{x}{t}\right) d\left(\frac{x}{t}\right). \tag{3.78}$$

Differentiating (3.78) with respect to t yields

$$\frac{\partial u}{\partial t} = \int F\left(\frac{x}{t}\right) d\left(\frac{x}{t}\right) + \frac{tF(x/t)d(x/t)}{dt}, \tag{3.79}$$

but, since $d(x/t)/dt = -(x/t)/t$,

$$\frac{\partial u}{\partial t} = \int F\left(\frac{x}{t}\right) d\left(\frac{x}{t}\right) - \left(\frac{x}{t}\right) F\left(\frac{x}{t}\right). \tag{3.80}$$

Forming the second derivative gives

$$\frac{\partial^2 u}{\partial t^2} = \frac{(x/t)^2 dF(x/t)}{td(x/t)}. \tag{3.81}$$

It is much simpler to show that

$$\frac{\partial^2 u}{\partial x^2} = \frac{dF(x/t)}{td(x/t)}, \tag{3.82}$$

and substitution of these results in (3.77) results in

$$\frac{\{\rho(x/t)^2 - S\}dF(x/t)}{td(x/t)} = 0. \tag{3.83}$$

There are two solutions of interest, and these divide the cable into three regions. In the first place the requirement

$$\frac{dF(x/t)}{d(x/t)} = 0 \tag{3.84}$$

means that $F(x/t)$ is constant, which implies a region of constant strain. Thus a plastic wave front advances into the cable with a velocity $c_p =$

$(S_p/\rho)^{1/2}$, where S_p has the value relevant to the plastic strain behind the front. The region from $x = 0$ to $c_p t$ delineates the plastic region.

In the region from $x = c_p t$ to $x = ct$ we have from (3.83) the relation

$$x = \left(\frac{S}{\rho}\right)^{1/2} t, \tag{3.85}$$

where $E < S < S_p$. Ahead of the elastic wave front, which is the fastest moving portion, no influence can be felt. But at $x = ct$ there must be a discontinuity in strain. The chain of events is illustrated by the curve in figure 3.15a. The imposed end velocity, which causes propagation of the plastic wave, is the area under that curve, namely,

$$\dot{u}_{0p} = \int_0^{\varepsilon_p} \left(\frac{x}{t}\right) d\varepsilon = \int_0^{\varepsilon_p} \left(\frac{S}{\rho}\right)^{1/2} d\varepsilon, \tag{3.86}$$

where ε is the strain [23].

For high-strength steels a plausible form for the stress-strain curve is (see figure 3.15b)

$$\frac{\sigma}{\sigma_u} = \tanh\left(\frac{E\varepsilon}{\sigma_u}\right), \tag{3.87}$$

where σ_u is the ultimate tensile strength. Therefore

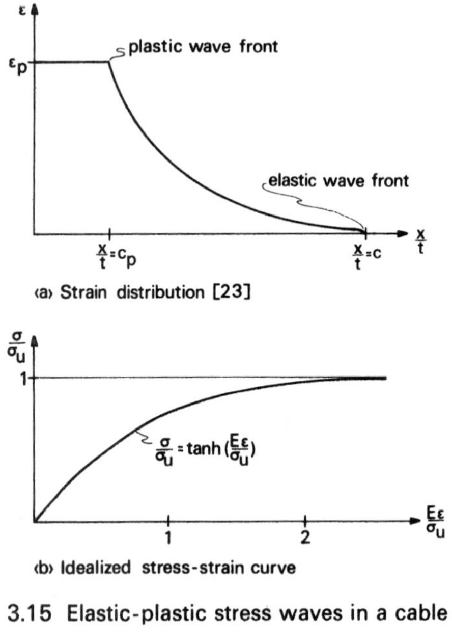

3.15 Elastic-plastic stress waves in a cable

$$S = E\,\mathrm{sech}^2\left(\frac{E\varepsilon}{\sigma_u}\right), \tag{3.88}$$

and so

$$\dot{u}_{0p} = \frac{\sigma_u}{\rho c}\int_0^{E\varepsilon_p/\sigma_u} \mathrm{sech}\left(\frac{E\varepsilon}{\sigma_u}\right) d\left(\frac{E\varepsilon}{\sigma_u}\right) = \frac{2\sigma_u}{\rho c}\left[\tan^{-1}\left\{\exp\left(\frac{E\varepsilon_p}{\sigma_u}\right)\right\} - \frac{\pi}{4}\right]. \tag{3.89}$$

When the strain is near ultimate, $E\varepsilon_u \gg \sigma_u$, and (3.89) reduces to the attractively simple form

$$\dot{u}_{0u} = \frac{\pi\sigma_u}{2\rho c}, \tag{3.90}$$

which is independent of ε_u. In high-strength steels the elastic limit is reached at about half of the ultimate tensile strength; so in theory it might be possible for an elastic wave to pass to the far end and cause fracture there as the wave is reflected. This velocity would be

$$\dot{u}_{0r} \simeq \frac{\sigma_u}{2\rho c}, \tag{3.91}$$

and therefore $\dot{u}_{0u} \simeq \pi\dot{u}_{0r}$. We conclude that, if fracture by reflection at the support is unlikely, it is even more unlikely in the initial grabbing process.

Example 3.7 Airplane Impact on a Barrage Balloon Cable

This example was originally solved by Housner, also during World War II [17]. The problem is nonlinear, the nonlinearity here stemming from geometry. Since the war several more complicated methods of solution have been proposed, but that due to Housner is the neatest, and we shall use it. Housner writes:

> I solved the problem during the Second World War when I was in North Africa with the 9th Bomber Command and the low-level raid on the Ploesti oil refineries was planned. The solution is for an infinitely long cable, in a vacuum, with stress wave velocity of propagation $c = (E/\rho)^{1/2}$. The barrage balloons were flown at about 2,000 ft altitude and were tethered by a wire of diameter equal to 0.1 inches, approximately. This high-strength, drawn wire behaves elastically until it reaches a stress of about 200,000 psi at which point it breaks.
>
> The method of solution is of the inverse type, that is, the solution is postulated by intuition, and then it is shown that it satisfies Hooke's law, Newton's law, and the conditions of compatibility. If the solution satisfies these three, then it obviously would also satisfy the partial differential equations of dynamic elasticity, since they are derived from those three statements.
>
> The aircraft velocity of 500 ft/sec required to break the cable was greater than airplanes in the Second World War were capable of. Therefore the stress wave traveled down to the anchor point and reflected back, traveling past the airplane wing, causing the wire to saw past the wing. When the triangular cable wave reaches the base anchor point, the cable will certainly break. Seven of the aircraft returned with evidence of having run through a cable. There were neat cuts through the leading edge of the wing (thin aluminum sheet) which clearly were caused by the sawing action of the cable. I had concluded that the total time,

before breaking, that the cable exerted a force on the wing was so small that it would not affect appreciably the flight of the bomber plane; and the returning pilots said that they had not felt any action of the cable on the plane.

Figure 3.16a shows an unstressed cable which is impacted with velocity v. The impact causes propagation of a stress wave and a disturbance wave, with velocities of propagation c and kv, respectively. The geometric factor k is substantially greater than unity, but we expect that $kv \ll c$, that is, the stress wave moves much faster than the disturbance wave. Figure 3.16b shows a force balance in which F is the reaction of the wing on the cable and $\rho c^2 \varepsilon A$ is the tensile force generated in the cable, equal to σA. The other force components, shown dotted, represent inertia terms.

From Newton's second law these forces must equal the rate of change of momentum of the kinked portion. Thus in the vertical direction

$$\rho c^2 \varepsilon A = \frac{d}{dt}\left(\frac{M}{2} kv\right) \simeq (\rho A k v) kv,$$

while in the horizontal direction

$$F = \frac{d}{dt}(Mv) \simeq (2\rho A k v) v,$$

where ε is the (constant) strain over the whole portion of the cable up to the stress wave fronts.

During a time t the additional length generated by stretching is $2\varepsilon ct$, and the length needed, if the requirements of compatibility are to be met, is

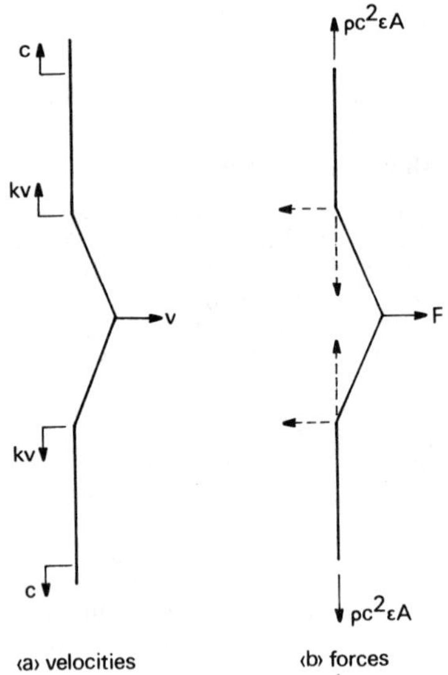

3.16 Definition diagrams for impact of an airplane wing on a barrage balloon cable (courtesy G. W. Housner)

$2[\{(vt)^2 + (kvt)^2\}^{1/2} - kvt]$. Equating the two, and retaining only the most important terms, leads to

$$\frac{v}{c} \simeq 2k\varepsilon.$$

But k may be found from the first equilibrium equation, namely,

$$k \simeq \frac{\sqrt{\varepsilon}}{v/c}.$$

Therefore

$$\frac{v}{c} \simeq \sqrt{2}\,\varepsilon^{3/4},$$

which gives an upper limit on v if a specified strain is not to be exceeded. The exact expression is $v/c = (2\varepsilon\sqrt{\varepsilon} - \varepsilon^2)^{1/2}$, which is not much different.

For a cold drawn high-strength steel wire the breaking strain is about 0.007, and the barrage cable will break on impact if the airplane speed is greater than about 3 percent of the stress wave velocity, or greater than about 150 m/sec. The airplane speed that causes immediate fraction is independent of cable size although the force on the wing certainly isn't!

Exercises

3.1 The variation in tension of a chain rotated about one of its ends with a constant angular velocity Ω is readily shown to be

$$\frac{dH}{dx} = -\Omega^2 m(x)x,$$

where x measures distance from the axis of rotation and $m(x)$ is the distribution of mass along its length. The effects of gravity are considered negligible. For small vibrations of the chain perpendicular to the plane of rotation show that regardless of the distribution of mass the frequency of the fundamental mode of vibration is

$$\omega_1 = \Omega$$

and its form is

$$\tilde{w}_1(x) = Ax,$$

where A is an arbitrary constant. This mode of vibration involves just a tilting of the plane of rotation [41].

3.2 The use of Fourier series permits the vertical components of the symmetric in-plane modes of vibration of a flat-sag suspended cable to be approached from a different viewpoint. Noting that the righthand side of (3.15) may be written as

$$\frac{mg\tilde{h}}{H} \sum_{n,\,\text{odd}} \frac{4}{n\pi} \sin\left(\frac{n\pi x}{l}\right),$$

show that the solution

$$\tilde{\mathbf{w}}(\mathbf{x}) = \tilde{\mathbf{h}} \sum_{n,\,\text{odd}} \frac{4}{n\pi(\omega^2 - n^2\pi^2)} \sin n\pi\mathbf{x}$$

gives the vertical component of each symmetric mode when ω^2 is known. Show further that the natural frequencies are found from the roots of the transcendental equation

$$\sum_{n,\,\text{odd}} \frac{1}{n^2\pi^2(\omega^2 - n^2\pi^2)} = \frac{1}{8\lambda^2}.$$

As is evident, the roots corresponding to the classical taut string are recovered when λ^2 is small, while, when λ^2 is large, we obtain the infinite series first given by Rohrs. As a matter of fact use of the first two terms in the series gives an accurate estimate of the first symmetric mode for all values of λ^2. The estimate is

$$\omega_1^2 = \frac{(10\pi^2 + 80\lambda^2/9\pi^2) - \{(10\pi^2 + 80\lambda^2/9\pi^2)^2 - 4(9\pi^4 + 656\lambda^2/9)\}^{1/2}}{2},$$

and it may easily be shown that the radical is always positive.

3.3 The vertical component of the ith in-plane mode of a flat-sag suspended cable is governed by

$$\frac{d^2\tilde{\mathbf{w}}_i}{d\mathbf{x}^2} + \omega_i^2\tilde{\mathbf{w}}_i = \lambda^2\mathbf{q}_i,$$

where $\mathbf{q}_i = \int_0^1 \tilde{\mathbf{w}}_i\, d\mathbf{x}$ and $\tilde{\mathbf{w}}_i$ is antisymmetric or symmetric, depending on whether $\mathbf{q}_i = 0$ or $\mathbf{q}_i \neq 0$, respectively. Similarly for the jth mode

$$\frac{d^2\tilde{\mathbf{w}}_j}{d\mathbf{x}^2} + \omega_j^2\tilde{\mathbf{w}}_j = \lambda^2\mathbf{q}_j.$$

Show that, because $\tilde{\mathbf{w}}_i$ and $\tilde{\mathbf{w}}_j$ are zero at each end of the cable, multiplication of these equations by $\tilde{\mathbf{w}}_j$ and $\tilde{\mathbf{w}}_i$, respectively, and integration by parts yields

$$(\omega_i^2 - \omega_j^2) \int_0^1 \tilde{\mathbf{w}}_i\tilde{\mathbf{w}}_j\, d\mathbf{x} = 0.$$

In most cases $i \neq j$ implies $\omega_i \neq \omega_j$, and we obtain the familiar expression

$$\int_0^1 \tilde{\mathbf{w}}_i \tilde{\mathbf{w}}_j \, d\mathbf{x} = 0, \quad i \neq j.$$

However, as has been demonstrated in section 3.2, there are situations where $\omega_i = \omega_j$, even though the associated modes are different. In these situations, which arise only for certain values of λ^2, the natural frequency of an antisymmetric mode may be identical to that of a symmetric mode. Nevertheless this equation still holds because the product of a symmetric mode and an antisymmetric mode is an antisymmetric quantity, the line integral of which must be zero.

3.4 For a flat-sag suspended cable:

(a) Show that the symmetric in-plane modes of vibration and the participation factors for additional tension are connected by the relation

$$\sum_n \alpha_n \left(1 - \tan\frac{\omega}{2}\sin\omega\mathbf{x} - \cos\omega\mathbf{x}\right) = 1,$$

and hence by integration show that

$$\sum_n \alpha_n \omega_n^2 = \lambda^2.$$

(b) Show that, if damping is ignored, the additional tension generated by the sudden application of a uniformly distributed load is given by

$$\mathbf{h(t)} = \mathbf{p}\sum_n \alpha_n (1 - \cos\omega_n \mathbf{t}),$$

where ω_n is the frequency of the nth symmetric mode. Damping eventually removes the harmonic modulation, and we are left with the static solution. Thus

$$\sum_n \alpha_n = \frac{1}{(1 + 12/\lambda^2)}.$$

As expected, the dynamic amplification doubles at most the static response.

3.5 Show that the steady-state additional tension generated in a flat-sag suspended cable responding to symmetric vertical acceleration at its ends of $\mathbf{a}\sin\omega\mathbf{t}$ is

$$\mathbf{h(t)} = \frac{\mathbf{a}\,\{\tan(\omega/2) - \omega/2\}}{\{\tan(\omega/2) - \omega/2 + 4(\omega/2)^3/\lambda^2\}}\sin\omega\mathbf{t}.$$

The interesting feature of this equation is that, for a sequence of forcing frequencies that satisfy $\tan(\omega/2) = \omega/2$, no additional tension is generated. The reason is that at these frequencies the spatial mean of the additional vertical deflection is zero.

3.6 A tugboat of mass m pulls a barge of mass M. If the hawser goes slack, after which the tug's velocity increases by δV, show that the length of hawser necessary to absorb adequately the jerk transmitted as the hawser tightens is given by

$$l = \frac{EA}{H_a^2}\frac{m(\delta V)^2}{(1 + m/M)},$$

where H_a is the allowable tension in the rope.

3.7 A long cable is stretched taut and flat, the tension being H_0. If one end is suddenly released, but then is equally suddenly arrested after it has moved a short distance, show that the peak dynamic tension registered is likely to be about $2\sqrt{3}H_0$.

This particular exercise has a direct application in the problem of the dynamic tension generated in an electrical transmission line when the conductor fractures suddenly in a span. The insulator bundles from which the conductor is hung in that span will swing outward and upward from the vertical, and the cable movements in other spans will start to be arrested when the insulator bundles have reached the horizontal. If the conductor is not firmly attached to each bundle, a cascading line collapse will result. In addition, if the conductor is overstressed by the dynamic tension, line fracture may occur elsewhere.

Our simplified result will tend to give an upper bound to the true and exceedingly complicated situation. It is probably an upper bound because the presence of sag in each span will reduce the peak dynamic tension. The results of field fracture tests indicate peak tensions in the range between about H_0 and $4H_0$, depending on circumstances. Large displacement dynamic analysis programs have proved effective in matching the measured time history of tension in, say, the insulator bundle.†

The implication is that a factor of safety of at least four is required if further fracture is to be avoided. Of course with such a factor of safety the probability of line fracture from any cause recedes.

For a different type of transmission line problem see problem 4.5.

3.8 A tension leg platform is anchored to the sea floor by four vertical riser-cables, each under a tension T_0. If one cable suddenly fractures, show that the peak dynamic tension recorded in each of the two closest cables will be $3T_0$.

Wave swell causes the cable tension to change in an essentially periodic fashion (in responding to wave forcing, the lateral stiffness of the riser group will be time varying, which makes analysis complicated). The peak tension could well approach $2T_0$ (although this must be an upper bound

† Personal communication, Alain H. Peyrot.

for the minimum would then be negative, which is inadmissible). Therefore a factor of safety of 6 on riser fracture, or foundation pull-out, is indicated.

References

1. Abdel-Ghaffar, A. M. 1976. *Earthqu. Eng. Res. Lab.*, California Institute of Technology: EERL 76-01.

2. Abdel-Ghaffar, A. M., and Housner, G. W. 1978. *J. Eng. Mech. Div., Proc. ASCE*, 104:983–999.

3. Baron, F. 1979. *ASCE Ann. Conv.*, Boston, Preprint 3590.

4. Biggs, J. M. 1964. *Introduction to Structural Dynamics.* New York: McGraw-Hill, ch. 1.

5. Bleich, F., et al. 1950. *The Mathematical Theory of Vibration in Suspension Bridges.* Bureau of Public Roads, U.S., Dept. of Commerce. Washington, D.C.: Government Printing Office.

6. Buckland, P. G., et al. 1979. *J. Struct. Div., Proc. ASCE*, 105:859–874.

7. Carrier, G. F. 1945. *Quart. Appl. Math.*, 3:157–165.

8. Carrier, G. F. 1949. *Quart. Appl. Math.*, 7:97–101.

9. Chikwendu, S. C., and Kevorkian, J. 1972. *SIAM J. Appl. Math.*, 22:235–258.

10. Clough, R. W., and Penzien, J. 1975. *Dynamics of Structures.* New York: McGraw-Hill, chs. 15, 21.

11. Coulson, C. A., 1968. *Waves.* 7th ed. Edinburgh: Oliver & Boyd, ch. 2.

12. Davenport, A. G., and Steels, G. N. 1965. *J. Struct. Div., Proc. ASCE*, 91: 43–70.

13. Dean, D. L. 1961. *J. Struct. Div., Proc. ASCE*, 87:1–21.

14. Dempsey, K. M., and Irvine, H. M. 1978. *Int. J. Earthqu. Eng. Struct. Dyn.*, 6:511–515.

15. Goodey, W. J. 1961. *Q. J. Mech. Appl. Math.*, 14:118–127.

16. Hartmann, A. J., and Davenport, A. G. 1966. *Eng. Sci. Res. Rep.*, University of Western Ontario: ST-4-66.

17. Housner, G. W. 1979. Personal communication.

18. Irvine, H. M., and Caughey, T. K. 1974. *Proc. R. Soc.* (Lond.), A341:299–315.

19. Irvine, H. M. 1976. *J. Struct. Div., Proc. ASCE*, 102:1286–1288.

20. Irvine, H. M., and Griffin, J. H. 1976. *Int. J. Earthqu. Eng. Struct. Dyn.*, 4:389–402.

21. Irvine, H. M. 1978. *J. Struct. Div., Proc. ASCE*, 104:343–347.

22. Irvine, H. M. 1980. *Q. J. Mech. Appl. Math.*, 33:227–234.

23. Kolsky, H. 1963. *Stress Waves in Solids.* New York: Dover, chs. 3, 7.

24. Lindsay, R. B. 1945. Introduction to *The Theory of Sound.* Vol. 1. By Lord Rayleigh. New York: Dover.

25. Luongo, A., and Rega, G. 1978. *Instituto di Scienza delle Costruzioni*, Università di Roma: Pubb. 11–237.

26. Meirovitch, L. 1975. *Elements of Vibration Analysis*. New York: McGraw-Hill, ch. 10.

27. Miles, J. W. 1965. *J. Acoust. Soc. Am.*, 38:855–861.

28. Møllmann, H. 1965. PhD dissertation, Technical University of Denmark.

29. Nigam, N. C., and Jennings, P. C. 1969. *Bull. Seis. Soc. Am.*, 59:909–922.

30. Pugsley, A. G. 1949. *Q. J. Mech. Appl. Math.*, 2:412–418.

31. Ramberg, S. E., and Griffin, O. M. 1977. *J. Struct. Div., Proc. ASCE*, 103: 2079–2092 (also *J. Struct. Div., Proc. ASCE*, 104:1926–1927).

32. Rannie, W. D. 1941. *The Failure of the Tacoma Narrows Bridge*. Board of engineers: Amman, O. H., von Karman, T., Woodruff, G. Append. 6. Washington D.C.: Federal Works Agency (also personal communication).

33. Rayleigh, Lord 1945. *The Theory of Sound*. Vol. 1. New York: Dover, ch. 6.

34. Rohrs, J. H. 1851. *Trans. Camb. Phil. Soc.*, 9:379–398.

35. Routh, E. J. 1955. *Advanced Dynamics of Rigid Bodies*. 6th ed. New York: Dover, ch. 13, pp. 278, 410–412.

36. Saxon, D. S., and Cahn, A. S. 1953. *Q. J. Mech. Appl. Math.*, 6:273–285.

37. Segedin, C. M., 1978. Personal communication.

38. Simpson, A. 1966. *Proc. Instn. Elect. Engrs.* 113:870–878.

39. Smith, C. E., and Thompson, R. S. 1973. *J. Appl. Mech., Trans. ASME*, 40:624 626.

40. Soler, A. I. 1970. *J. Franklin Inst.* 290:377–387.

41. Southwell, R. V. 1969. *Theory of Elasticity*. New York: Dover, pp. 481–482.

42. Tyndall, T. 1869. *On Sound*. 2nd ed. London: Longmans, Green, p. 104.

43. Vincent, G. S. 1965. *Proc. Conf. Wind Effects Build. Struct.*, 2:488–515, Teddington, England: HMSO (also a discussion 1945, *Trans. ASCE*, 110:512–522).

44. Watson, G. N. 1966. *Theory of Bessel Functions*. 2nd ed. Cambridge: Cambridge University Press, pp. 3–5.

45. Yu, A-T. 1952. *Proc. Soc. Exptl. Stress Anal.*, 9:141–157.

4

Applications

4.1 Introduction

In this chapter we present some direct applications of the theory of the first three chapters. To this end, sections are included on selected aspects of the response of clusters of guy cables, cable trusses and suspension bridges. Further applications are given as occasionally lengthy exercises at the end of the chapter.

4.2 Statics and Dynamics of a Cluster of Guy Cables

We start with the calculation of the horizontal and vertical forces induced at the common point of attachment of a cluster of three guys when that point moves horizontally and vertically and the guys are, and remain, taut and straight. These expressions are then linearized: they are extended to include the situation when n guys are arranged symmetrically in plan. Finally, a correction for self-weight is applied. Following examples on stability calculations for simple guyed towers, the linearized dynamic stiffness of a cluster is calculated, and an example is worked on the natural frequencies and modes of vibration of a multilevel guyed mast. The final subsection deals with the step-by-step calculation of guy response, using the exact equations of the elastic catenary.

Linearized Static Response

Consider the cluster of three guys shown in figure 4.1. Besides being tensioned to the same level, the guys are all identical in their geometric and material properties. Since their self-weight is ignored, the guys are assumed to be taut and straight and to remain so subsequently. Initially the axial load supplied by the cluster is $3T(b/c)$, where T is the pretension in each guy and (b/c) is the sine of the angle subtended to the horizontal.

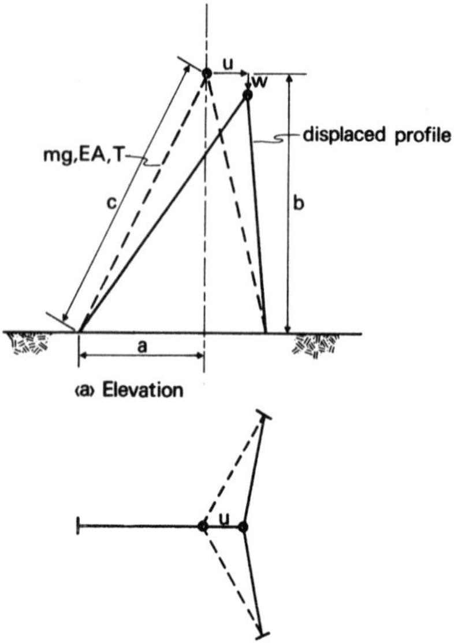

4.1 Definition diagrams for a symmetric cluster of three guys

Suppose that the cluster's junction moves horizontally and vertically by u and w, respectively, both displacements being small. The vertical plane containing one guy is typical, so these displacements may be assumed to occur in that plane. Since the displacements usually arise from the rigid body rotation and flexing of a mast, it is reasonable to assume that $w \sim O(u^2)$.

An analysis of the geometry of the displacements yields the following results: the increase in tension of the windward guy is

$$\Delta T' = EA\left\{\frac{au}{c^2} + \frac{a^2u^2}{8c^4} - \frac{u^2}{2c^2} + \frac{bw}{c^2}\right\} + O(u^3), \tag{4.1}$$

while the decrease in tension in each of the leeward guys is

$$\Delta T'' = -EA\left\{\frac{au}{2c^2} + \frac{a^2u^2}{8c^4} - \frac{u^2}{2c^2} + \frac{bw}{c^2}\right\} + O(u^3), \tag{4.2}$$

where EA is the axial stiffness of one guy and (a/c) is the cosine of the angle subtended to the horizontal by a guy.

The horizontal component of tension in the windward guy is

$$F'_u = T\left(\frac{a}{c} + \frac{u}{c}\right) + (EA - T)\left\{\frac{a^2 u}{c^3} + \frac{3a}{2c}\left(1 - \frac{a^2}{c^2}\right)\frac{u^2}{c^2} - \frac{abw}{c^3}\right\} + O(u^3), \tag{4.3}$$

while its vertical component is

$$F'_w = T\left(\frac{b}{c} - \frac{w}{c}\right) + (EA - T)\left\{\frac{abu}{c^3} - \frac{3a^2bu^2}{2c^5} + \frac{b}{c}\left(\frac{u^2}{2c^2} - \frac{bw}{c^2}\right)\right\} + O(u^3). \tag{4.4}$$

The horizontal component of tension in a leeward guy (in the u direction) is

$$F''_u = T\left(\frac{a}{2c} - \frac{u}{c}\right) - (EA - T)\left\{\frac{a^2 u}{4c^3} - \frac{3a}{4c}\left(1 - \frac{a^2}{4c^2}\right)\frac{u^2}{c^2} + \frac{abw}{2c^3}\right\} + O(u^3), \tag{4.5}$$

and its vertical component is

$$F''_w = T\left(\frac{b}{c} - \frac{w}{c}\right) - (EA - T)\left\{\frac{abu}{2c^3} + \frac{3a^2bu^2}{8c^5} + \frac{b}{c}\left(\frac{u^2}{2c^2} - \frac{bw}{c^2}\right)\right\} + O(u^3). \tag{4.6}$$

Therefore the expressions for the overall response of the cluster are

$$F_u (= F'_u - 2F''_u) = 3T\left(\frac{u}{c}\right) + \frac{3}{2}\left(\frac{a}{c}\right)^2 (EA - T)\left(\frac{u}{c}\right)\left\{1 - \frac{3}{4}\left(\frac{a}{c}\right)\left(\frac{u}{c}\right)\right\} + O(u^3), \tag{4.7}$$

$$\begin{aligned} F_w (= F'_w + 2F''_w) = {} & 3T\left(\frac{b}{c}\right)\left\{1 - \left(\frac{c}{b}\right)\left(\frac{w}{c}\right)\right\} \\ & + \frac{3}{2}(EA - T)\left\{\left(\frac{b}{c}\right)\left(1 - \frac{3}{2}\left(\frac{a}{c}\right)^2\right)\left(\frac{u}{c}\right)^2 \right. \\ & \left. - 2\left(\frac{b}{c}\right)^2\left(\frac{w}{c}\right)\right\} + O(u^3). \end{aligned} \tag{4.8}$$

These equations may be simplified considerably. It is clear that (u/c) will be small in practice (a large value would be 0.01), so $3(a/c)(u/c)/4$ may be neglected in comparison to unity in (4.7). By the same token, $(c/b)(w/c)$ may also be neglected in (4.8), and the second-order terms to the right of the righthand side of this equation may be dropped when it is realized that they would provide a contribution of $O(u^3)$ in the differential equation of equilibrium of, say, the mast of a guyed mast (see example 4.1). Thus (4.7) and (4.8) may be replaced with

$$F_u = \left\{3T + \frac{3}{2}\left(\frac{a}{c}\right)^2 (EA - T)\right\}\left(\frac{u}{c}\right),$$
$$F_w = 3T\left(\frac{b}{c}\right), \tag{4.9}$$

and the axial load supplied to the point of attachment is, to first order, independent of cluster displacements.

The first equation of (4.9) can obviously be simplified further as $T \ll EA$ for the structural materials typically used for guys, such as steel. (But we require $T > (a/c)EA(u/c)/2$ if the leeward guys are not to lose all their pretension.) The equation has been left this way temporarily to illustrate the point (of solely academic interest) that, when the guys are horizontal, the lateral resistance provided by the guy pretension is $3T(u/c)/2$. The other limiting case is, when the guys are vertical, the lateral resistance of the cluster is $3T(u/c)$. Nevertheless in the practical case, when (a/c) is not vanishingly small, the term $3(a/c)^2EA/2c$ dominates the horizontal stiffness to the exclusion typically of the contribution provided by the pretensions. Therefore for simplicity (4.9) is written in the form

$$F_u = \left\{3T + \frac{3}{2}\left(\frac{a}{c}\right)^2 EA\right\}\left(\frac{u}{c}\right),$$
$$F_w = 3T\left(\frac{b}{c}\right), \tag{4.10}$$

and these are the linearized equations for a cluster of three guys that remain straight.

In general a cluster of n guys arranged symmetrically around the mast gives rise to

$$F_u = n\left\{T + \frac{1}{2}\left(\frac{a}{c}\right)^2 EA\right\}\left(\frac{u}{c}\right),$$
$$F_w = nT\left(\frac{b}{c}\right). \tag{4.11}$$

This result, which holds for all $n \geq 3$, has an intuitive appeal. It may be established by summation of the trigonometric series (in the plan angle) that arise [31].

In reality the guys are not straight but, owing to their self-weight, dip slightly. To first order this does not change the expression for F_w, but it leads to a reduction in F_u—the equivalent spring is softer. Following an analysis similar to that presented in chapter 2, F_u may be modified to read

$$F_u = n\left[T + \frac{(a/c)^2\, EA}{2\{1 + (mga/T)^2\, EA/12T\}}\right]\left(\frac{u}{c}\right), \quad n \geq 3, \tag{4.12}$$

where mg is the weight per unit length of one guy. This correction for self-weight is well known and can be significant [13, 52], although, if the guys sag appreciably, geometric nonlinearity will feature in the response, and step-by-step methods will have to be used. In general the guys of land-based masts sag little, so (4.12) is probably sufficient for static purposes. On the other hand, in offshore applications such as the very tall guyed oil production platforms presently on oil industry drawing boards, the guys are noticeably curved, and the more sophisticated step-by-step methods are required.

Subject to these qualifications the results given here are general and apply to the static response of any symmetric cluster of n guys. Consequently, if a mast is stayed at several points, suitably modified forms of these equations determine local cluster actions. However, in linearized dynamic analyses it is the dynamic stiffness of the cluster that is of interest and this, as we shall see in the next subsection, may be readily established with little sacrifice on accuracy.

Example 4.1 Stability of Simple Guyed Masts

The mast shown in figure 4.2 is stayed at its top by three guys. From the point of view of mast stability the guys have two opposing effects. They supply an axial force to the mast tip, thereby giving rise to the possibility of instability, but they also provide a measure of lateral restraint with part of the lateral stiffness being directly proportional to the guy pretension. Although this latter effect is usually insignificant, it does play a part in the limiting trends in behavior. In addition for tall towers the destabilizing effect of tower self-weight may materially affect the calculations.

The differential equation for shear force equilibrium (see figure 4.3) is

$$EI\frac{d^3u}{dz^3} + F_w\frac{du}{dz} + (mg)_m z\frac{du}{dz} = F_u,$$

where z is measured from the top of the mast as is u (therefore u_1 is the relative displacement between top and bottom), EI is the flexural rigidity of the mast,

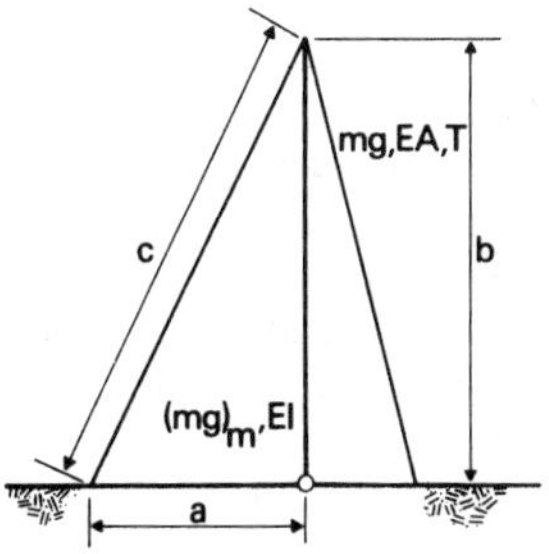

4.2 Definition diagram for a guyed mast with a pinned base

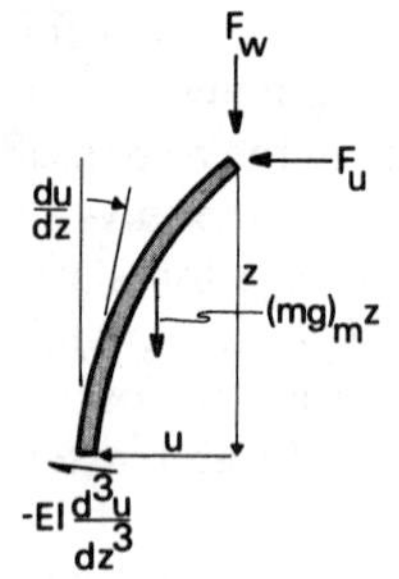

4.3 Portrayal of terms in the differential equation governing shear force equilibrium in the mast

$(mg)_m$ is its weight per unit length, and F_u and F_w are given by (4.10). This equation may be recast in the following dimensionless form

$$\frac{d^3\mathbf{u}}{d\mathbf{z}^3} + \alpha^2(1 + \beta^2\mathbf{z})\frac{d\mathbf{u}}{d\mathbf{z}} = \alpha^2\left(1 + \frac{\lambda^2}{2}\right)\mathbf{u}_1,$$

where $\mathbf{z} = z/b$, $\mathbf{u} = u/(3T(b/c)b^3/EI)$, $\alpha^2 = 3T(b/c)b^2/EI$, $\beta^2 = (mg)_m b/(3T(b/c))$, and $\lambda^2 = (a/c)^2[(EA/T)/\{1 + (mga/T)^2 EA/(12T)\}]$. The smallest value of λ is zero, which occurs when the guys are vertical, but it is likely to be large when a/c takes on values more representative of actual clusters.

Boundary conditions may be selected from either

1. a pinned base

$$\mathbf{u}(0) = 0, \quad \mathbf{u}(1) = \mathbf{u}_1,$$

$$\frac{d^2\mathbf{u}}{d\mathbf{z}^2}(0) = \frac{d^2\mathbf{u}}{d\mathbf{z}^2}(1) = 0;$$

2. a fixed base

$$\mathbf{u}(0) = 0, \quad \mathbf{u}(1) = \mathbf{u}_1,$$

$$\frac{d^2\mathbf{u}}{d\mathbf{z}^2}(0) = 0, \quad \frac{d\mathbf{u}}{d\mathbf{z}}(1) = 0.$$

Because of the form of the nondimensionalization, either α or β may be taken as the dependent variable (with either EI or $(mg)_m b$ being the dependent dimensional variable), while λ is considered to be the independent variable.

Analytical solutions to the problem are not straightforward on account of the presence of the term for mast self-weight. By a suitable transformation a solution in terms of integrals of Bessel functions of one-third order may be found for $d\mathbf{u}/d\mathbf{z}$ [4], but this is of little help since a further integration is required.

On the other hand, solutions are readily obtained if mast self-weight is ignored (that is, $\beta = 0$). In this case the solution of the reduced governing equation for a pinned base mast results in the eigenvalue equation

$$\sin\alpha = 0,$$

the only relevant solution of which is

$$\alpha = \pi.$$

This result is independent of λ since the mast tip cannot move if the condition of zero moment at the base is to be met. This case is simply that of a pinned-pinned Euler strut [31] (see figure 4.4a).

With full fixity provided at the base of the mast, the eigenvalue equation may be shown to be [31]

$$\tan \alpha = \frac{\alpha}{1 + 2/\lambda^2},$$

the first nonzero root of which lies in the range

$$\pi \leq \alpha < 1.43\pi.$$

The upper limit applies when $\lambda \to \infty$, and the guys offer almost complete lateral resistance to mast tip displacement. The mast behaves as a pinned-fixed Euler strut (see figure 4.4b). The lower limit applies when $\lambda = 0$ (when the guys are vertical, a situation of academic interest only). The mast tip displacement is nonzero, but no moment is induced at the base because the line of action of guy pretension passes through the base. This limiting case is simply a pinned-pinned Euler strut. (The problem of a cantilever column acted on by a compressive force of fixed line of action—tilt buckling—is discussed by Simitses [48].)

The inclusion of mast self-weight necessitates the numerical solution of the governing equation. Figure 4.5 shows plots of the dependent variable β for a range of values of α and λ for the pinned base mast. Several points are worthy of note. When $\alpha \to \pi$, $\beta \to 0$ for all λ; when $\alpha \to 0$, $\beta \to \lambda$ for all λ (this is a rigid body mode that can be established from simple statics, see example 4.2); finally, when $\lambda \to \infty$, the envelope is very closely given by the upper bound

$$\left(\frac{\alpha}{\pi}\right)^2 \left(1 + \frac{\beta^2}{2}\right) = 1,$$

which may be established by using a one-term Rayleigh solution with a shape function $\mathbf{u}(\mathbf{z}) = \sin \pi \mathbf{z}$.

Results for fixed base masts may also be found, although fixed bases tend not to be favored in practice for a variety of reasons. An excellent upper limit to the general solution in this case useful when the guys provide a large measure of lateral

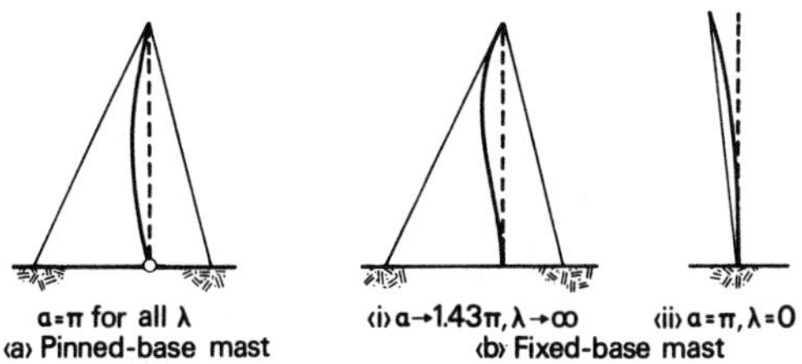

4.4 Modes of buckling when tower self-weight may be ignored

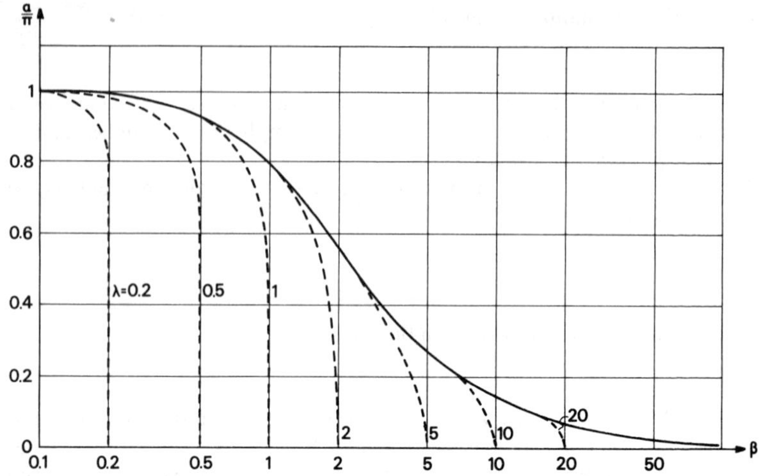

4.5 Semilogarithmic plots of the full solution for a pinned-base mast. The dotted lines represent numerical results and the full line represents the envelope given by

$$\left(\frac{\alpha}{\pi}\right)^2\left(1 + \frac{\beta^2}{2}\right) = 1.$$

restraint is

$$\left(\frac{\alpha}{\varepsilon}\right)^2(1 + \delta\beta^2) = 1,$$

where

$$\delta = \frac{1}{2} - \frac{2}{\varepsilon^2} - \frac{4}{\varepsilon^4}\{(1 + \varepsilon^2)^{1/2} + 1)\},$$

and ε is the first nonzero root of $\tan \varepsilon = \varepsilon$; that is, $\varepsilon = 1.43\pi$. This result may be obtained (after some labor) by employing a one-term Rayleigh solution with a shape function corresponding to the first mode of buckling of a pinned-fixed Euler strut. The envelope is finally

$$\left(\frac{\alpha}{1.43\pi}\right)^2(1 + 0.35\beta^2) = 1.$$

Reference [31] shows how these results may be extended to include an external axial load applied to the tower top.

In actual guyed masts design axial loads are frequently a factor of four or five, or more, less than the elastic critical loads. This margin usually means that it is sufficiently accurate to ignore the effect of axial load in lateral response calculations, provided that displacements are small in the accepted structural sense. (This means that mast stiffnesses and fixed end moments do not need to be modified to account for the presence of axial load. The influence of the $P - \Delta$

effect on mast bending moments then enters as a final correction to them.) But, as the next example shows, tall towers that undergo large displacements in response to design loads can be in a precarious position.

Example 4.2 Large Displacement Stability of a Tall Guyed Offshore Oil Production Platform

In the last few years, as economic constraints have dictated that exploration move farther and farther offshore, there has been considerable interest in the concept of using compliant guyed towers to support production platforms. Several investigations into the feasibility of such systems have been conducted, and some of the major oil companies are now preparing to proceed with detailed design of structures that will operate in depths of 500 m to 1,000 m, which is far outside the range of presently installed conventional steel jacket platforms and reinforced concrete gravity platforms.

Needless to say, major problems are posed by operating in such depths. The offshore environment is harsher, foundation conditions are of considerable concern, and the structural behavior is complicated by the fact that these compliant structures may have fundamental periods two or three times longer than predominant wave periods (compared to conventional platforms in shallow water that operate at the opposite end of the response spectrum with periods that may be a factor of four or five less than the wave period). As a result the platform undergoes large displacements in response to extreme design storms.

A compliant platform consists of a multi-tube, cross-braced truss tower, stayed at or near the top by a radiating array of bridge strands, and is supported at the mud line either by a piled foundation (in which case flexural restraint is provided there) or by a spud can which in burying itself provides the necessary vertical restraint. In the latter case the tower is effectively pinned at its base and will behave like a rigid body; design lateral loads, which are concentrated near the level of guy attachment, do not then cause significant flexural stresses in the tower (at least if first mode response predominates). At the mud line the guys are first attached to a long line of heavy clump weights before being anchored off some distance away. The purpose of this is to produce a force-limiting mode in the windward guys which are required to lift off progressively in large storms. Under more frequent small storms the guys are stiffer [18, 19].

Figure 4.6 shows a schematic of a tower for which the following data are

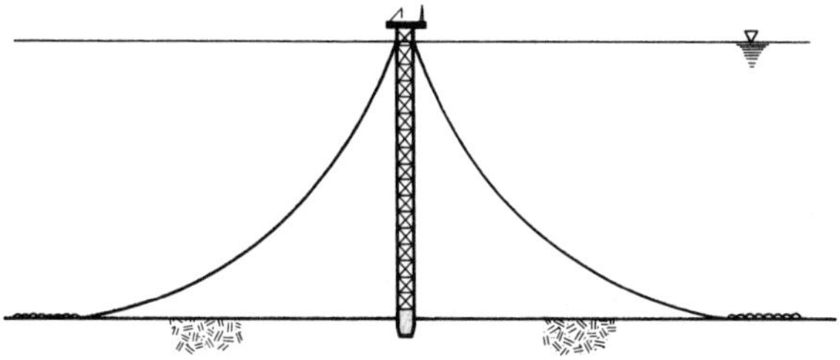

4.6 Schematic of a guyed offshore oil production platform

believed to be representative: Height, l = 500 m, platform load (plus axial component of guy tensions), $P = 9 \times 10^4$ kN, submerged weight of tower $(mg)'l = 4 \times 10^4$ kN, initial stiffness of equivalent horizontal spring (provided by the guys), $k = 1.5 \times 10^3$ kN/m, number of guys (9 cm bridge strand) = 20, inclination to vertical = 60°, initial tension = 1.5×10^3 kN. Due to the softening associated with lift-off of the clump weights, the lateral force supplied by the guys is approximately as shown in figure 4.7.

Elastic stability is assured if

$$\left(\frac{(mg)'l}{2} + P\right) < kl.$$

The lefthand side is 11×10^4 kN, while the righthand side is 75×10^4 kN, which is some seven times greater. There is therefore no problem with ensuring elastic stability of the structure.

However, under a design storm (such as a 30 m wave (crest-to-trough), a 150 km/hr wind plus a 5 km/hr current), the situation is dramatically different. If this equivalent static force is labeled F_s (as in figure 4.8), while the resistance of the guys is F_r, equilibrium can be maintained when

$$\left(\frac{(mg)'l}{2} + P\right)\Delta + F_s l < F_r l,$$

where Δ is the lateral deflection at the top. Referring to figure 4.7, we note that a reasonable representation of the force-deflection characteristics of the guy cluster is given by

$$F_r = F_{rm}\left\{1 - \exp\left(\frac{-\Delta}{\Delta_0}\right)\right\},$$

where F_{rm} is the maximum allowable resisting force ($\simeq 1.8 \times 10^4$ kN) and $\Delta_0 = F_{rm}/k$. Therefore we require

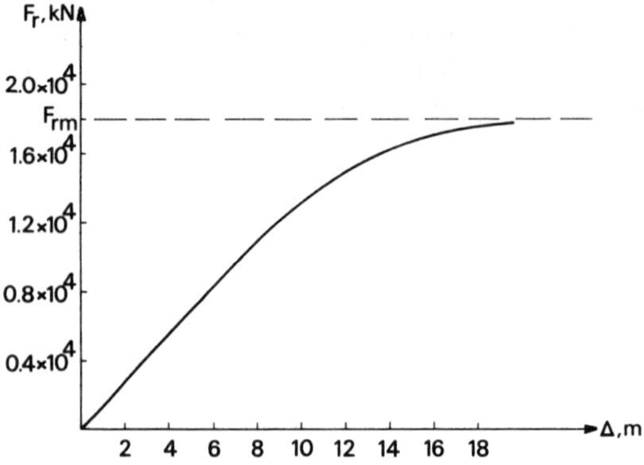

4.7 Force-deflection relation for a cluster of twenty bridge strands when allowance is made for lift-off of clump weights on the windward guys

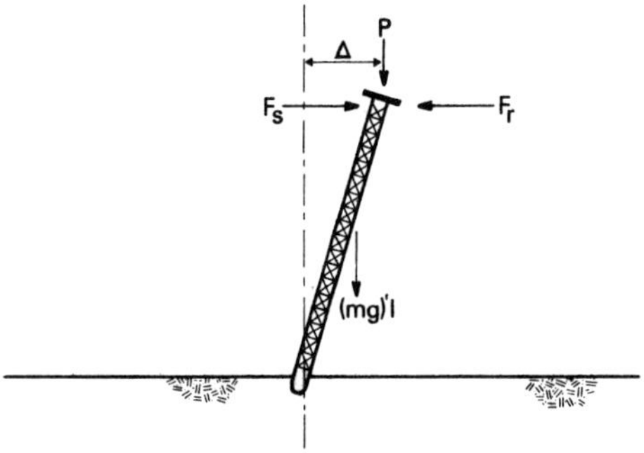

4.8 Equilibrium of the tower under the design storm

$$F_{rm}\left\{1 - \exp\left(\frac{-\Delta}{\Delta_0}\right)\right\} > F_s + \left(\frac{(mg)'l}{2} + P\right)\frac{\Delta}{l}.$$

Plotting this, as in figure 4.9, indicates that stability is no longer possible when

$$\Delta' = \Delta_0 \ln\left\{\frac{kl}{((mg)'l/2 + P)}\right\}$$

and

$$F_s = F_{rm}\left\{1 - \left(\frac{(mg)'l/2 + P}{F_{rm}}\right)\left(\frac{\Delta_0 + \Delta'}{l}\right)\right\}.$$

Hence in the present case

$$\Delta' = 23 \text{ m}, \quad \text{and} \quad \frac{F_s}{F_{rm}} = 0.57.$$

The potential difficulty highlighted by these results is that the design storm is likely to involve values of F_s and Δ approaching these limits and that an adequate margin of safety is obviously lacking. However, the structure may be dynamically stable because, with a period of approximately 30 seconds compared to a wave period of, say, 12 seconds, the response lags the forcing by nearly 180°. So overall this static analysis points to a problem area rather than make a definitive state-

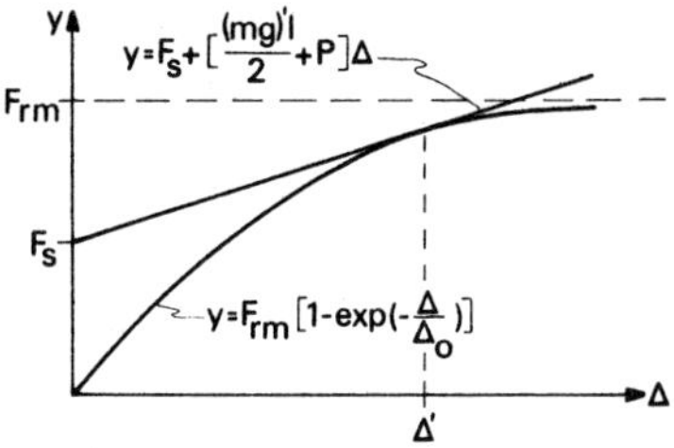

4.9 Large displacement stability plot for the tower

ment about it. It may be noted in passing that if extra buoyancy tanks are fitted below the region of wave influence, and the displaced volume is sufficiently large, no problem of stability will be encountered. These buoyancy tanks may be arranged around the tower to counter torsional components in the response that arise because the platform load is usually eccentric and the cable cluster cannot easily provide the necessary torsional resistance.

Linearized Dynamic Stiffness of a Cluster of Guys

For the purposes of dynamic response the cluster of n guys will be judged equivalent to $n/2$ single cables. Embodied in this is the assumption that the majority of the dynamic stiffness of the cluster arises from additional tension induced in the guys. Thus we are interested only in the symmetric cable modes—the antisymmetric modes do not induce additional tension, and we need not concern ourselves with them. Neither are we concerned with pendulum modes in which the guys swing to and fro out of their vertical planes.

Therefore we consider an inclined cable with movements of $u_1 \cos\theta$ and $u_1 \sin\theta$ imposed at its far end (see figure 4.10). It is noted that $\cos\theta = a/c$ and $\sin\theta = b/c$ in conformity with figure 4.1. We need only consider the displacement $u_1 \cos\theta$, as is justified by subsequent comparison with the static stiffness.

The equation of motion of this guy is

$$\frac{d^2\tilde{\mathbf{w}}_*}{d\mathbf{x}_*^2} + \boldsymbol{\omega}_*^2\tilde{\mathbf{w}}_* = \tilde{\mathbf{h}}_*, \tag{4.13}$$

where $\mathbf{x}_* = x_*/c$, $\tilde{\mathbf{w}}_* = \tilde{w}_*/(mg\,c^2 \cos\theta/T)$, $\tilde{\mathbf{h}}_* = \tilde{h}_*/T$, and $\boldsymbol{\omega}_* = \omega_* c(m/T)^{1/2}$. The cable equation is

$$\tilde{\mathbf{h}}_* = \lambda_*^2\left\{\tilde{\mathbf{u}}_* + \int_0^1 \tilde{\mathbf{w}}_*\,d\mathbf{x}_*\right\}, \tag{4.14}$$

where $\tilde{\mathbf{u}}_* = \tilde{u}_1 \cos\theta/\{(mg\,c^2\cos\theta/T)\,(mgc\cos\theta/T)\}$ and $\lambda_*^2 =$

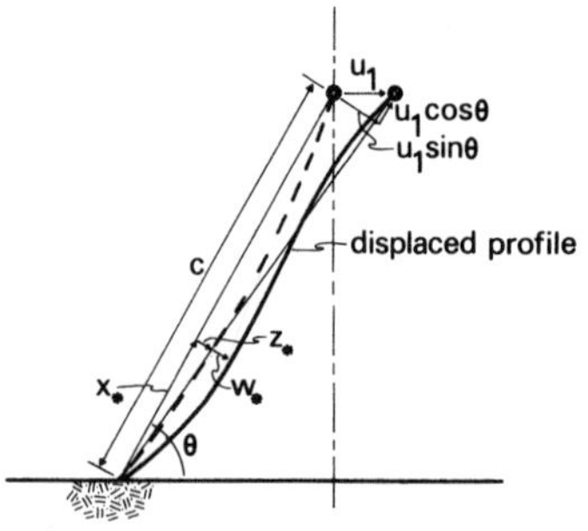

4.10 Definition diagram for the dynamic response of a guy

$(mgc\cos\theta/T)^2\,(EA/T)$. This notation represents a compromise between the work of examples 2.1 and 3.1 and the previous subsection. The solution, subject to boundary conditions written about the displaced chord, is

$$\tilde{\mathbf{w}}_*(\mathbf{x}_*) = \frac{\tilde{\mathbf{h}}_*}{\omega_*^2}\left(1 - \tan\frac{\omega_*}{2}\sin\omega_*\mathbf{x}_* - \cos\omega_*\mathbf{x}_*\right). \tag{4.15}$$

After rearrangement we have from the cable equation

$$\tilde{\mathbf{h}}_* = \frac{4\,\tilde{\mathbf{u}}_*(\omega_*/2)^3}{\{\tan(\omega_*/2) - (\omega_*/2 - 4(\omega_*/2)^3/\lambda_*^2)\}}. \tag{4.16}$$

Therefore the dynamic displacement of the guy relative to the displaced chord is

$$\tilde{\mathbf{w}}_*(\mathbf{x}_*) = \frac{\tilde{\mathbf{u}}_*(\omega_*/2)(1 - \tan(\omega_*/2)\sin\omega_*\mathbf{x}_* - \cos\omega_*\mathbf{x}_*)}{\{\tan(\omega_*/2) - (\omega_*/2 - 4(\omega_*/2)^3/\lambda_*^2)\}}. \tag{4.17}$$

Consistent with the approximations made, the total restoring force of the n guys is

$$\frac{n}{2}T\tilde{\mathbf{h}}_*\cos\theta.$$

Thus in dimensional terms the dynamic lateral stiffness of the cluster is

$$k_{\text{dyn}} = \frac{nT\{2/mgc/T)^2\}(\omega_*/2)^3}{c\{\tan(\omega_*/2) - (\omega_*/2) - 4(\omega_*/2)^3/\lambda_*^2\}}. \tag{4.18}$$

When the frequency of the combined systems of clusters plus mast is low (which is after all what ω_* refers to),

$$\tan\frac{\omega_*}{2} - \left\{\frac{\omega_*}{2} - \frac{4}{\lambda_*^2}\left(\frac{\omega_*}{2}\right)^3\right\} \simeq \left(\frac{\omega_*}{2}\right)^3\left(\frac{1}{3} + \frac{4}{\lambda_*^2}\right), \tag{4.19}$$

and

$$k_{\text{dyn}} \simeq \frac{nEA\cos^2\theta}{2c}\cdot\frac{1}{(1 + \lambda_*^2/12)}, \tag{4.20}$$

which is precisely the static stiffness in (4.12) minus the term nT/c which, as we have mentioned previously, is of little importance for guys of normal inclination.

The first work on the calculation of the dynamic stiffness of a guy was that by Kolousek [34], and it has been simplified and extended by Davenport [15] and Davenport and Steels [16]. Their work was based on a Fourier series approach and lacks the conciseness of the present direct method, although both approaches appear to give essentially similar expressions. The advantages of our analytical solution may be listed as,

1. A cluster of n guys is replaced by one expression for its dynamic stiffness, equation (4.18).

2. The associated dynamic displacements in the guys are found from substitution in a single equation, namely, equation (4.17).

Thus a saving in computational effort results when a dynamic analysis of a multilevel guyed mast is undertaken. The number of degrees of freedom necessary for reasonable accuracy is no greater than $2i$, where i is the number of staying levels because, if axial displacements are judged insignificant, a knowledge of just the lateral displacement and rotation of the mast at each stay level is necessary. The next example illustrates the utility of this approach.

Example 4.3 Natural Frequencies and Modes of Vibration of a Multilevel Guyed Mast

The mast shown in figure 4.11 is a three-legged braced truss of tubular steel construction. For the purposes of this exercise mast and guy properties are assumed to be uniform. The pertinent data are

1. for the mast

$$E = 2.1 \times 10^5 \text{ MPa},$$

$$I = 1.5 \times 10^{-3} \text{ m}^4,$$

$$(mg)_m = 0.6 \text{ kN/m};$$

2. for each guy (three per cluster)

$$T = 30 \text{ kN},$$

$$E = 1.5 \times 10^5 \text{ MPa},$$

$$A = 2 \times 10^{-4} \text{ m}^2,$$

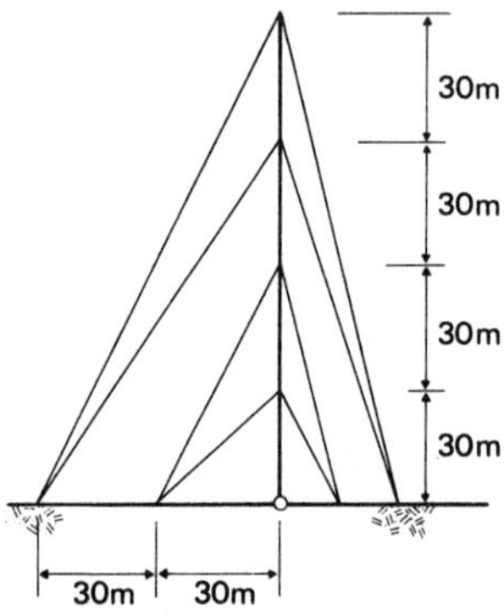

4.11 Guy arrangements for a 120 m tall multilevel guyed mast

$(mg)_c = 0.025$ kN/m.

The static stiffnesses of each cluster are calculated to be (see equation 4.12)

$k_1 \simeq 500$ kN/m (bottom cluster),

$k_2 \simeq 130$ kN/m,

$k_3 \simeq 90$ kN/m,

$k_4 \simeq 60$ kN/m (top cluster).

These spring stiffnesses may be used to obtain initial estimates of the natural frequencies of the mast-guy system, using standard plane-frame eigenvalue routines—the mast is modeled as a beam on a discrete elastic foundation. The values of the natural frequencies so obtained may then be used to find dynamic stiffnesses (see 4.18) of each cluster for each mode. The eigenvalue analysis is then run again to obtain improved estimates of the frequencies and modes, and this process is continued until a satisfactory result is obtained. (In using equation 4.18, some care has to be exercised; otherwise spurious negative dynamic stiffnesses may be calculated. This did not occur with the present problem, and there are indications that it will not occur if the static stiffnesses are used to start the iterative process. Nevertheless, it is important that one be aware of this possibility.)

In the present example only three such iterations were required to achieve acceptable answers. The results for the first three modes are shown in figure 4.12. In each case the dynamic guy stiffnesses were not greatly different from the static stiffnesses, because the periods of the combined system were somewhat shorter than the fundamental periods of the guys themselves. A fundamental period of 0.43 seconds indicates that this stayed mast is quite stiff. The fundamental mode is in this instance similar to a rigid body mode, and in fact, if hand calculations are performed on this basis, one obtains a value of 0.3 seconds for its period. This is shorter than the true value, as it should be, because the mast is flexible. The

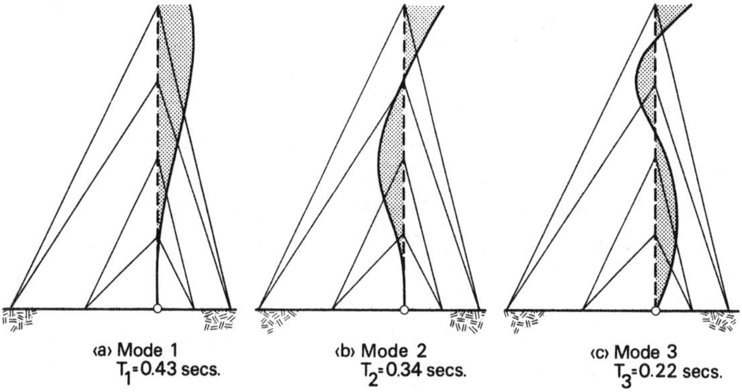

4.12 First three modes of vibration (guy movements have not been shown—they may be found from equation 4.17)

modal participation factors of 1.25, -0.57, and 0.79, respectively, suggest that the first mode is likely to predominate in the dynamic response, although appreciable contributions from the higher modes may be present if the loading has significant higher frequency components. (Pursuing this rigid body analogy is not a bad way to start an assessment of mast behavior. For example, under a uniformly distributed wind load of 1.5 kN/m the mast tip deflection is about 0.5 m, the peak bending stress in the mast legs is about 200 MPa (these stresses usually highest in the vicinity of the lower stay points), and the peak guy stress is about 500 MPa, although the pretensions in the leeward guys are all but lost.)

One of the earlier studies of the dynamic response of guyed masts to wind loading was done by Shears [46]. A more recent account of their free vibrational characteristics is given in the thesis by Howson [23]. Descriptions of particular designs and design methods have appeared in the journals of various engineering societies, and there exist several general-purpose computer programs for guyed mast analysis.

A vexing question still not completely resolved concerns the interaction of the guys and the mast when the guys are ice-laden. Such guys may gallop under wind loading, and these galloping oscillations usually occur at the frequency of the fundamental inplane mode of the guy, which is often somewhat lower than the fundamental frequency of the combined system. Some expensive, hotly contested mast failures have occurred in which proponents of the galloping guy theory (failure is due to the tail wagging the dog) and their opponents, claiming that the failure may be attributed to the mast itself, have been unable to reach a consensus. Be that as it may, galloping oscillations of ice-laden guys do occur, and they can be of appreciable magnitude. The provision of damping devices to control these low frequency oscillations appears to be worthwhile, and some work in this direction is being done [43].

Step-by-Step Calculation of Guy Response

When the sag is appreciable, the linearized expressions developed in the earlier subsection are no longer adequate, and other methods of determining guy response must be sought. One possibility is to use the finite element method by dividing the cable up into several elements, writing approximate equilibrium and compatibility equations for each element, combining these into a global system for the complete guy, and so proceeding to determine its response. Acceptable accuracy will be assured if several elements are used and suitably small calculation steps are chosen. This approach has been thoroughly explored in the literature. An alternative approach is to use step-by-step methods based on approximate analytical equations for the complete guy—this has been advocated by, for example, Livesley [35].

However, recall that in chapter 1 we obtained exact analytical solutions for the elastic catenary. If we use these results we have available a powerful means of expeditiously determining cluster response. In contrast to the multi-element techniques, the guy may be represented by a single element.

The potential savings in computer time make the method attractive for static response calculations.

The guy shown in figure 4.13a is fixed at its lower end. Its upper end is attached to a point that undergoes displacements dl and dh, thereby inducing changes in the horizontal and vertical reactions there of dH and dV (see figure 4.13b). (No allowance is made here for the alteration in profile due to the effects of wind or current drag on the cable. Except perhaps for very long guys, it will be sufficient to transfer an appropriate portion of the total drag on the guy to its upper stay level and thereafter ignore the problem.) For convenience the notation adopted in chapter 1 has been retained.

We assume that the two components H and V of the tension at the upper end of the initial profile have been determined from the implicit equations

$$l = \frac{HL_0}{EA} + \frac{HL_0}{W}\left[\sinh^{-1}\left(\frac{V}{H}\right) - \sinh^{-1}\left(\frac{V-W}{H}\right)\right], \tag{4.21}$$

$$h = \frac{WL_0}{EA}\left(\frac{V}{W} - \frac{1}{2}\right) + \frac{HL_0}{W}\left[\left\{1 + \left(\frac{V}{H}\right)^2\right\}^{1/2} - \left\{1 + \left(\frac{V-W}{H}\right)^2\right\}^{1/2}\right]. \tag{4.22}$$

It is further assumed that the lower portion of the guy never lies on the ground, although this assumption may be relaxed as and when necessary.

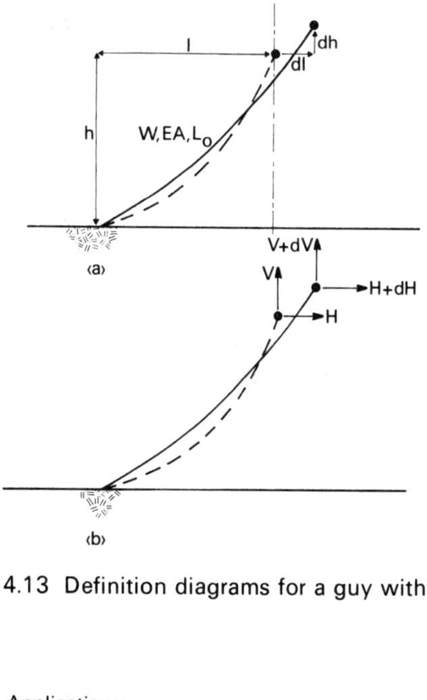

4.13 Definition diagrams for a guy with appreciable sag

Equations (4.21) and (4.22) may be written as

$$\begin{aligned} l &= f(H, V), \\ h &= g(H, V), \end{aligned} \tag{4.23}$$

so that

$$\begin{aligned} dl &= \frac{\partial f}{\partial H} dH + \frac{\partial f}{\partial V} dV, \\ dh &= \frac{\partial g}{\partial H} dH + \frac{\partial g}{\partial V} dV. \end{aligned} \tag{4.24}$$

In matrix notation

$$\begin{Bmatrix} dl \\ dh \end{Bmatrix} = \mathbf{F} \begin{Bmatrix} dH \\ dV \end{Bmatrix}, \tag{4.25}$$

where

$$\mathbf{F} = \begin{bmatrix} \dfrac{\partial f}{\partial H} & \dfrac{\partial f}{\partial V} \\ \dfrac{\partial g}{\partial H} & \dfrac{\partial g}{\partial V} \end{bmatrix} = \begin{bmatrix} f_{11} & f_{12} \\ f_{21} & f_{22} \end{bmatrix}$$

Therefore

$$\begin{Bmatrix} dH \\ dV \end{Bmatrix} = \mathbf{K} \begin{Bmatrix} dl \\ dh \end{Bmatrix}, \tag{4.26}$$

where

$$\mathbf{K} = \mathbf{F}^{-1} = \begin{bmatrix} f_{22} & -f_{12} \\ -f_{21} & f_{11} \end{bmatrix} \frac{1}{\det \mathbf{F}}$$

and

$$\det \mathbf{F} = f_{11} f_{22} - f_{12} f_{21} \cdot$$

Because of geometric nonlinearity the flexibility matrix **F** and the stiffness matrix **K** are nonsymmetric, and the elements need to be constantly updated. (From (4.3) and (4.4) the linearized equations for a straight guy are

$$\begin{Bmatrix} \Delta F'_u \\ \Delta F'_w \end{Bmatrix} = \frac{1}{c} \begin{bmatrix} T\dfrac{b^2}{c^2} + EA\dfrac{a^2}{c^2} & (EA - T)\dfrac{ab}{c^2} \\ (EA - T)\dfrac{ab}{c^2} & T\dfrac{a^2}{c^2} + EA\dfrac{b^2}{c^2} \end{bmatrix} \begin{Bmatrix} u \\ -w \end{Bmatrix}.$$

The matrix is symmetric. This result is contained in equation 4.26 and is an upper limit to the stiffness obtained when most of the sag has been pulled out.)

From (4.21) and (4.22) we see that

$$f_{11} = \frac{\partial f}{\partial H} = \frac{L_0}{EA} + \frac{L_0}{W}\left\{\sinh^{-1}\left(\frac{V}{H}\right) - \sinh^{-1}\left(\frac{V-W}{H}\right)\right\}$$
$$+ \frac{L_0}{W}\left[-\frac{V/H}{\{1+(V/H)^2\}^{1/2}} + \frac{(V-W)/H}{\{1+((V-W)/H)^2\}^{1/2}}\right], \tag{4.27}$$

$$f_{12} = \frac{\partial f}{\partial V} = \frac{L_0}{W}\left[\left\{1+\left(\frac{V}{H}\right)^2\right\}^{-1/2} - \left\{1+\left(\frac{V-W}{H}\right)^2\right\}^{-1/2}\right], \tag{4.28}$$

$$f_{21} = \frac{\partial g}{\partial H} = \frac{L_0}{W}\left[\left\{1+\left(\frac{V}{H}\right)^2\right\}^{1/2} - \left\{1+\left(\frac{V-W}{H}\right)^2\right\}^{1/2}\right]$$
$$+ \frac{2L_0}{W}\left[-\frac{V^2/H^2}{\{1+(V/H)^2\}^{1/2}} + \frac{((V-W)/H)^2}{\{1+((V-W)/H)^2\}^{1/2}}\right], \tag{4.29}$$

and

$$f_{22} = \frac{\partial g}{\partial V} = \frac{L_0}{EA} + \frac{L_0}{W}\left[\frac{V/H}{\{1+(V/H)^2\}^{1/2}} - \frac{((V-W)/H)}{\{1+((V-W)/H)^2\}^{1/2}}\right]. \tag{4.30}$$

(For purposes of calculation it is best to replace the inverse hyperbolic sine by its logarithmic representation, namely, $\sinh^{-1} x = \ln\{x + (1 + x^2)^{1/2}\}$.)

Equations (4.27) to (4.30) are useful for response calculations for various types of floating structures. Comprehensive accounts of the analysis and design of the mooring systems of ocean engineering structures and oceanographic buoys are to be found in the books by Evans and Adamchak [17] and Berteaux [8]; in addition both of these works contain extensive reference lists.

In those situations where the mass of the moored structure is significantly greater than that of the mooring system, the dynamic response of the structure may, given the very probable uncertainty in the fluid loading, be assessed using these static stiffnesses. The nature of wave forces, and fluid-structure interaction in general, is obviously a crucial item in the calculations, but it is outside both the scope and the purpose of this work to pursue it here. Among works that treat the subject mention may be made of the books by Newman [39] and Wiegel [57].

When the guy stays a mast or a tower, $dh \sim O(dl)^2$, and we may set dh

to zero with little likelihood of appreciable error. Then

$$dH = \frac{f_{22}}{f_{11}f_{22} - f_{12}f_{21}} dl, \tag{4.31}$$

$$dV = \frac{-f_{21}}{f_{11}f_{22} - f_{12}f_{21}} dl, \tag{4.32}$$

and the springs are uncoupled, although it is difficult to show this by a single diagram.

If an increment in displacement is prescribed, several iterations within that step are necessary to arrive at the correct increment in the reaction. A schematic of this procedure based on the tangent modulus method is shown in figure 4.14a. After each iteration new values of l are found by direct substitution of the new values of H and V in (4.21). Iterations proceed until the difference between the updated value of l and its initial value match the prescribed displacement.

If an increment in the horizontal reaction is prescribed, several iterations within that step are again required to arrive at the correct increment in displacement (see figure 4.14b). The disadvantage here is that at the end of each iteration new values of H and V have to be found implicitly from (4.21) and (4.22). This necessitates iterations within that iteration

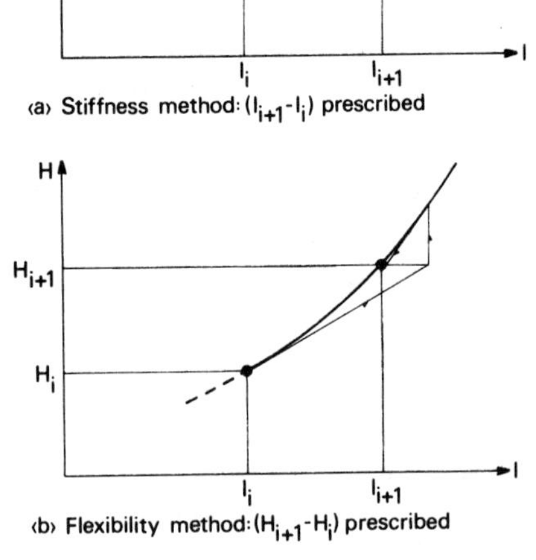

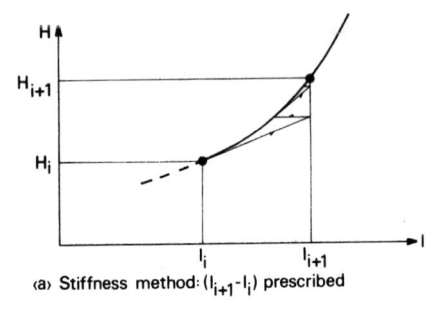
H
H_{i+1}
H_i
l_i
l_{i+1}
l
(a) Stiffness method: $(l_{i+1}-l_i)$ prescribed
H
H_{i+1}
H_i
l_i
l_{i+1}
l
(b) Flexibility method: $(H_{i+1}-H_i)$ prescribed

4.14 Iteration within a calculation step

using, for example, the Newton-Raphson scheme

$$\mathbf{F}_k \begin{Bmatrix} \Delta H \\ \Delta V \end{Bmatrix}_{k+1} + \begin{Bmatrix} f \\ g \end{Bmatrix}_k = \begin{Bmatrix} l \\ h \end{Bmatrix}. \tag{4.33}$$

Even though the guy response is in general nonlinear, it is nonetheless elastic, so loading and unloading follow the same curve. Therefore these procedures are unaffected by the signs of the imposed increments.

4.3 Cable Trusses

Cable trusses are composed of top and bottom chords of continuous prestressed cables, anchored at each end, between which numerous vertical, light, rigid spacers are placed. Since the cable prestress is usually high, the geometry of the truss is determined in large part by the span and the lengths of the spacers (see figure 4.15). The cable truss has several advantages as a means of supporting the roofs of large-span buildings. Because of the structural efficiency of the truss, the roof is light and yet possesses considerable rigidity.

Although the cable truss appears to be a relatively recent invention, there is a growing body of technical literature on the subject. The report by the ASCE Subcommittee on Cable-Suspended Structures includes a section on the cable truss [51]. The report also cites works (up to 1971) on the analysis, design and construction of cable trusses.

One of the first structures utilizing cable trusses is Zetlin's Municipal Auditorium in Utica, New York [59]. This well-known structure, completed in 1959, is circular in plan, and the roof is supported by a radial array of trusses spanning a distance of 76 m. The Workers' Gymnasium of Beijing (People's Republic of China) was built in 1961, and the diameter of its roof is 90 m (see figure 4.16). In 1962 a roof was placed over the

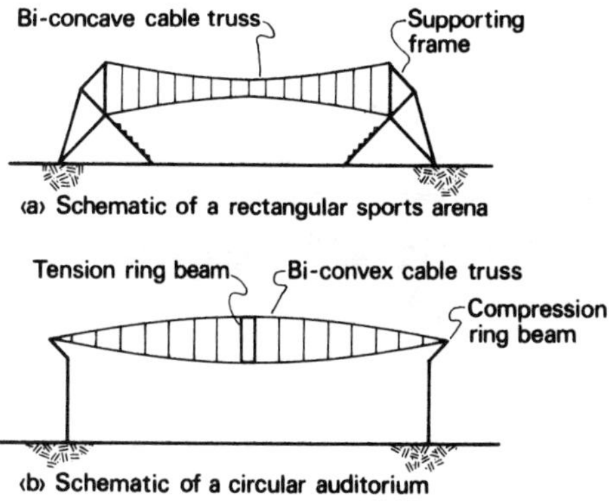

4.15 Possible uses for cable trusses

4.16 The Workers' Gymnasium of Beijing, 1961: diameter 90 m (courtesy Ye Yaoxian)

Johanneshov Ice Stadium in Stockholm using a system of roof trusses known as the Jawerth system, after the designer. The Jawerth cable truss is biconcave, and the whole truss is prestressed by adjusting the lengths of the diagonal ties that form the web members. Since then many other structures have been built incorporating the cable truss.

An analysis of the engineering principles upon which the cable truss design is based has been presented by Zetlin [59], and a book by Ye gives an account of Chinese practice [58]. Comprehensive analytical treatments have been given by Schleyer [45] and by Møllmann [38].

The analyses presented here are taken from the work described in references [26] and [27]. Here, however, we extend the analysis in reference [27] to cover static and dynamic responses of trusses symmetric only about a vertical axis at midspan rather than the simpler, less common case of trusses initially symmetric about both axes at midspan. The results then approach, in breadth of applicability, those due originally to Møllmann [38]. Reference [26] discusses the lateral stability of cable trusses, and part of it will be presented as example 4.4. So far as can be determined, the question of lateral stability has not been addressed by others.

Static Response

The cable truss will be analyzed as a single entity, and the analysis will proceed on the assumption that its top and bottom chords are either biconvex or biconcave (refer back to figure 4.15). The relatively small weight of the cables and the spacers will be ignored, so that the initial free-hanging geometry will be specified by the cable pretensions, the lengths of the spacers, and the span. In the analysis the spacers will be replaced by a continuous diaphragm whose adjacent vertical elements may slide freely with respect to each other. Each vertical element of the diaphragm is considered inextensible. It will be assumed that the slopes of the chords are, and remain, small. As a rule of thumb the maximum difference between spacer length and the spacing of the chords at the supports should be less than one-quarter of the span. In practical situations this requirement will always be met. The small longitudinal movements of the chords associated with the vertical movements of the truss under load must be allowed to occur freely. The analyses will not hold for biconcave systems (under loads not symmetrical about midspan) if, for example, the chords are clamped together at midspan. (Møllmann considers this case [38].) Finally, only trusses with vertical spacers will be considered. The use of inclined spacers may stiffen the truss significantly, but analytical solutions are not really feasible, and it is best to resort directly to numerical methods.

Consider a *biconvex cable truss*, anchored on rigid supports, as shown in figure 4.17a. Suppose that under applied vertical loading the shear force at some cross section x along the span is S. Then from vertical equilibrium

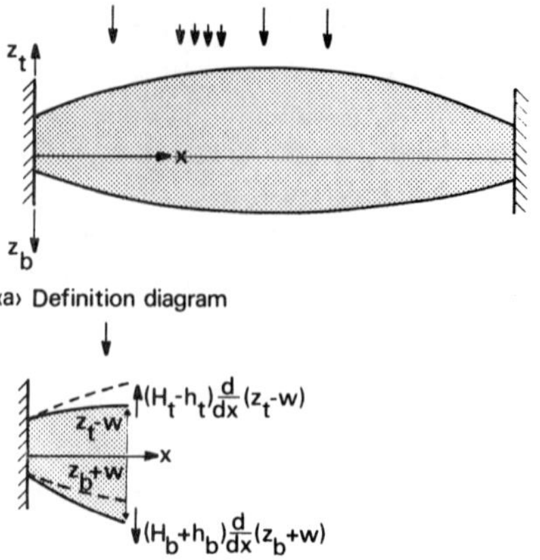

(a) Definition diagram

(b) Vertical equilibrium at a cross-section

4.17 Equilibrium of a loaded biconvex cable truss

it is clear that (see figure 4.17b)

$$(H_b + h_b)\frac{d}{dx}(z_b + w) - (H_t - h_t)\frac{d}{dx}(z_t - w) = S, \tag{4.34}$$

where H_b and H_t are the horizontal components of the pretensions in the bottom and top chords, respectively, h_b and h_t are the additional horizontal components of cable tension owing to applied load, z_b and z_t are the initial profiles of the chords, and w is the additional vertical deflection.

The internal equilibrium of the unloaded truss is expressed by

$$H_b\frac{dz_b}{dx} = H_t\frac{dz_t}{dx}. \tag{4.35}$$

This permits calculation of the axial forces that the spacers must resist in the initial geometry. When the spacers are struts, as they are for biconvex trusses, the possibility of buckling must be guarded against.

Substitution of (4.35) allows (4.34) to be reduced to

$$(H_b + H_t)\frac{dw}{dx} + (h_b - h_t)\frac{dw}{dx} + h_b\frac{dz_b}{dx} + h_t\frac{dz_t}{dx} = S. \tag{4.36}$$

The cable equations are

$$\begin{aligned} \frac{h_b L_{eb}}{EA_b} &= \int_0^l \left(\frac{dz_b}{dx}\right) \frac{dw}{dx}\, dx + \frac{1}{2}\int_0^l \left(\frac{dw}{dx}\right)^2 dx, \\ \frac{h_t L_{et}}{EA_t} &= \int_0^l \left(\frac{dz_t}{dx}\right)\left(\frac{dw}{dx}\right) dx - \frac{1}{2}\int_0^l \left(\frac{dw}{dx}\right)^2 dx, \end{aligned} \tag{4.37}$$

where A_b and A_t are the areas of the chords and L_{eb} and L_{et} are specified by $L_{eb,t} = \int_0^l \left(\frac{ds_{b,t}}{dx}\right)^3 dx$.

These equations are sufficient to obtain closed-form solutions for the dependent variables w, h_b and h_t by following the general approach given in chapter 2. Within reason, numerous different truss geometries could be chosen to yield a variety of architecturally pleasing structural solutions. In general a separate solution is required for each of these. The results are not as tidy as for a single cable because h_b and h_t must be found from a pair of coupled cubic equations, which can be separated only with the assumption of linear response. That assumption is plausible for normally encountered roof loads on trusses of appreciable curvature, and we shall exploit it here.

First, let us confine our attention to those trusses in which

$$H_b = H_t = H, \quad A_b = A_t = A. \tag{4.38}$$

The first implies that

$$z_b = z_t = z, \tag{4.39}$$

so that initially the truss is symmetrical about the longitudinal axis. Then, as may be seen from (4.37), h_b and h_t differ by a term of the second order in w, and the equilibrium equation (4.36) is linear, except for a term of the third order in w. Since this nonlinearity is of minor significance, we are justified in writing (4.36) as

$$2H\frac{dw}{dx} + (h_b + h_t)\frac{dz}{dx} \simeq S, \tag{4.40}$$

Adding the results in (4.37), we have

$$\frac{(h_b + h_t)L_e}{2EA} = \int_0^l \left(\frac{dz}{dx}\right)\left(\frac{dw}{dx}\right) dx. \tag{4.41}$$

Equations (4.40) and (4.41) are linear, and they accurately describe the displacement response of the structure. However, significant nonlinearities could be present in the individual additional tension responses in the chords, even though this nonlinearity is largely absent from their sum. This facet of cable behavior is not widely recognized, although the same phenomenon arises in the torsional analysis of suspension bridges where as with the cable truss the geometry of the cables changes by equal and opposite amounts [24]. It is a phenomenon well known in other branches of mechanics and in physics. For example, the linear range of an amplifier can be extended considerably if two amplifiers are used, which are connected in a push-pull arrangement.

In proceeding now to consider particular results for trusses whose chord properties and profiles differ, we shall assume linear behavior. There is less justification for this than for symmetric trusses, but it is still a reasonable assumption. Deflection response will be slightly overestimated, which is conservative, while additional tension in the hanging chord will be underestimated. If this is thought to be significant, a correction can always be found by substituting the linearized solution into (4.37). The profiles of the chords are assumed to be parabolic, as shown in figure 4.18, and given by

$$z_b = b_b\left\{1 + 4\left(\frac{d_b - b_b}{b_b}\right)\frac{x}{l}\left(1 - \frac{x}{l}\right)\right\},$$
$$z_t = b_t\left\{1 + 4\left(\frac{d_t - b_t}{b_t}\right)\frac{x}{l}\left(1 - \frac{x}{l}\right)\right\}. \tag{4.42}$$

If a point load, $\mathbf{P}$, is applied at $\mathbf{x}_1$, the linearized solution of (4.36) is

$$\mathbf{w} = (1 - \mathbf{x}_1)\mathbf{x} - \frac{(\mathbf{h}_b + \mathbf{h}_t)}{4\mathbf{P}}\mathbf{x}(1 - \mathbf{x}), \tag{4.43}$$

for $0 \le \mathbf{x} \le \mathbf{x}_1$, and

$$\mathbf{w} = (1 - \mathbf{x})\mathbf{x}_1 - \frac{(\mathbf{h}_b + \mathbf{h}_t)}{4\mathbf{P}}\mathbf{x}(1 - \mathbf{x}), \tag{4.44}$$

for $\mathbf{x}_1 \le \mathbf{x} \le 1$, where $\mathbf{x} = x/l$, $\mathbf{x}_1 = x_1/l$, $\mathbf{w} = w/\{Pl/(H_b + H_t)\}$, $\mathbf{h}_b = h_b/H_b$, $\mathbf{h}_t = h_t/H_t$, and $\mathbf{P} = Pl/\{16(d_b - b_b)H_b\}$.

Substitution of (4.43) and (4.44) into linearized versions of (4.37) yields

$$\mathbf{h}_b = \lambda_b^2\left\{\frac{\mathbf{P}}{2}\mathbf{x}_1(1 - \mathbf{x}_1) - \frac{(\mathbf{h}_b + \mathbf{h}_t)}{24}\right\},$$
$$\mathbf{h}_t = \lambda_t^2\left\{\frac{\mathbf{P}}{2}\mathbf{x}_1(1 - \mathbf{x}_1) - \frac{(\mathbf{h}_b + \mathbf{h}_t)}{24}\right\}. \tag{4.45}$$

The simultaneous solution of these produces the results

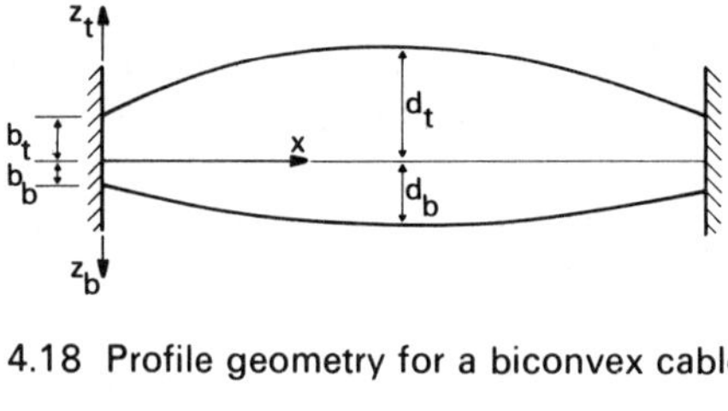

4.18 Profile geometry for a biconvex cable truss (see equation 4.42)

$$\mathbf{h}_b = \frac{1}{\{(1 + \lambda_t^2/\lambda_b^2)/2 + 12/\lambda_b^2\}} 6\mathbf{P}\,\mathbf{x}_1(1 - \mathbf{x}_1),$$

$$\mathbf{h}_t = \frac{1}{\{(1 + \lambda_b^2/\lambda_t^2)/2 + 12/\lambda_t^2\}} 6\mathbf{P}\,\mathbf{x}_1(1 - \mathbf{x}_1), \tag{4.46}$$

where $\lambda_b^2 = \{8(d_b - b_b)/l\}^2\,[l/\{(H_b + H_t)L_{eb}/2EA_b\}]$
and $\lambda_t^2 = \{8(d_t - b_t)/l\}^2\,[l/\{(H_b + H_t)L_{et}/2EA_t\}]$.
There is obviously a strong similarity here with the linearized solution for the single cable.

Under a uniformly distributed load $\mathbf{p}$, extending from $\mathbf{x}_2$ to $\mathbf{x}_3$ along the span, the results are

$$\mathbf{w} = \left[\left\{(\mathbf{x}_3 - \mathbf{x}_2) - \frac{1}{2}(\mathbf{x}_3^2 - \mathbf{x}_2^2)\right\}\mathbf{x} - \frac{(\mathbf{h}_b + \mathbf{h}_t)}{4\mathbf{p}}\mathbf{x}(1 - \mathbf{x})\right], \tag{4.47}$$

for $0 \le \mathbf{x} \le \mathbf{x}_2$,

$$\mathbf{w} = \left[\left\{-\frac{1}{2}\mathbf{x}_2^2 + \mathbf{x}_3\mathbf{x} - \frac{1}{2}\mathbf{x}^2 - \frac{1}{2}(\mathbf{x}_3^2 - \mathbf{x}_2^2)\mathbf{x}\right\} - \frac{(\mathbf{h}_b + \mathbf{h}_t)}{4\mathbf{p}}\mathbf{x}(1 - \mathbf{x})\right], \tag{4.48}$$

for $\mathbf{x}_2 \le \mathbf{x} \le \mathbf{x}_3$, and

$$\mathbf{w} = \left[\frac{1}{2}(\mathbf{x}_3^2 - \mathbf{x}_2^2)(1 - \mathbf{x}) - \frac{(\mathbf{h}_b + \mathbf{h}_t)}{4\mathbf{p}}\mathbf{x}(1 - \mathbf{x})\right], \tag{4.49}$$

for $\mathbf{x}_3 \le \mathbf{x} \le 1$, where $\mathbf{x}_2 = x_2/l$, $\mathbf{x}_3 = x_3/l$, $\mathbf{w} = w/\{pl^2/(H_b + H_t)\}$, and $\mathbf{p} = pl^2/\{16(d_b - b_b)H_b\}$.
The addition tensions are

$$\mathbf{h}_b = \frac{1}{\{(1 + \lambda_t^2/\lambda_b^2)/2 + 12/\lambda_b^2\}} 6\mathbf{p}\left\{\frac{1}{2}(\mathbf{x}_3^2 - \mathbf{x}_2^2) - \frac{1}{3}(\mathbf{x}_3^3 - \mathbf{x}_2^3)\right\},$$

$$\mathbf{h}_t = \frac{1}{\{(1 + \lambda_b^2/\lambda_t^2)/2 + 12/\lambda_t^2\}} 6\mathbf{p}\left\{\frac{1}{2}(\mathbf{x}_3^2 - \mathbf{x}_2^2) - \frac{1}{3}(\mathbf{x}_3^3 - \mathbf{x}_2^3)\right\}. \tag{4.50}$$

All these results apply equally to *biconcave trusses*. There $b_b > d_b$ and $b_t > d_t$, so $\mathbf{P}$ and $\mathbf{p}$ are negative and $\mathbf{h}_b$ and $\mathbf{h}_t$ are negative. The additional deflection $\mathbf{w}$ is of course still positive, and the interpretation to be placed on the change in sign of $\mathbf{h}_b$ and $\mathbf{h}_t$ is that the bottom chord now suffers a reduction in tension as against the increase recorded in the top chord. All this is as it should be.

In the case of circular roofs supported by a radial array of cable trusses an important example to consider is when a *triangular load block* of maximum intensity $\mathbf{p}$ acts at each end of the span. The additional deflection is

$$\mathbf{w} = \frac{1}{3}\left\{\frac{1}{8} - \left(\frac{1}{2} - \mathbf{x}\right)^3\right\} - \frac{(\mathbf{h}_b + \mathbf{h}_t)}{4\mathbf{p}}\mathbf{x}(1 - \mathbf{x}), \tag{4.51}$$

for $0 \leq \mathbf{x} \leq 1/2$, where $\mathbf{w} = w/\{pl^2/(H_b + H_t)\}$
and $\mathbf{p} = pl^2/\{16(d_b - b_b)H_b\}$. The additional tensions are

$$\begin{aligned} \mathbf{h}_b &= \frac{1}{\{(1 + \lambda_t^2/\lambda_b^2)/2 + 12/\lambda_b^2\}}\frac{3}{8}\mathbf{p}, \\ \mathbf{h}_t &= \frac{1}{\{(1 + \lambda_b^2/\lambda_t^2)/2 + 12/\lambda_t^2\}}\frac{3}{8}\mathbf{p}. \end{aligned} \tag{4.52}$$

If λ_b^2 and λ_t^2 are sufficiently large, the midspan will rise under this loading. In the symmetrical truss this occurs when $\lambda_b^2(\equiv \lambda_t^2) > 96$.

Placing the roof corresponds to imposing a uniformly distributed load over the whole area, and, if live loading is applied in addition to this dead load, vertical equilibrium across a section of a truss requires that

$$(H_b + h_b + h_b')\frac{d}{dx}(z_b + w + w') - (H_t - h_t - h_t')\frac{d}{dx}(z_t - w - w')$$

$$= S + S', \tag{4.53}$$

where the primes denote live load quantities. Upon expanding (4.53), and substituting for (4.34), we arrive at

$$(H_b + H_t)\frac{dw'}{dx} + h_b'\frac{dz_b}{dx} + h_t'\frac{dz_t}{dx} = S', \tag{4.54}$$

correct to first order. This is similar in every respect to (4.36). Thus the total horizontal reaction at the support never differs much from the pretension load $(H_b + H_t)$. In other words the addition of live loading causes another rearrangement in the distribution of this reaction between the top and bottom chords, but the sum remains constant.

A cable truss consists of a drape cable and a hanging cable. The area of the hanging cable is usually larger, but its stress in the initial profile is usually less than the stress in the drape cable. Under dead load and live load the difference in these stresses reduces, but it is important to check that the tension can never be lost in the drape cable.

Example 4.4 Lateral Stability of Cable Trusses

A single cable truss may be laterally unstable under the influence of applied vertical loading, although this eventuality is of limited practical importance in a well-designed roof system because of the presence of the roofing material and cross bracing between adjacent trusses. Nonetheless the theory is of some interest, as it appears to open up a new field of buckling problems [26].

The lateral stability of the cable truss is not directly analogous to any of the well-known classes of buckling problems. The nearest analogy is with the lateral buckling of a thin deep beam, but even here it is incomplete.

In this example attention will be confined to those cable trusses that are symmetrical about the longitudinal axis. Variational methods are used to set up the governing differential equation, and a solution is given for a triangular chord truss under a point load. Other cases are explored in reference [26].

As we have seen previously, a truss with a vertical load will displace vertically, with the horizontal components of chord pretension changing as well. In this vertically displaced position, the application of a small torsional disturbance to the truss will be resisted at each cross section by a torque that depends on the chord tensions, the chord spacing, and the out-of-plane rotation θ (see figure 4.19a). We choose to ignore the additional tensions generated by the vertical movement and assume in addition that θ is sufficiently small for it to cause no change in the tensions either. (The latter assumption is more plausible than the first. In reference [28] the effects of additional tension due to the vertical displacement, prior to the possibility of lateral instability, are considered.) The chords then move laterally by equal and opposite amounts, and the equivalent system of figure 4.19b describes the situation.

If in the original configuration the length of an element of the chord is ds, and its length in the laterally displaced position is ds', then

$$ds^2 = dx^2 + dz^2,$$

and

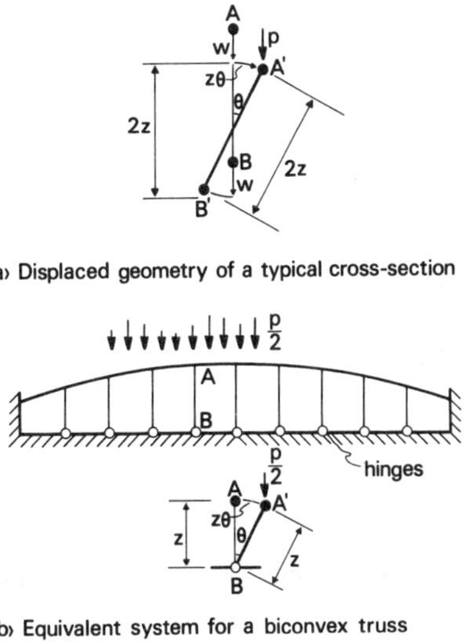

(a) Displaced geometry of a typical cross-section

(b) Equivalent system for a biconvex truss

4.19 Definition diagrams for the lateral stability of a cable truss

$$ds'^2 = dx^2 + \{(z + dz)\cos(\theta + d\theta) - z\cos\theta\}^2$$
$$+ \{(z + dz)\sin(\theta + d\theta) - z\sin\theta\}^2.$$

Because θ is small,

$$ds' - ds \simeq \frac{1}{2}z^2\left(\frac{d\theta}{ds}\right)^2 ds.$$

The pretension in the chord is T, and the potential energy stored in the element is

$$\frac{1}{2}Tz^2\left(\frac{d\theta}{ds}\right)^2 ds.$$

However, since $T = H\,ds/dx$, this may be replaced by

$$\frac{1}{2}Hz^2\left(\frac{d\theta}{dx}\right)^2 dx,$$

and the total for both chords is

$$\int_0^l Hz^2\left(\frac{d\theta}{dx}\right)^2 dx.$$

On the other hand, the work done by the load is

$$\int_0^l pz(1 - \cos\theta)dx \simeq \frac{1}{2}\int_0^l pz\theta^2 dx.$$

The difference between the potential energy stored and the work done is

$$\int_0^l \left\{-Hz^2\left(\frac{d\theta}{dx}\right)^2 + \tfrac{1}{2}pz\theta^2\right\} dx.$$

Performing the variation with respect to θ yields

$$\int_0^l \left\{-2Hz^2\frac{d\theta}{dx}\frac{d(\delta\theta)}{dx} + pz\theta\delta\theta\right\} dx = 0,$$

Upon integration (by parts where necessary) the following equation is obtained

$$-2Hz^2\frac{d\theta}{dx}\delta\theta\Big|_0^l + \int_0^l\left\{2H\frac{d}{dx}\left(z^2\frac{d\theta}{dx}\right) + pz\theta\right\}\delta\theta dx = 0.$$

If the boundary conditions are such that

$$2Hz^2\frac{d\theta}{dx}\delta\theta\Big|_0^l = 0,$$

then, because $\delta\theta$ is arbitrary, it follows that the differential equation governing the lateral stability of the truss is

$$2H\frac{d}{dx}\left(z^2\frac{d\theta}{dx}\right) + pz\theta = 0.$$

This equation may equally well be written in terms of the lateral displacement

$\varepsilon(= z\theta)$, namely,

$$2H\frac{d}{dx}\left(z\frac{d\varepsilon}{dx} - \varepsilon\frac{dz}{dx}\right) + p\varepsilon = 0.$$

In this form the statement of torsional equilibrium is more clearly demonstrated, and it proves convenient to use it for the one case we consider here, which is of a triangular chord truss under a point load P acting at a distance x_1 from the left-hand support. The apex of the truss occurs at $x_0(\geq x_1)$ (see figure 4.20).

When the truss is displaced laterally, the out-of-plane rotations of the apex and the point of load application are θ_0 and θ_1, respectively. A relationship between these two quantities must be established, and for this an energy argument is best.

If the initial lengths of the components of a chord are a, c, and e, and b and d are the half-spacing of the chords at the support and at the apex, then the increments in the chord length are

$$\Delta a \simeq \frac{1}{2}\frac{z_1 b\theta_1^2}{a}, \quad \Delta c \simeq \frac{1}{2}\frac{z_1 d(\theta_0 - \theta_1)^2}{c}, \quad \Delta e \simeq \frac{1}{2}\frac{bd\theta_0^2}{e},$$

where z_1 is the half-spacing of the chords at the point of load application. The potential energy stored in the chords is

$$H\left\{\frac{z_1 b\theta_1^2}{x_1} + \frac{z_1 d(\theta_0 - \theta_1)^2}{(x_0 - x_1)} + \frac{bd\theta_0^2}{(l - x_0)}\right\}.$$

For given θ_1, the correct θ_0 is that which minimizes this potential energy, giving

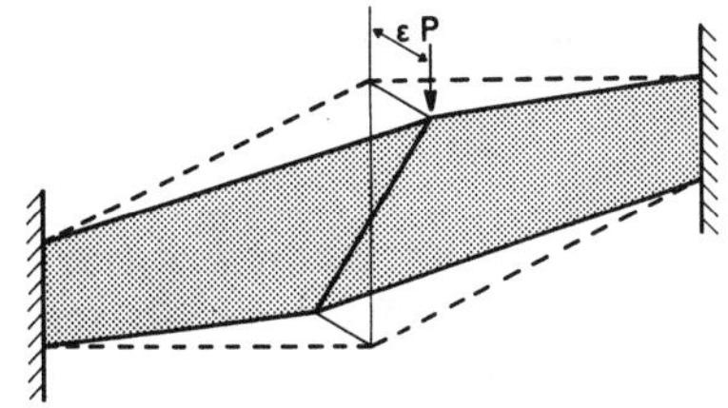

(a) Three dimensional view of a buckled triangular chord truss

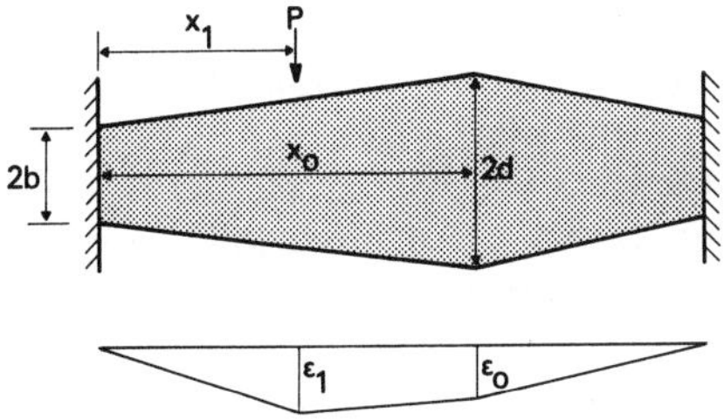

(b) A general triangular chord truss

4.20 A triangular chord cable truss buckled by a point load

$$\frac{\theta_0}{\theta_1} = \frac{z_1(l - x_0)}{\{z_1(l - x_0) + b(x_0 - x_1)\}} \leq 1,$$

The lateral displacements are in the ratio

$$\frac{\varepsilon_0}{\varepsilon_1} = \frac{d(l - x_0)}{\{z_1(l - x_0) + b(x_0 - x_1)\}}.$$

The governing differential equation may be integrated to

$$2H\left(z\frac{d\varepsilon}{dx} - \varepsilon\frac{dz}{dx}\right)\Bigg|_{x_0^-}^{x_0^+} + 2H\left(z\frac{d\varepsilon}{dx}\right)\Bigg|_{x_1^-}^{x_1^+} + P\varepsilon_1 = 0,$$

so that

$$P_{\text{crit}} = 2bH\left\{\frac{1}{x_1} + \frac{\varepsilon_0/\varepsilon_1}{(l - x_0)}\right\}.$$

But $z_1 = b\left\{1 + \frac{(d - b)}{b}\frac{x_1}{x_0}\right\}$, and we have finally

$$P_{\text{crit}} = \frac{2bH}{x_1}\left\{\frac{(1 - x_1/x_0)b + (x_1/x_0)d}{(1 - x_1/x_0)b + x_1(1 - x_0/l)d/x_0}\right\}$$

This result holds regardless of whether the truss is biconvex or biconcave. Numerous special cases are contained within it. When the geometry and/or loading are more complicated, the effort required to effect solutions is a somewhat daunting prospect, although a few cases have been presented in reference [26]. A general solution for a truss with the point of load application and the apex both at midspan is given as an exercise (see problem 4.7).

Dynamic Response

Because cable trusses are principally used as structural members supporting the roofs of buildings of large span, the dynamic characteristics of such roof systems under the dynamic effects of wind or earthquakes must be assessed.

Two different situations will be discussed. The first, relates to rectangular buildings where the weight of the roof associated with each truss is assumed to be uniformly distributed along the span. The second relates to circular buildings where the weight of the roof associated with each truss is assumed to be distributed as a triangular load block.

Only the vertical components of the vibrations will be considered, since these are the most important. For simplicity the term mode will be applied to the vertical components, although in reality an in-plane mode consists of two components.

The equilibrium equation for free vibrations for the *uniformly distributed* roof load is

$$(H_b + h_b + h_b')\frac{\partial^2}{\partial x^2}(z_b + w + w') - (H_t - h_t - h_t')\frac{\partial^2}{\partial x^2}(z_t - w - w')$$

$$= -p + m\frac{\partial^2 w'}{\partial t^2}, \tag{4.55}$$

where the primes denote dynamic quantities and m is the mass of the roof per unit length, thus $p \equiv mg$. The equation of static equilibrium for the truss under roof load is

$$(H_b + H_t)\frac{d^2 w}{dx^2} + h_b\frac{d^2 z_b}{dx^2} + h_t\frac{d^2 z_t}{dx^2} = -p. \tag{4.56}$$

Hence

$$(H_b + H_t)\frac{\partial^2 w'}{\partial x^2} + h_b'\frac{d^2 z_b}{dx^2} + h_t'\frac{d^2 z_t}{dx^2} = m\frac{\partial^2 w'}{\partial t^2} \tag{4.57}$$

is the linearized equation of motion.

The important feature of this equation is that h_b, h_t, and w are absent, and so it differs from the single cable considered in the previous chapter. (There are, however, similarities with the hyperbolic paraboloidal network considered in the next chapter.) The cable equations are

$$\frac{h_b' L_{eb}}{EA_b} = \int_0^l \frac{dz_b}{dx}\frac{\partial w'}{\partial x}\,dx,$$
$$\frac{h_t' L_{et}}{EA_t} = \int_0^l \frac{dz_t}{dx}\frac{\partial w'}{\partial x}\,dx. \tag{4.58}$$

After separating time from these equations, they may be recast in the dimensionless form

$$\frac{d^2\tilde{\mathbf{w}}'}{d\mathbf{x}^2} + \omega^2\tilde{\mathbf{w}}' = \frac{1}{2\mathbf{p}}(\tilde{\mathbf{h}}_b' + \tilde{\mathbf{h}}_t'), \tag{4.59}$$

and

$$\frac{\tilde{\mathbf{h}}_b'}{\mathbf{p}} = \lambda_b^2\int_0^1 \tilde{\mathbf{w}}'\,d\mathbf{x},$$
$$\frac{\tilde{\mathbf{h}}_t'}{\mathbf{p}} = \lambda_t^2\int_0^1 \tilde{\mathbf{w}}'\,d\mathbf{x}, \tag{4.60}$$

where $\mathbf{x} = x/l$, $\tilde{\mathbf{w}}' = \tilde{w}'/\{mgl^2/(H_b + H_t)\}$, $\boldsymbol{\omega}^2 = m\omega^2 l^2/(H_b + H_t)$, $\mathbf{p} = mgl^2/\{16(d_b - b_b)H_b\}$, $\tilde{\mathbf{h}}_b' = \tilde{h}_b'/H_b$, $\tilde{\mathbf{h}}_t' = \tilde{h}_t'/H_t$, and $\lambda_b^2 = \{8(d_b - b_b)/l\}^2 [l/\{(H_b + H_t)L_{eb}/2EA_b\}]$, $\lambda_t^2 = \{8(d_t - b_t)/l\}^2 [l/\{(H_b + H_t)L_{et}/2EA_t\}]$.

If we set $\lambda_e^2 = (\lambda_b^2 + \lambda_t^2)/2$ and $\tilde{\mathbf{h}}_e' = (\tilde{\mathbf{h}}_b' + \tilde{\mathbf{h}}_t')/2\mathbf{p}$, we have

$$\frac{d^2\tilde{\mathbf{w}}'}{d\mathbf{x}^2} + \omega^2\tilde{\mathbf{w}}' = \tilde{\mathbf{h}}'_e, \tag{4.61}$$

with

$$\tilde{\mathbf{h}}'_e = \lambda_e^2 \int_0^1 \tilde{\mathbf{w}}'\, d\mathbf{x}, \tag{4.62}$$

which is identical in form to the equations for a single cable. Therefore the whole of the theory pertaining to the free and forced vibrations of the single cable is directly applicable to this cable truss. For this reason we need consider the matter no further.

If the roof load is distributed as a *triangular block*, as when a radial array of trusses support a circular roof, we start by positioning the axes at midspan. The dynamic equation of equilibrium for a representative truss is

$$(H_b + H_t)\frac{\partial^2 w'}{\partial x^2} + h'_b\frac{d^2 z_b}{dx^2} + h'_t\frac{d^2 z_t}{dx^2} = m\frac{2x}{l}\frac{\partial^2 w'}{\partial t^2}, \quad 0 \le x \le l/2, \tag{4.63}$$

where m is the mass per unit length of the roof at the supports.

The antisymmetric modes are of the form

$$\tilde{\mathbf{w}}' = \mathbf{A}\,\omega^{1/3}\mathbf{x}^{1/2} J_{1/3}\left(\frac{2}{3}\omega\mathbf{x}^{3/2}\right), \quad 0 \le \mathbf{x} \le 1, \tag{4.64}$$

where $\tilde{\mathbf{w}}' = \tilde{w}'/\{mgl^2/(H_b + H_t)\}$, $\omega^2 = m\omega^2(l/2)^2/(H_b + H_t)$, and $\mathbf{x} = 2x/l$. The associated natural frequencies are found from the roots of

$$J_{1/3}\left(\frac{2}{3}\omega\right) = 0, \tag{4.65}$$

where $J_{1/3}(\ \ )$ is the Bessel function of the first kind and one-third order. The natural frequency of the first antisymmetric mode is $\omega_1 = 4.35$.

In an actual roof these modes will depend to some degree on the plan angle. However, since there is no stiffness in that direction, the plan-wise modulation of the roof mode, provided that it is compatible with the antisymmetric nature of the individual truss modes, will not affect these results.

The modes in which additional tensions are generated are radially symmetric and are of more interest. If parabolic profiles are used, the solutions become unnecessarily involved. To avoid this complication, we shall assume that the initial chord profiles are given by the cubics

$$z_b = d_b\left\{1 - \left(\frac{d_b - b_b}{d_b}\right)\left(\frac{2x}{l}\right)^3\right\}, \tag{4.66}$$

$$z_t = d_t\left\{1 - \left(\frac{d_t - b_t}{d_t}\right)\left(\frac{2x}{l}\right)^3\right\},$$

for $0 \le 2x/l \le 1$. The results that follow can for practical purposes be assumed to apply to trusses with parabolic profiles.

In dimensionless form the equation of motion reads

$$\frac{d^2\tilde{\mathbf{w}}'}{d\mathbf{x}^2} + \omega^2\mathbf{x}\tilde{\mathbf{w}}' = \frac{1}{2\mathbf{p}}(\tilde{\mathbf{h}}_b' + \tilde{\mathbf{h}}_t'), \quad 0 \le \mathbf{x} \le 1, \tag{4.67}$$

with

$$\frac{\tilde{\mathbf{h}}_b'}{2\mathbf{p}} = 4\lambda_b^2\int_0^1 \tilde{\mathbf{w}}'\mathbf{x}\,d\mathbf{x},$$
$$\frac{\tilde{\mathbf{h}}_t'}{2\mathbf{p}} = 4\lambda_t^2\int_0^1 \tilde{\mathbf{w}}'\mathbf{x}\,d\mathbf{x}. \tag{4.68}$$

The solution of (4.67) that satisfies boundary conditions of $d\tilde{\mathbf{w}}'(0)/d\mathbf{x} = 0$ and $\tilde{\mathbf{w}}'(1) = 0$ is

$$\tilde{\mathbf{w}}'(\mathbf{x}) = \frac{1}{2\mathbf{p}}\left(\frac{\tilde{\mathbf{h}}_b' + \tilde{\mathbf{h}}_t'}{\omega^2}\right)\left\{1 - \frac{\mathbf{x}^{1/2}J_{-1/3}(2\omega\mathbf{x}^{3/2}/3)}{J_{-1/3}(2\omega/3)}\right\}. \tag{4.69}$$

(The presence of a tension ring beam at midspan will modify the first boundary condition. If the weight of the ring beam is substantial, its effect will be felt both in the static profile and dynamic response.) The transcendental frequency equation is

$$\frac{J_{2/3}(2\omega/3)}{J_{-1/3}(2\omega/3)} = \frac{3}{4}\left(\frac{2}{3}\omega\right) - \frac{27}{16(\lambda_b^2 + \lambda_t^2)}\left(\frac{2}{3}\omega\right)^3. \tag{4.70}$$

The dimensionless quantities are $\mathbf{x} = 2x/l$, $\tilde{\mathbf{w}}' = \tilde{w}/\{mgl^2/(H_b + H_t)\}$, $\mathbf{p} = mgl^2/\{12(d_b - b_b)H_b\}$, $\omega^2 = m\omega^2 l^2/\{4(H_b + H_t)\}$, $\tilde{\mathbf{h}}_b' = \tilde{h}_b'/H_b$, $\tilde{\mathbf{h}}_t' = \tilde{h}_t'/H_t$, and $\lambda_b^2 = \{6(d_b - b_b)/l\}^2\,[l/\{(H_b + H_t)L_{eb}/EA_b\}]$, $\lambda_t^2 = \{6(d_t - b_t)/l\}^2\,[l/\{(H_b + H_t)L_{et}/EA_t\}]$. Notice the differences with the definitions listed after (4.60).

It may readily be shown that [4]

$$\lim_{x\to 0}\frac{J_{2/3}(x)}{J_{-1/3}(x)} = \frac{3}{4}x, \tag{4.71}$$

which indicates that the first root of (4.70) is the trivial result $\omega = 0$. In this respect (4.70) is analogous to (3.17).

When $(\lambda_b^2 + \lambda_t^2)$ is very small, (4.70) reduces to

$$J_{-1/3}\left(\frac{2}{3}\omega\right) = 0, \tag{4.72}$$

the first nonzero root of which is $\omega_1 = 2.81$. On the other hand, when λ^2 is large, the transcendental equation reduces to

$$\frac{J_{2/3}(2\omega/3)}{J_{-1/3}(2\omega/3)} = \frac{3}{4}\left(\frac{2}{3}\omega\right), \tag{4.73}$$

On account of a recurrence relation for the Bessel functions [4], this equation may be further reduced to

$$J_{5/3}\left(\frac{2}{3}\omega\right) = 0, \tag{4.74}$$

the first nonzero root of which is $\omega_1 = 7.08$. Therefore $2.81 < \omega_1 < 7.08$.

When $(\lambda_b^2 + \lambda_t^2) < 26$, the first symmetric mode has no internal nodes. When $(\lambda_b^2 + \lambda_t^2) = 26$, the first symmetric mode is tangential to the static profile at its ends. When $(\lambda_b^2 + \lambda_t^2) > 26$, the first symmetric mode has two internal nodes. The apparent difference in this cutoff, compared to the value $(\lambda_b^2 + \lambda_t^2)/2 = 4\pi^2$ for the uniformly loaded truss, is mainly due to the differing definitions of λ^2 in each case.

Example 4.5 Calculations for a Cable Truss Roof of Rectangular Plan

The biconcave cable truss shown in figure 4.21 is typical of a parallel array used to support a rectangular roof. The following data are specified: $l = 80$ m, $b_b = 5$ m, $d_b = 1$ m, $b_t = 6$ m, $d_t = 1$ m, $H_b = 600$ kN, $H_t = 480$ kN, $A_b = 2 \times 10^{-3}$ m^2, $A_t = 3 \times 10^{-3}$ m^2, $E = 1.5 \times 10^5$ MPa. The horizontal stiffness of each backstay to the top, or hanging, cable is 1.5×10^4 kN/m, while that for each backstay to the bracing cable is calculated to be 1×10^4 kN/m. The supporting columns are flexible but are axially inextensible.

We are required to find the dead load profile under roof cladding equivalent to 2 kN/m and also to find the periods of the first antisymmetric mode and the first symmetric mode. To do this, we need first to calculate various dimensionless quantities.

The chords are parabolas, and internal equilibrium in the initial profile is satisfied if

$$\frac{H_b(d_b - b_b)}{8l^2} = \frac{H_t(d_t - b_t)}{8l^2},$$

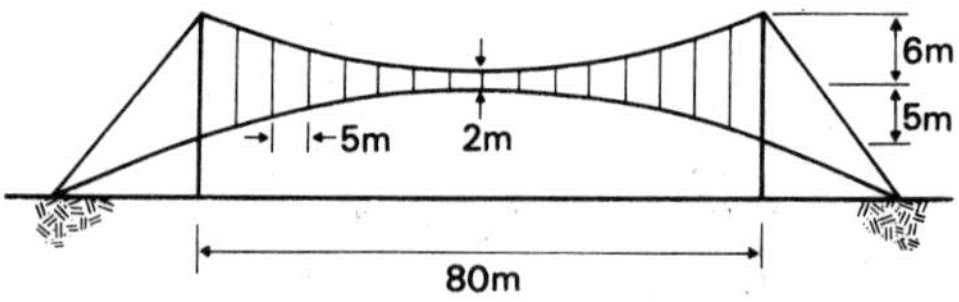

4.21 A typical member of a parallel array of biconcave cable trusses supporting a roof rectangular in plan which is certainly the case. The tensile force in each of the ties connecting the chords is

$$\frac{H_b(b_b - d_b)}{8l^2} \times \text{spacing} = \frac{600 \times 4}{8 \times 80 \times 80} \times 5 = 0.23 \text{ kN}.$$

For the dead load profile (see equations 4.47 through 4.50)

$$\mathbf{p} = \frac{pl^2}{16(d_b - b_b)H_b} = -\frac{2 \times 80 \times 80}{16 \times 4 \times 600} = -0.33.$$

The negative sign indicates that we are dealing with a biconcave truss. Before we calculate λ_b^2 and λ_t^2, we must modify the values of EA_b and EA_t to account for longitudinal support flexibility. Following the method described in the footnote prior to (2.20)

$$(EA_b)_{\text{equiv}} = \frac{2EA_b}{1 + (2EA_b/(1 \times 10^4\ L_{eb}))} = 1.7 \times 10^5 \text{ kN},$$

$$(EA_t)_{\text{equiv}} = \frac{2EA_t}{1 + (2EA_t/(1.5 \times 10^4\ L_{et}))} = 2.6 \times 10^5 \text{ kN}.$$

In each case support flexibility has caused a significant reduction in the equivalent axial stiffness of the cable, the amount being about 43 percent. Hence

$$\lambda_b^2 = \left(\frac{8(d_b - b_b)}{l}\right)^2 \frac{l}{(H_b + H_t)L_{eb}/(EA_b)_{\text{equiv}}} = 25,$$

$$\lambda_t^2 = \left(\frac{8(d_t - b_t)}{l}\right)^2 \frac{l}{(H_b + H_t)L_{et}/(EA_t)_{\text{equiv}}} = 87.$$

Since the dead load covers the whole span, (4.50) gives the additional tensions as

$$\mathbf{h}_b = \frac{\mathbf{p}}{1/2(1 + \lambda_t^2/\lambda_b^2) + 12/\lambda_b^2} = -0.12,$$

$$\mathbf{h}_t = \frac{\mathbf{p}}{1/2(1 + \lambda_b^2/\lambda_t^2) + 12/\lambda_t^2} = -0.42.$$

The minus signs indicate that the tension in the bottom chord has reduced to $0.88 \times 600 = 530$ kN, while the tension in the top chord has increased to $1.42 \times 480 = 680$ kN.

Under pretension plus dead load the stress in the bottom chord is 270 MPa, and the stress in the top chord is 230 MPa. (A snow load of, say, 4 kN/m will give total stresses of approximately 200 MPa and 370 MPa, respectively. The factor of safety on the top chord is then about four, which is adequate.) The ends of the clear span of the top chord will have moved together by 0.027 m, and the ends of the bottom chord will have moved apart by 0.015 m.

From (4.48) the additional vertical deflection at midspan is

$$\mathbf{w}_{\max} = \frac{1}{8}\left\{1 - \frac{(\mathbf{h}_t + \mathbf{h}_t)}{2\mathbf{p}}\right\} = 0.023.$$

In other words

$$w_{max} = 0.023 \times \frac{pl^2}{(H_b + H_t)} = 0.27 \text{ m}.$$

If support flexibility had been ignored, the additional deflection would have been 0.16 m—the difference is obviously significant.

Turning now to the free vibrational response, we see from (4.62) that

$$\lambda_e^2 (= (\lambda_b^2 + \lambda_t^2)/2) = 56.$$

Because this is greater than $4\pi^2$, it follows that the frequency of the first anti-symmetric mode will be the lowest. The frequency of this mode is

$$\omega_1 = 2\pi,$$

so the period is

$$T_1 = \left(\frac{ml^2}{H_b + H_t}\right)^{1/2} = 1.1 \text{ sec}.$$

The frequency of the first symmetric mode is (see table 3.1)

$$\omega_1 = 2.25\,\pi,$$

and the period is

$$T_1 = 1.0 \text{ sec}.$$

All these results suggest that, notwithstanding its appreciable span, the truss possesses a marked deg ee of rigidity.

4.4 Elements of the Theory of Suspension Bridges

The first work that attempted to account for the interplay under live load between the deck and the cables of a suspension bridge was produced by Rankine around the middle of the nineteenth century. By assuming that the cables provided a uniformly distributed uplift equal in magnitude to the live loading, extimates could be made of additional cable forces and deck bending moments. There are of course any number of statically admissible solutions, but only one at the same time satisfies compatibility of displacements between deck and cables. The classical elastic theory (based on Castigliano's theorem) restores some measure of compatibility. The elastic theory is in fact correct to the first order for self-anchored bridges. However, when independent anchorages are provided, as is usual, it is necessary to use a full deflection theory. The deflection theory was established independently by Müller-Breslau and by Melan in the latter part of the nineteenth century, and all modern practice is based on this work. New work always takes some time to filter through, and many of the major bridges built in that period were designed without much in the way of analysis. The Roeblings' towering accomplishment, the Brooklyn bridge, completed in 1884, is a case in point [36]. (As a piece of engineering history McCullough's account of the building of the Brooklyn bridge is well worth reading.) Chronologically the Brooklyn bridge was too early to have been influenced by Melan's work. The deflection theory was not utilized until 1904, when Moisseiff applied the analysis in his design of the Manhattan bridge [14].

According to the deflection theory, the equilibrium of a slice of the deck and the cables is expressed in the form of

$$-\frac{d^2}{dx^2}\left(EI\frac{d^2w}{dx^2}\right) + 2(H + h)\frac{d^2w}{dx^2} = -p - 2h\frac{d^2z}{dx^2},$$

with

$$\frac{hL_e}{EA} = \int\left(\frac{dz}{dx}\right)\left(\frac{dw}{dx}\right)dx + \frac{1}{2}\int\left(\frac{dw}{dx}\right)^2 dx,$$

the various terms of which are self-explanatory. Analytical solutions abound for bridges with single and multiple suspended spans (see, for example, the standard works by Timoshenko [54] and by Pugsley [42]). Naturally enough, the solutions increase in complexity as account is taken of variations in span properties and tower interaction, and, if geometric nonlinearity is also included, a point is quickly reached at which an analytical solution is not worth obtaining, even if it can be found.

Therefore current practice in the consulting firms and companies engaged in the design and construction of major suspension bridges is to make extensive use of numerical techniques. These approaches have been widely reported in the technical literature. One method for the full static analysis of a bridge, which has been used on some recent European bridges, is described in a paper by Jennings and Mairs [33]. Another method for the dynamic analysis of a bridge, which has been checked out against the results of field tests, is contained in the report by Abdel-Ghaffar [1], to mention just two. The interested reader will come across other works in journals such as the publications of the American Society of Civil Engineers, the Japan Society of Civil Engineers, and European engineering organizations.

We shall consider two aspects of behavior in which a knowledge of cable theory is sufficient to give some indication of the predominant physical characteristics. Both relate to dynamic behavior. The first represents an introduction to aeroelastic stability of a bridge and, in the second, a simple formula is produced for earthquake-generated additional tension in the cables.

Aeroelastic Stability of Suspension Bridges

The story of the aeroelastic failure in 1940 of the first Tacoma Narrows bridge is too well known to bear retelling here. The demise of that bridge represented something of a watershed, for in hindsight it should have been clear that wind action is very much a dynamic phenomenon—there had been numerous wind-related failures of smaller bridges in the previous century. (For an interesting account of the problems experienced with some of these early bridges, see reference [9].)

Even though aeroelastic theory had been established in the previous decade, its principal field of application was to the lifting surfaces of aircraft. The failure of the Tacoma Narrows bridge made engineers realize that unless adequate precautions were taken, adverse performance could be expected in many flexible structures exposed to a moving fluid environment. An extensive series of windtunnel tests were carried out, to both determine the cause of the original failure and suggest improvements, primarily in fairing of the cross section and adding torsional rigidity in the deck. Such windtunnel tests on scaled models are today an essential ingredient in the design of major bridge structures. (For a review of the state of the art see Scanlan [44].)

Currently research is active in many areas of aeroelasticity, such as vortex shedding, galloping, buffeting, and flutter. All have different connotations, but they are more or less linked by a common notion that the motion is generated by forces that themselves depend on that motion. A wide range of structures are susceptible to flow-induced vibrations, which includes suspension bridges, mooring systems, electrical transmission lines, pipes containing flowing fluids, lifting surfaces, rotating blades, chimney stacks and even slender multistory buildings. It is an area in which theory and experiment often go hand in hand, because the analysis usually depends on experimentally determined aerodynamic and fluid dynamic coefficients. Books on the subject have been written by Fung [20], Blevins [10], and Simiu and Scanlan [49].

The theory that we now present is flawed by the impossibility of using precise aerodynamic information. This is an inevitable consequence of the necessity of treating individual bridges as individual bridges: bridge properties and locations are too variable for any one theory to be successfully defended as general. Nonetheless, the present approach represents a first pass at the problem, and the simple formulas that result enable preliminary assessments to be made.

Consider the unstiffened single suspended-span bridge shown in elevation and in enlarged cross section in figure 4.22. The dead load profile of the cables is

$$z = \frac{mgl^2}{4H}\frac{x}{l}\left(1 - \frac{x}{l}\right), \tag{4.75}$$

where mg is the weight of the bridge per unit length (two cables, their hangers, and the deck), and H is the horizontal component of tension in one cable.

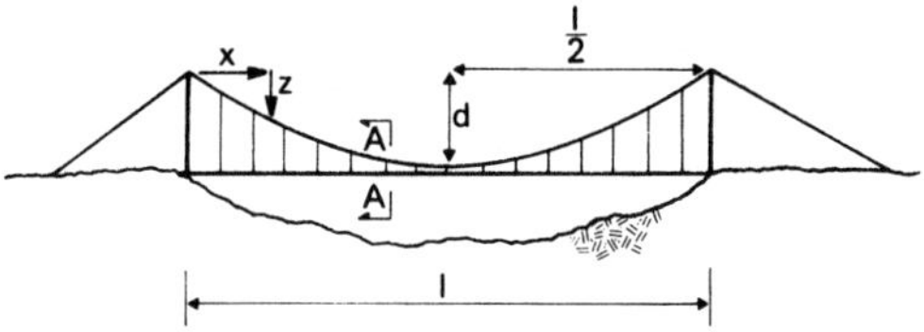

(a) Elevation on bridge

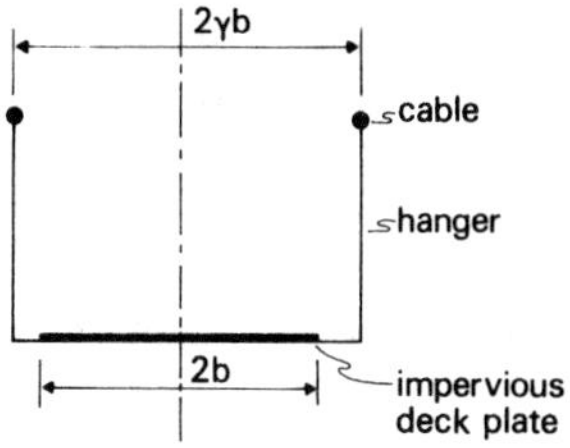

(b) Section A-A

4.22 Definition diagram for an unstiffened single suspended-span suspension bridge

Consider now the free vibrations after a small torsional disturbance is applied to the deck, over which a steady uniform wind of speed V is blowing. The angle of twist about the longitudinal centerline of the deck θ and the additional vertical displacement (of the two cables and deck together) w are assumed to be small (see figure 4.23). Torsional equilibrium of a slice of the deck and cables is given by

$$2H\gamma^2 b^2 \frac{\partial^2\theta}{\partial x^2} + 2h_1\gamma b \frac{d^2 z}{dx^2} = I_m \frac{\partial^2\theta}{\partial t^2} - \pi\rho V^2 b^2 \theta, \tag{4.76}$$

with

$$\frac{h_1 L_e}{EA} = \int_0^l \left(\frac{dz}{dx}\right) \frac{\partial(\gamma b\theta)}{\partial x} dx.$$

The statement of vertical equilibrium is

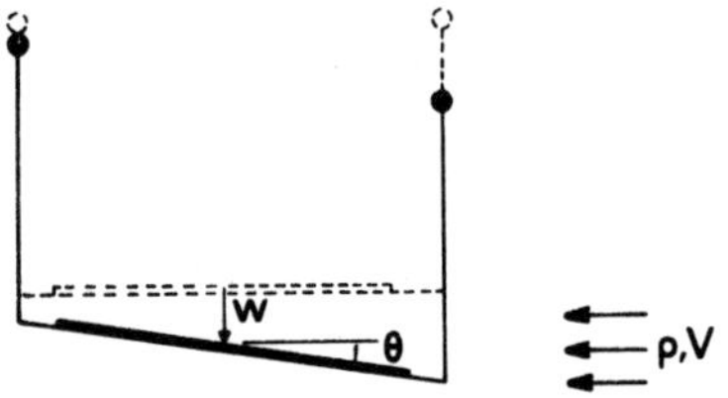

4.23 Twist and vertical displacement under wind action

$$2H\frac{\partial^2 w}{\partial x^2} + 2h_2\frac{d^2 z}{dx^2} = m\frac{\partial^2 w}{\partial t^2} - 2\pi\rho V^2 b\theta, \tag{4.77}$$

with

$$\frac{h_2 L_e}{EA} = \int_0^l \left(\frac{dz}{dx}\right)\left(\frac{\partial w}{\partial x}\right) dx.$$

Here h_1 is the additional component of tension generated in one cable due to the θ motion (the righthand cable according to figure 4.23, whereas the lefthand cable suffers a reduction in tension of $-h_1$), I_m is the moment of inertia of the deck and cables about the longitudinal centerline, ρ is the density of air, and h_2 is the additional component of tension generated in one cable by the w motion. The longitudinal centerline is the elastic axis of the bridge.

The term $2\pi\rho V^2 b\theta$ is the theoretical aerodynamic lift that a flat plate experiences when slightly inclined to the wind and may be derived from ideal fluid flow [37]. Wind tunnel measurements yield smaller values. The center of lift acts at the quarter point of the chord, producing a torque of $\pi\rho V^2 b^2\theta$. These displacement-dependent aerodynamic forces result in coupling between the two components of motion.

In this formulation any damping of structural and aerodynamic origin has been ignored, and likewise the inertia forces caused by added mass effects. Lateral displacements due to wind drag are not coupled with the other displacements and maybe ignored too. Finally, the temporal and spatial characteristics of the wind, which are obviously site-dependent, receive no consideration.

We could include aerodynamic damping which is perhaps the most important of the omitted terms. However, this refinement, although frequently of significance, yields a messy flutter determinant that requires a numerical solution. (See Fung [20], pp. 187–193 for the damping terms and pp. 235–238 for methods of solving the flutter determinant.) Notwithstanding these qualifications, the essence of the problem is as outlined. The fundamental features of flutter in unstiffened suspension bridges are similar in many respects to the flutter of the wing and aileron of an aircraft given in the book by von Kármán and Biot [56].

Therefore in dimensionless form (4.76) and (4.77) become

$$\frac{\partial^2 \boldsymbol{\theta}}{\partial \mathbf{x}^2} - \mathbf{I}_m\frac{\partial^2 \boldsymbol{\theta}}{\partial \mathbf{t}^2} + \frac{\mathbf{V}^2}{\gamma^2}\boldsymbol{\theta} = \mathbf{h}_1, \tag{4.78}$$

with

$$\mathbf{h}_1 = \lambda^2 \int_0^1 \boldsymbol{\theta}\, d\mathbf{x},$$

and

$$\frac{\partial^2 \mathbf{w}}{\partial \mathbf{x}^2} - \frac{\partial^2 \mathbf{w}}{\partial \mathbf{t}^2} = \mathbf{h}_2 - \frac{2\mathbf{V}^2}{\gamma}\boldsymbol{\theta}, \tag{4.79}$$

with

$$\mathbf{h}_2 = \lambda^2 \int_0^1 \mathbf{w}\, d\mathbf{x},$$

where $\mathbf{x} = x/l$, $\mathbf{h}_{1,2} = h_{1,2}/H$, $\mathbf{t} = t/\{l(m/2H)^{1/2}\}$, $\boldsymbol{\theta} = \theta\gamma b/(mgl^2/2H)$, $\mathbf{w} = w/(mgl^2/2H)$, $\mathbf{V}^2 = \pi\rho V^2 l^2/2H$, $\mathbf{I}_m = I_m/(m\gamma^2 b^2)$, and $\lambda^2 = (mgl/2H)^2\{l/(HL_e/EA)\}$.

When $\mathbf{V} = 0$ (4.78) and (4.79), describe free uncoupled vibrations. Let

$$\boldsymbol{\theta}(\mathbf{x}, \mathbf{t}) = \tilde{\boldsymbol{\theta}}(\mathbf{x})e^{i\omega_1 \mathbf{t}},\ \mathbf{h}_1(\mathbf{t}) = \tilde{\mathbf{h}}_1 e^{i\omega_1 \mathbf{t}}, \tag{4.80}$$

and

$$\mathbf{w}(\mathbf{x}, \mathbf{t}) = \tilde{\mathbf{w}}(\mathbf{x})e^{i\omega_2 \mathbf{t}},\ \mathbf{h}_2(\mathbf{t}) = \tilde{\mathbf{h}}_2 e^{i\omega_2 \mathbf{t}},$$

where $\boldsymbol{\omega}_1$ and $\boldsymbol{\omega}_2$ are the dimensionless natural frequencies of torsional and vertical vibration, respectively. Therefore

$$\frac{d^2\tilde{\boldsymbol{\theta}}}{d\mathbf{x}^2} + \mathbf{I}_m \boldsymbol{\omega}_1^2 \tilde{\boldsymbol{\theta}} = \tilde{\mathbf{h}}_1, \tag{4.81}$$

with

$$\tilde{\mathbf{h}}_1 = \lambda^2 \int_0^1 \tilde{\boldsymbol{\theta}}\, d\mathbf{x},$$

and

$$\frac{d^2\tilde{\mathbf{w}}}{d\mathbf{x}^2} + \boldsymbol{\omega}_2^2 \tilde{\mathbf{w}} = \tilde{\mathbf{h}}_2, \tag{4.82}$$

with

$$\tilde{\mathbf{h}}_2 = \lambda^2 \int_0^1 \tilde{\mathbf{w}}\, d\mathbf{x}.$$

We know that if $\lambda^2 > 4\pi^2$, which is invariably true for suspension bridges, the first torsional and first vertical modes of vibration are in each case the first antisymmetric modes. These modes exhibit S-shaped curves, with nodes at the ends and midspan, so it is clear that $\tilde{\mathbf{h}}_1 = \tilde{\mathbf{h}}_2 = 0$. Thus

$$\begin{aligned} \mathbf{I}_m \boldsymbol{\omega}_1^2 &= (2\pi)^2, \\ \boldsymbol{\omega}_2^2 &= (2\pi)^2. \end{aligned} \tag{4.83}$$

Only the first modes of vibration need be considered, because only the lowest critical velocity is of importance.

An expression for I_m may be derived if it is assumed that the mass of the deck is uniformly distributed across the cable spacing $2\gamma b$. This is usually not far from the truth. If the mass per unit length of two cables is denoted by m_c,

$$I_m = \frac{m\gamma^2 b^2}{3} + \frac{2}{3} m_c \gamma^2 b^2, \tag{4.84}$$

or in dimensionless terms

$$\mathbf{I}_m = \frac{1}{3} + \frac{2}{3}\frac{m_c}{m}. \tag{4.85}$$

But by definition $0 < m_c/m < 1$, so that $1/3 < \mathbf{I}_m < 1$ and $1 < \boldsymbol{\omega}_1/\boldsymbol{\omega}_2 < \sqrt{3}$. Therefore the torsional frequency is always greater than its vertical counterpart.

Now attention may be turned to the case when the wind is blowing. The cable parameter λ^2 remains unchanged, and the antisymmetric modes are still the first modes. As is evident from (4.78), (4.79), and the argument leading to (4.83), the frequency of the first torsional mode is now given by

$$\mathbf{I}_m \boldsymbol{\omega}_*^2 = (2\pi)^2 - \frac{\mathbf{V}^2}{\gamma^2}, \tag{4.86}$$

while that of the first vertical mode is as before

$$\boldsymbol{\omega}_2^2 = (2\pi)^2. \tag{4.87}$$

The wind causes a lowering of the torsional frequency. As $\mathbf{V} \to 2\pi\gamma$, $\boldsymbol{\omega}_* \to 0$, and if $\mathbf{V} > 2\pi\gamma$, $\boldsymbol{\omega}_*$ is imaginary, which means that any torsional disturbance will be amplified and grow exponentially with time. The *divergence speed* of the deck is thus given by

$$\mathbf{V}_d = 2\pi\gamma. \tag{4.88}$$

However, this is not the full story.

It was shown that in the absence of wind the torsional frequency was always greater than its vertical counterpart. With the wind blowing, the torsional frequency is reduced, while the vertical frequency remains unchanged. As the wind speed is increased, a situation is reached in which $\boldsymbol{\omega}_* = \boldsymbol{\omega}_2$. In this case the righthand side of equation (4.79) contains a forcing term from the torsional motion, the frequency of which is identical to the frequency of the vertical motion. This gives rise to resonance in the vertical motion. The deck undergoes sustained coupled vibrations (of large amplitude) at a common frequency (at $\boldsymbol{\omega}_2$). This is called flutter.

The *flutter speed* is given by

$$\mathbf{I}_m(2\pi)^2 = (2\pi)^2 - \frac{\mathbf{V}_f^2}{\gamma^2}, \tag{4.89}$$

which after some rearrangement reduces to

$$\mathbf{V}_f = \left(\frac{2}{3}\right)^{1/2}\left(1 - \frac{m_c}{m}\right)^{1/2} 2\pi\gamma. \tag{4.90}$$

Hence the greatest speed at which flutter is possible is $(2/3)^{1/2}$, or 0.82 times the divergence speed. The flutter speed is the critical one. Therefore in dimensional terms

$$V_{\text{crit}} = \left\{\frac{\pi g}{3\rho(d/l)}\right\}^{1/2} \gamma(m_d/l)^{1/2}, \tag{4.91}$$

where $m_d(= m - m_c)$ is the mass of the deck.

The first term on the right is essentially a constant independent of the particular bridge (the ratio of sag to span varies little). Consequently (4.91) shows that the critical speed is increased as the spacing of the cables is increased in relation to the width of the deck, as the mass of the deck is increased, or as the span is reduced.

We have in (4.91) an explanation of why some of the early suspension bridges fared badly in the wind. The weight of the chains formed a significant part of the total weight of the bridge, so the critical wind speeds were accordingly low. In fact, if appreciable deck stiffness were indeed absent, modern bridges would not do much better, because none of the variables at the right of (4.91) is able to be changed significantly. However, while modern long-span bridges may have decks in which, relative to the cables, flexural rigidity is low, the relative torsional rigidity is high. This concept has arisen from considerations of aeroelastic theory, backed up by wind-tunnel tests where it was found that, if the natural frequencies in torsion and in flexure could be well spaced, the critical speeds were increased. This in turn has led to the use of box trusses and more recently to box girders.

If one assumes that the torsional resistance is derived largely from shear flow, as might be appropriate in a box girder section, the dimensionless parameters that indicate the relative importance of flexure and torsion compared to cable action are

$$\alpha^2 = \frac{EI}{2Hl^2} \quad \text{and} \quad \beta^2 = \frac{GI_t}{2H\gamma^2 b^2},$$

where GI_t is the torsional stiffness of the deck. Since $GI_t \sim EI$ (their ratio is almost unity for a thin-walled tube), $\beta^2 \gg \alpha^2$, and in long-span bridges where typically $\alpha^2 \ll 1$ we may plausibly ignore flexural stiffness and concentrate on the correction obtained when allowance is made for torsional rigidity.

Equation (4.78) is then [25]

$$(1 + \beta^2)\frac{\partial^2 \boldsymbol{\theta}}{\partial \mathbf{x}^2} - \mathbf{I}_m\frac{\partial^2 \boldsymbol{\theta}}{\partial \mathbf{t}^2} + \frac{\mathbf{V}^2}{\lambda^2}\boldsymbol{\theta} = \mathbf{h}_1 \tag{4.92}$$

with

$$\mathbf{h}_1 = \lambda^2 \int_0^1 \boldsymbol{\theta} d\mathbf{x},$$

and (4.79) remains as it is. If now

$$\lambda^2 > 4\pi^2(1 + \beta^2), \tag{4.93}$$

which is usually the case, the critical speed is given by

$$\mathbf{V}_{\text{crit}} = \left(\frac{2}{3}\right)^{1/2}\left(1 - \frac{m_c}{m} + \frac{3}{2}\beta^2\right)^{1/2} 2\pi\gamma. \tag{4.94}$$

Other things being equal, this formula will give some indication of the likely critical speed in a single suspended-span bridge possessing a streamlined deck.

Example 4.6 The Aeroelastic Failure of the Matukituki Suspension Footbridge

The Matukituki river drains a remote catchment in the south of New Zealand. In April 1977 a small suspension footbridge that crosses the river blew down in a gale, just twelve days after it was completed. By chance observers were on hand, and their description of the failure shows that it behaved very much like the other bridges that have come down in the wind. The deck persisted in lurching and twisting wildly until failure occurred, and for part of the time a node was noticeable at midspan.

The collapse is believed to have been aeroelastic in origin. Other similar foot-bridge in similar sites have stood for many years, and, although they have exhibited a certain amount of liveliness under wind loading, none has failed. The only discernible difference was that the deck planks in the Matukituki bridge butted together, thereby rendering the deck relatively impervious—the planks on other bridges were always gapped. This oversight is suspected to be the root cause of the failure.

The original bridge was unstiffened and had the following approximate properties: $l = 100$ m, $d = 8$ m, $mg = 1$ kN/m, $m_c g = 0.125$ kN/m. The width of the planking was 1 m, and the cables were spaced at 1.33 m; therefore $\gamma = 1.33$. The density of air is $\rho = 1.2$ kg/m^3.

The cable parameter λ^2 is calculated to be about 400, so the first modes of vibration are the antisymmetric modes. The flutter speed is

$$V_{\text{crit}} = \left(\frac{\pi \times 875}{3 \times 1.2 \times 8}\right) \times 1.33 = 13 \text{ m/sec} = 47 \text{ km/hr} (\equiv 34 \text{ knots}),$$

and the divergence speed is

$$V_d = \left(\frac{3}{2}\right)^{1/2} \left(1 - \frac{m_c}{m}\right)^{1/2} V_{\text{crit}}$$

$$= 61 \text{ km/hr} (\equiv 44 \text{ knots}).$$

The gale that caused failure was assessed as a Beaufort Scale Force 9 Strong Gale, having a ten-minute mean wind speed of 41 to 47 knots, with the possibility of a three-second gust of 70 to 80 knots. This storm not only exceeded the estimated flutter speed of the bridge but also its divergence speed. It is therefore safe to conclude that the initial cause of the failure was aeroelastic instability.

Gaps between adjacent planks allow air to bleed through, and the critical velocity is much higher because lift and torque coefficients are substantially reduced. The planks of the replacement structure were gapped, and, although gales of similar magnitude occur several times each year in that part of New Zealand, no further problems have arisen.

Earthquake-Generated Additional Tension in a Suspension Bridge Cable

We can derive a simple formula for the peak additional cable tension that can be expected in a suspension bridge with a single suspended span undergoing earthquake excitation. The method involves application of the response spectrum technique and rests on several plausible assumptions. The formula gives a reasonable upper bound, suitable for preliminary design estimates, irrespective of whether the ground motions at each end are in-phase or out-of-phase, as is probable with a long-span bridge [32].

For the preliminary details to be adequately covered, reference should be made to section 3.3 and example 3.4. Suffice it to say here that the bridge shown in figure 4.24 (in which the backstays are considered unloaded) is replaced by a single cable fixed to rigid supports at the same

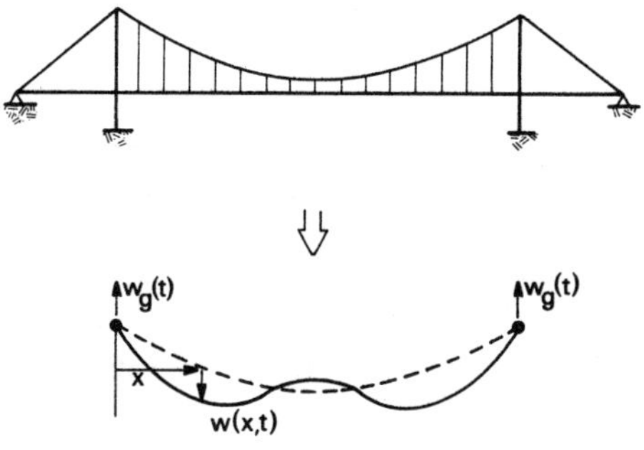

4.24 Equivalent single cable undergoing vertical excitation

level. The influence of the stiffening truss is neglected; it is probably conservative to ignore its effect on cable response in general. The effects of tower compliance, which ensure that tower top movements will differ from motions fed in at the foundation level, are also not specifically addressed. It is clear then that we are concentrating on but one (important) component of a highly complicated structural system. Sight of this fact should not be lost in the ensuing analysis.

Under vertical ground accelerations that are the same at each end of the cable, the normal coordinate equations may be shown to be (refer to example 3.4)

$$\frac{d^2\phi_n}{dt^2} + 2\omega_n\zeta_n\frac{d\phi_n}{dt} + \omega_n^2\phi_n = \frac{\alpha_n\omega_n^2}{\mathbf{h}_n}\ddot{\mathbf{w}}_g, \tag{4.95}$$

where the additional tension is

$$\mathbf{h(t)} = \sum_n \mathbf{h}_n\phi_n(\mathbf{t}). \tag{4.96}$$

(Because the ground motion at each end is assumed to be in-phase, just the vertical motion gives rise to additional cable tension: longitudinal motion does not then enter the problem.) The modal participation factor for additional tension is (recall equation 3.36)

$$\alpha_n = \frac{2/3}{[1 + \lambda^2\{\tan(\omega_n/2)/(\omega_n/2)\}^2/12]}. \tag{4.97}$$

The modal maximum for additional tension is

$$\mathbf{h}_{n,\,\max} = \mathbf{h}_n\phi_n(\mathbf{t})\big|_{\max} = \alpha_n\omega_n^2\mathbf{S}_{d,\,n}, \tag{4.98}$$

where $\mathbf{S}_{d,\,n}$ is the displacement response spectrum value for the nth symmetric mode under the acceleration $\ddot{\mathbf{w}}_g$. If we use a flat-hyperbolic spectrum for acceleration, for which some justification exists [22], and also use the concept of the pseudovelocity spectrum, namely, $\mathbf{S}_{pv}$ $(= \omega\mathbf{S}_d)$, we have

$$\mathbf{h}_{n,\,\max} = \alpha_n\omega_n\mathbf{S}_{pv,\,n}, \tag{4.99}$$

where the nondimensionalization is $\mathbf{S}_{pv} = S_{pv}/\{(mgl/H)(gl)\}^{1/2}$. We further assume that $\mathbf{S}_{pv,n} = \mathbf{S}_{pv}$—the pseudovelocity is independent of mode number (which tends to overestimate the higher mode contributions), and damping is for convenience taken to be the same in all the modes. Finally, because the natural frequencies are well spaced, we may apply the SRSS (square root of the sum of the squares) combination rule to give

$$\mathbf{h}_{\max} = \mathbf{S}_{pv}\left\{\sum_n (\alpha_n\omega_n)^2\right\}^{1/2}. \tag{4.100}$$

Values of λ^2 for suspension bridges lie broadly and typically in the range $100 < \lambda^2 < 400$. Therefore, using data from tables 3.1 and 3.4 for a value of $\lambda^2 = 36\pi^2$, we obtain

$$\mathbf{h}_{max} \simeq \left\{(0.035 \times 2.82\pi)^2 + (0.134 \times 4.78\pi)^2 + \left(\frac{2}{3} \times 6\pi\right)^2 + (0.11 \times 7.14\pi)^2 + \ldots\right\}^{1/2} \mathbf{S}_{pv},$$

or

$$\mathbf{h}_{max} \simeq 4.1\pi\,\mathbf{S}_{pv}. \tag{4.101}$$

In dimensional terms this rearranges to

$$\frac{h_{max}}{H} \simeq \frac{4.1\pi S_{pv}}{(8d/l)^{1/2}(gl)^{1/2}}. \tag{4.102}$$

Values of the ratio of sag to span of 1:10 to 1:12 are common, so incorporating the smaller of these and rounding the answer yields

$$\frac{h_{max}}{H} \simeq \frac{15 S_{pv}}{(gl)^{1/2}}. \tag{4.103}$$

We could have obtained the answer directly in terms of $(gd)^{-1/2}$, but this was not done because the clear span seems a more obvious variable to work with. Equation (4.103) is conservative: if we had used a value of $\lambda^2 = 16\pi^2$, the result would have been about two-thirds as great because the major contribution would then have come from the second mode and not the third mode. (The result, $h_{max}/H \simeq 10S_{pv}/(gl)^{1/2}$, could be used for cable roofs where smaller values of λ^2 occur.)

The last item to require discussion concerns the possible influence of out-of-phase motion. In the extreme case, when strong ground shaking occurs at one end before any occurs at the other support (see figure 4.25), the forcing term in (4.95) may be replaced by

$$\frac{\ddot{\mathbf{w}}_g}{2} + \lambda^2 \mathbf{u}_g,$$

where $\mathbf{u}_g(= u_g/\{(mgl^2/H)(mgl/H)\})$ is the horizontal component of ground displacement at the end. To estimate the importance of this effect, we let $\ddot{\mathbf{w}}_g = 0.25$ (the peak vertical component of the ground acceleration at the support is 25 percent g), and we set the horizontal component of ground acceleration at 50 percent g, say. This could correspond to a peak longitudinal displacement at the support of perhaps $u_g \sim 0.5$ m. In dimensional terms

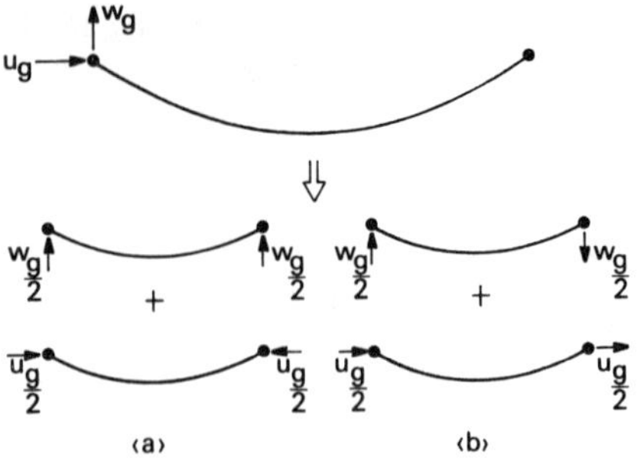

4.25 Definition diagrams for an extreme case of out-of-plane end motion: (a) additional tension generated; (b) no additional tension generated.

$$\lambda^2 \mathbf{u}_g \simeq \frac{u_g}{(Hl/EA)}, \tag{4.104}$$

and, given a value of $E \simeq 200{,}000$ MPa and a nominal dead load stress of $H/A \simeq 500$ MPa in a cable of 1,000 m span, we see that

$$\lambda^2 \mathbf{u}_g \simeq \frac{0.5 \times 200{,}000}{500 \times 1{,}000} \simeq 0.2.$$

With these data, $\ddot{\mathbf{w}}_g/2 + \lambda^2 \mathbf{u}_g \simeq 0.325$, which may be compared with 0.25 if in-phase behavior is not assumed at the outset. This represents an increase of 30 percent which could be significant.

Nevertheless we have endeavored to be conservative in this assessment, and, since the response is not markedly increased, it seems reasonable to let the original formula stand (which already has a rounding margin in it). It should be emphasized that phase effects are likely to be perceptible only for long-span bridges, and even there, because the towers may be founded on firm strata or bed rock, phase differences may be less pronounced than indicated by our computations here.

In addition Abdel-Ghaffar [3] has checked (4.103) against finite element, seismic response analyses for particular bridges and found good agreement for single suspended-span structures, but agreement was poor for triple suspended-span structures, partly because interaction with the side spans means that frequencies are not always well separated. It appears therefore that (4.103) is satisfactory as far as it goes. Improved estimates may always be had by resorting to more sophisticated techniques that consider the whole bridge. There are many excellent works available including references [9] and [50] on analytical techniques, references [6] and [1] on numerical techniques, references [2] and [11] on full-scale ambient vibration measurements, and reference [7] on design considerations for dynamic loads.

Example 4.7 Sample Calculations Using (4.103)

Application of the formula for a 1,000 m bridge, assuming a value of $S_{pv} = 0.25$ m/sec, yields

$$\frac{h_{\max}}{H} \simeq 0.05.$$

On the other hand, a much shorter, stiffer bridge (of around 100 m span) will respond more strongly under the same vertical input and then

$$\frac{h_{\max}}{H} \simeq 0.15.$$

In this case the more pronounced influence of the stiffening truss will tend to reduce this value.

In absolute terms the appearance of a large response for the shorter bridge is somewhat misleading, since the cables are often under relatively low dead load tensions and the total cable tension will remain low. For long-span bridges, where dead load tensions can be quite high (to make efficient use of the cables), the earthquake-generated component can be of some significance as a live load condition. This is especially true if the vertical spectral velocity is higher than the moderate value we have used. Design of a bridge for a seismically active region would necessitate the use of a local design spectrum calculated from the best available historical data. With these curves a more refined estimate of peak additional tension could be found by using the design vertical spectral velocity pertaining to the most strongly excited mode and the relevant value of damping for the structure as a whole.

Exercises

4.1 In work that could well lay claim to being the first for which a doctorate was awarded in cable theory, Grant in 1918 presented a thesis on the motion of a uniformly accelerated flexible cable [21]. An extract from the introduction to this work reads:

> The problem was suggested by observing the fluctuations in the movements of the steel cables used in hoisting loads of copper rock in the copper mines of Michigan. In some of these mines the shaft is in the shape of a common catenary, the inclination at the surface being about 54°, while about 4,200 feet farther down, the inclination has decreased to 34°. The accompanying diagram illustrates the conditions. . . . [see figure 4.26.]
> The question arose, what would be the shape of the curve assumed by the cable when subject to a uniform acceleration? The speed obtained is frequently as high as 50 miles per hour.

In attempting a solution, Grant assumed that the motion is steady, which requires that the velocity normal to the cable be zero. The normal velocity can only be zero if the tangential velocity of the cable varies directly with time—that is, the acceleration must be uniform. Armed with these assumptions, Grant was able to perform an elegant Lagrangian

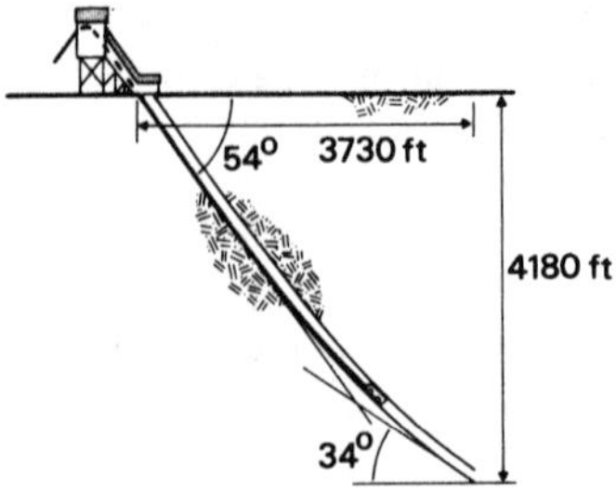

4.26 Schematic of the shaft of a copper mine typical of those in operation in Michigan at the turn of the century (from Grant [21], courtesy The University of Chicago Press)

analysis of the problem based on the classical catenary equation. He then compared the theory with observation.

The problem with such analysis is that it is completely implicit, and in any event the actual cable, when it is jerked into motion, will not be steady. As the loaded ore car begins its ascent, the cable straightens, and the amplitude of the movement of the cable will depend more or less directly on the tangential acceleration imparted to the car. Since it is inadvisable to have the cable tangle with the roof of the shaft, the amplitude of the cable movement must be limited, and this implies an upper limit on the allowable acceleration.

The geometry of the shaft shown in figure 4.26 indicates that the cable lies relatively close to the chord—a point worth exploiting. Show as a first approximation that the greatest acceleration, a, the car may start with is

$$\frac{a}{g} = \frac{D}{mg \cos \theta l_*^2/(8\, Mg \sin \theta)},$$

where g is the acceleration due to gravity, D is the tunnel diameter, mg is the weight per unit length of the cable, θ is the angle of inclination of the chord to the horizontal, l_* is the chord length, and Mg is the weight of the loaded car.

The rather gross assumption has been made that the cable tension is constant and equal to $Mg \sin \theta$. This assumption can easily be relaxed, and more refined estimates found (the solution then involves logarithmic terms), but it is debatable whether it is necessary. Data pertaining to the cable hoist [21] shown in figure 4.26 are $Mg = 30{,}000$ lbs, $mg = 3$ lb/ft, $\theta \simeq 45°$, $l_* = (3{,}730^2 + 4{,}180^2)^{1/2} = 5{,}600$ ft, $D = 10$ ft, which, on substitution in the given equation, yields $a \simeq 0.03g$. This is almost exactly the value found by Grant and is apparently typical of values used in practice.

4.2 The cable dragging a bottom-towed pipeline into place on a gently shelving sea bed must be paid out continuously if the end of the pipe is to avoid damage from excessive flexing. With reference to figure 4.27 show that

$$\frac{l_c}{l_p} = \frac{2S\mu(mg)_p}{(mg)_c}\left[\left\{1 + \frac{4}{2S\mu(mg)_p/(mg)_c}\right\}^{1/2} - 1\right].$$

In other words the ratio of cable length to pipe length should theoretically remain constant.

4.3 The single point mooring shown in figure 4.28 consists of a short scope riser chain anchored to a robust pile cap and attached to a mooring buoy at the surface. It might be thought that the sizing of the chain could be done directly once the design drag force (due to current and wind) H is known. This is, however, to overlook the fact that under a 180° change in the direction of the drag force the vessel will have acquired a kinetic energy of $\frac{1}{2}MV^2$ when it passes through the center of the mooring circle (of radius l). Unless this kinetic energy is transferred to gravitational potential energy in the chain, a dynamic impulse will be imparted to it, as the vessel comes to rest. This impulse could very well break the chain. The only way to increase the storage capability of the chain is to increase its weight mg—the corresponding chain area will usually be greatly in excess of that

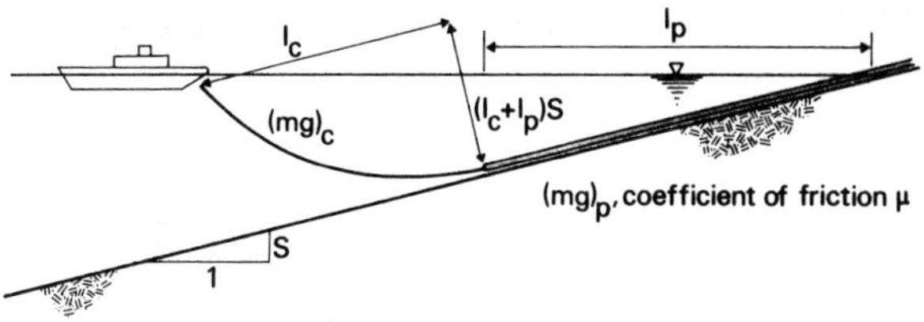

4.27 Definition diagram for a bottom-towed pipeline

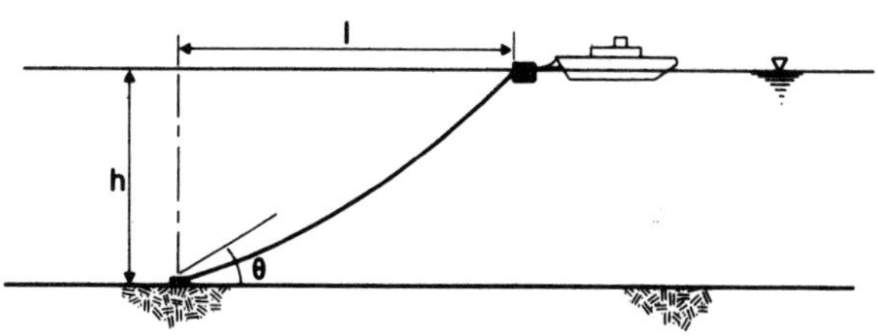

4.28 A short-scope riser chain used as a single-point mooring (from Shipp [47], courtesy ASCE)

required to resist the design drag force. For this reason the elasticity of the chain may be ignored.

Show then that the potential energy gained by the chain, given that it initially hangs limp and vertical at the center of the mooring circle, is

$$\frac{1}{2} mgl \sec \theta h - \frac{1}{2} mgh^2 - \frac{(mg)^2 \sec^2 \theta l^3}{12H},$$

where h is the water depth and $\sec \theta = (1 + h^2/l^2)^{1/2}$. The last term is a correction to allow for the slight dip that the chain has under the design drag force. Hence show that the necessary submerged weight of the chain is given to a very good degree of approximation by [29]

$$\frac{mgl}{H} = \frac{MV^2/Hl}{(h/l)\{(1 + h^2/l^2)^{1/2} - h/l\}} \left[1 + \frac{\frac{1}{2}(MV^2/Hl)(1 + h^2/l^2)}{3\{(h/l)^2((1 + h^2/l^2)^{1/2} - h/l)\}} \right].$$

This result compares favorably with calculations based on other less direct approaches [47].

4.4 A deeply submerged tri-moored navigational buoy is shown in figure 4.29. The guys are neutrally buoyant nylon ropes which are therefore straight. The initial tension in them is provided almost solely by the weight of the water displaced by the hollow spherical buoy. By treating the buoy as a point mass whose inertia is very much greater than the inertia of the guys, obtain expressions for the fundamental periods of pitching and heaving motion. Show that these periods are equal when the guys are inclined to the horizontal at

$$\theta = \tan^{-1} (1/\sqrt{2}) \simeq 35°.$$

The added mass of the water shifted by the buoy (the added mass coef-

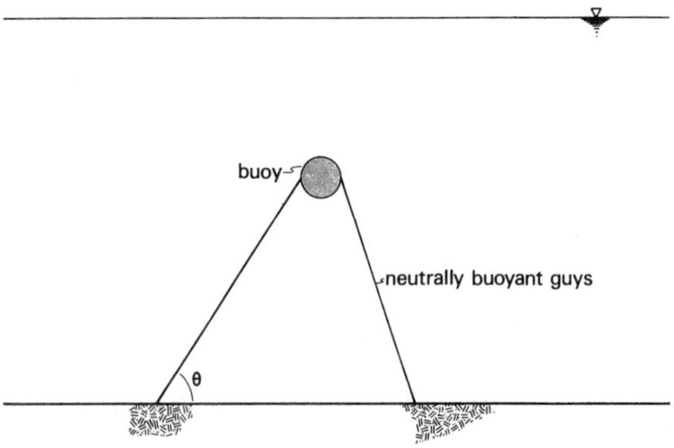

4.29 A submerged tri-moored navigational buoy

ficient is 0.5 in the case of a sphere) is the most significant mass component. If need be, corrections for guy mass can always be included. The elasticity of nylon ropes is not in general linear, but the pertinent tangent modulus is acceptable for stiffness calculations for small vibrations.

4.5 In Wisconsin in January 1975 64 mi of 345 kV electric transmission line came down in high winds. The 268 transmission towers supporting the line were of light aluminum construction and very flexible in the longitudinal direction, roughly coinciding with the wind direction on the day of the storm. Longitudinal stability was provided by a highly stressed top static, or lightning shield, wire. When this broke away, the unbalanced load on the towers caused a progressive cascading collapse that could not be arrested because there were no intermediate anchor spans. This type of along-wind failure has also occurred in other parts of the world. By contrast, a 2,000 tower line in Tewksbury, Massachusetts, which was of similar construction but had dead ends every 10 spans, suffered only minor damage in the great hurricane of 1938 [43].

Two design principles for along-wind response seem to be important: the first is that the allowable static wire tension in the span after a dead end should not be exceeded; the second, that the longitudinal tower top movement in the span adjacent to the next dead end should not be excessive. Because up to ten spans may be involved, calculations are difficult unless a systematic approach is adopted. The purpose of this problem is to suggest one way of tackling it.

Show that, if the static wire in the ith span from a dead end has a tension under longitudinal wind and dead load of H_i (while the original tension under dead load is H), and the longitudinal movements of the tower tops at each end of this span are u_{i+1} and u_i (see figure 4.30),

$$\mathbf{u}_{i+1} - \mathbf{u}_i = -\frac{1}{24}\left(1 - \frac{1}{\mathbf{H}_i}\right)^2 + \frac{1}{12}\left(1 - \frac{1}{\mathbf{H}_i}\right) + \frac{\mathbf{H}_i - 1}{\lambda_i^2},$$

where $\mathbf{u}_i = u_i/\{(mgl_i^2/H)(mgl_i/H)\}$, $\mathbf{H}_i = H_i/H$, and $\lambda_i^2 = (mgl_i/H)^2 \{l_i/(HLe_i/EA)\}$. Equilibrium at the ith tower requires

$$\mathbf{H}_{i+1} - \mathbf{H}_i = -\mathbf{P}_i + \mathbf{k}_i\mathbf{u}_i,$$

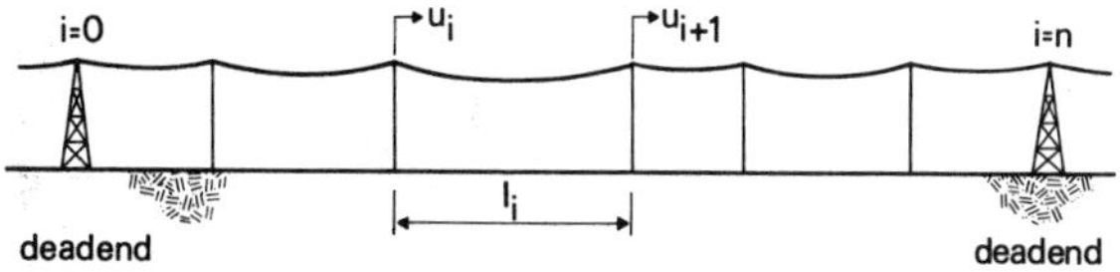

4.30 Definition diagram for the along-wind response of the static wire of a multispan electrical transmission line where $\mathbf{P}_i = P_i/H$, which is the dimensionless wind load to the tower top, with $\mathbf{k}_i = k_i(mgl_i/H)^2/mg$, which is the dimensionless longitudinal stiffness of the tower.

The calculation proceeds from $i = 0$ (the first anchor span) to $i = n - 1$ (the last span before the next dead end). Thus the boundary conditions are $\mathbf{u}_0 = \mathbf{u}_n = 0$. To start the calculation, let

$$\mathbf{H}_0 = \mathbf{H}_{\text{allow}},$$

where $\mathbf{H}_{\text{allow}}$ is the allowable tension in the static wire. This permits determination of $\mathbf{u}_1$ (from the first equation), and then $\mathbf{H}_1$ may be found from the second equation, and so on. When $\mathbf{u}_i$ reaches its assigned allowable value, a new dead end is called for.

It is appropriate at this juncture to mention that a great deal of work has of course been done on all aspects of electrical transmission-line performance. Transmission lines are by far the most prevalent form of cable structure, and methods for their design and construction are widely scattered throughout the technical literature. Peculiarities exist—for one, the conductor is a composite material—but the basic principles are much as indicated in the present work. Dynamic considerations, such as galloping of ice-laden conductors, are of extreme and continuing interest. In addition to the proceedings of specialty conferences and the journals of learned societies, reports are regularly produced by the electric power companies, the conductor manufacturers, and, in the United States, by the Electric Power Research Institute.

4.6 The multispan cable shown in figure 4.31 is anchored on rigid supports at each end and supported on rollers at intermediate points. Show that the symmetric in-plane modes are given by

$$\tilde{\mathbf{w}}_i(\mathbf{x}) = \tilde{\mathbf{h}}\frac{1}{\omega^2}\left\{1 - \tan\left(\frac{\alpha_i\omega}{2}\right)\sin(\alpha_i\omega\mathbf{x}) - \cos\alpha_i\omega\mathbf{x}\right\}, \quad 0 \le \mathbf{x} \le 1,$$

for the ith span (of length $\alpha_i l$) of the cable, where the natural frequencies are found from

$$\sum_{i=1}^{n} \tan\left(\frac{\alpha_i\omega}{2}\right) = \frac{\omega}{2} - \frac{4}{\lambda^2}\left(\frac{\omega}{2}\right)^3,$$

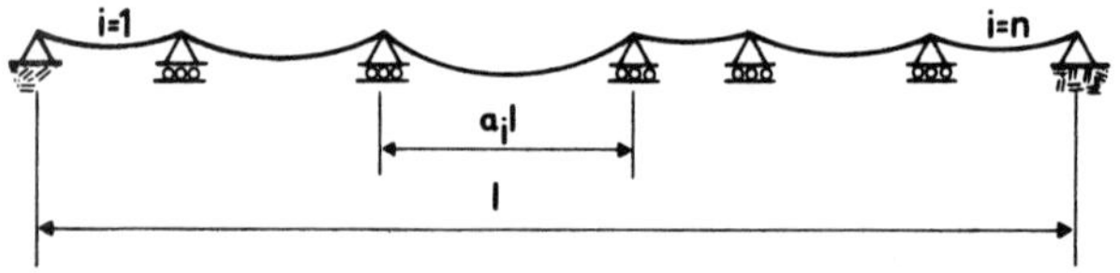

4.31 Definition diagram for free in-plane vibrations of a multispan cable where $\sum_{i=1}^{n} \alpha_i = 1$, $\lambda^2 = (mgl/H)^2 \{l/(HL_e/EA)\}$, $L_e = \sum_{i=1}^{n} L_{ei}$, and $\omega^2 = \omega^2 l^2 m/H$.

These results are useful for determining the natural frequencies of multispan transmission lines, where the conductors are hung from insulator bundles at their intermediate points, or they could be used to give some indication of the fundamental symmetric modes of a three-suspended-span suspension bridge, if tower interaction and deck stiffness are not appreciable.

4.7 In the case of a symmetric cable truss where the apex and the point of load application are both at midspan, prove that the critical load is

$$P_{\text{crit}} = \frac{4H}{d\int_0^{l/2} z^{-2}dx}.$$

(This result is due to Thomas K. Caughey.) If the truss is parabolic, that is, if

$$z = d\left\{1 - \left(\frac{d-b}{d}\right)\left(\frac{2x}{l}\right)^2\right\},$$

where coordinates are taken about midspan, then [26]

a for biconvex trusses

$$P_{\text{crit}} = \frac{8bH}{l}\,\frac{2}{\{1 + (b/d)(d/(d-b))^{1/2}\tanh^{-1}((d-b)/d)^{1/2}\}},$$

b for biconcave trusses

$$P_{\text{crit}} = \frac{8bH}{l}\,\frac{2}{\{1 + (b/d)(d/(b-d))^{1/2}\tan^{-1}((b-d)/d)^{1/2}\}}.$$

4.8 Sail theory falls within the realm of cable structures. The theory has been well worked over [55, 53, 40, 5, 12], the bulk of it being confined to studies of two-dimensional sails. The theory of three-dimensional sails is not well established, so that sail design for oceangoing yachts (for example) presently rests largely on the designer's skill and experience, with perhaps some recourse to numerical and experimental backup.

Even the development of a general linearized theory involving ideal fluid flow over a flexible two-dimensional sail is complicated, because the shape of the sail is determined by a pressure distribution, which is itself dependent on the profile of the sail, and numerical solution methods are generally necessary. Two exceptions exist. An approximate analytical solution may be found when the tension is high. There the sail profile is sufficiently shallow for the aerodynamic theory of the flat plate at incidence to provide a distribution of lift that allows the shape of the sail to be determined explicitly, see Thwaites [53].

At the other extreme Thwaites [53] and Nielsen [40] have shown that unstable situations may arise when the tension in the sail is insufficient to sustain the aerodynamic lift as curvature, or camber, begins to develop. In practice the instability manifests itself by flapping of the sail, in yachting terminology called luffing. It is in fact possible to provide a simple approximate result for the case when the sail is on the verge of instability [30].

By writing the equation of equilibrium for the whole profile shown in figure 4.32, with flow at zero incidence over a circular arc the lift being $4\pi(1/2\,\rho V^2)d$, show that the mid-chord displacement d is undetermined if

$$\frac{T}{\frac{1}{2}\rho V^2 l} = \frac{\pi}{2}.$$

This is in favorable agreement with the exact result of [53, 40]:

$$\frac{T}{\frac{1}{2}\rho\, V^2 l} = 1.73.$$

4.9 A simple and well-known way of stiffening a suspension bridge is to provide a rigid cross-braced frame between cables and deck at midspan. This suppresses the longitudinal cable motion associated with some of the antisymmetric modes. The antisymmetric modes affected in a single-suspended-span structure are the first, third, fifth and so forth. The alternate second, fourth and sixth antisymmetric modes are unaffected because the longitudinal components in these modes are zero at midspan (recall figure 3.2). For the affected antisymmetric modes, establish the results

$$\tilde{\mathbf{w}}(\mathbf{x}) = \frac{\tilde{\mathbf{h}}}{\omega^2}\left[1 - \tan\left(\frac{\omega}{4}\right)\sin\omega\mathbf{x} - \cos\omega\mathbf{x}\right], \quad 0 \le \mathbf{x} \le \frac{1}{2},$$

where $\boldsymbol{\omega}$ is found from the requisite nonzero root of [25]

$$\tan\left(\frac{\omega}{4}\right) = \left(\frac{\omega}{4}\right) - \frac{16}{\lambda^2}\left(\frac{\omega}{4}\right)^3,$$

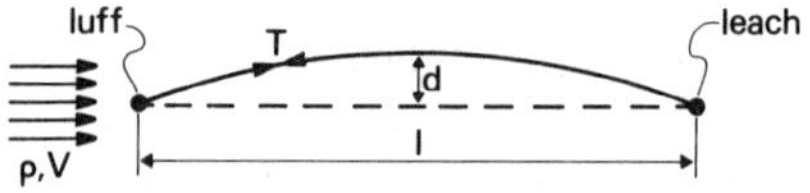

4.32 Luffing in a two-dimensional sail

We have deliberately chosen to ignore deck stiffness. Thus the cross-bracing at midspan will cause a rise in the lowest natural frequency for all bridges in which $\lambda^2 > 4\pi^2$—which, as we have frequently pointed out, is an inequality that is always met.

4.10 One of the important variables in the economics of a long-span suspension bridge is the ratio of sag to span of the cables. Given a fixed span and a total suspended weight which is essentially fixed with little variation in anchorage details, the larger ratios of sag to span will reduce cable forces and therefore cable areas. However, this occurs at the expense of taller towers. It may plausibly be assumed that the constructed costs of the cables and the towers bear directly on cable area on the one hand, and on tower height on the other.

Show that in this idealized situation an optimum ratio of sag to span exists, so that the cost of the two items together is minimized when their individual costs are equal. Actual cost investigations tend to confirm this conclusion and point to ratios of sag to span of about 1:10 to 1:12 [41].

4.11 Two different arrangements for a cable-stayed bridge with one suspended span are shown in figure 4.33. If in each case the strands pass freely over the tower tops and, because the individual strands are numerous, take all the dead load, show that the horizontal components of tension in the backstays are

$$H = \frac{mgl^2}{6d}$$

and

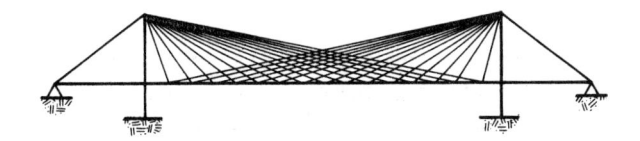

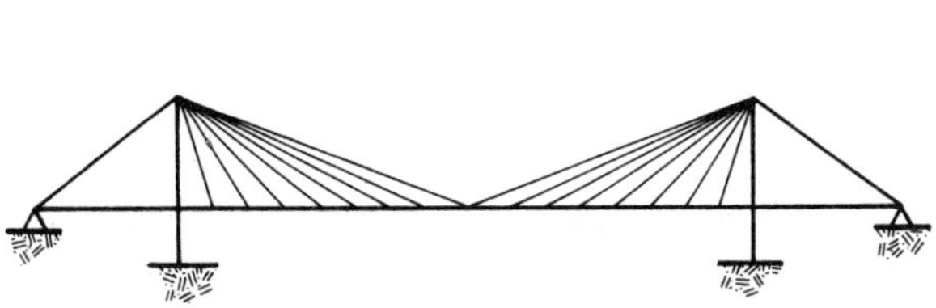

4.33 Two possible strand arrangements for a cable-stayed bridge with one suspended span

$$H = \frac{mgl^2}{8d},$$

respectively, where d is the tower height above the deck, and mg is the uniform dead load per unit length of the span l. In the first case it is assumed that the individual strands do not introduce axial forces into the deck.

Show further that, if the average stress in the backstays controls individual strand size, their area in the first case need only be 67 percent of the area of the individual strands in the second case. Nevertheless the total weight of cable used will be 33 percent higher in the first case.

4.12 A hyperboloid of revolution is a surface generated by a straight line that revolves about another nonparallel, nonintersecting line as an axis. It may also be generated by a line touching three circles whose planes are perpendicular to a common axis through their centers. Depending on the inclination of the generator, numerous different hyperboloids may be generated—the limiting cases being obviously a cylinder and a cone.

Suppose that the generators are taut flat cables, that the upper and lower circles are rigid plates, and that the longitudinal axis (taken as vertical) is an axially inextensible structural member. Show that the stiffness of this cable cluster, for a (horizontal) shearing displacement of one end relative to the other, is

$$n\left(\frac{T}{c} + \frac{1}{2}\cos^2\theta\,\frac{EA}{c}\right),$$

where n is the number of cables, T the tension in each, θ is the (constant) angle subtended to the horizontal by each cable of length c, and EA is its axial stiffness.

This result is identical to the first equation of (4.11) and is surprisingly simple, given the rather complicated geometry of the surface. Such staying systems have been used for elevated buildings, the central cores of which provide services and lifts. To eliminate the torque on the core, half the cables pass up the height of the structure from left to right, the other half passing from right to left.

References

1. Abdel-Ghaffar, A. M. 1976. Ph.D. dissertation report. Earthqu. Eng. Res. Lab., California Institute of Technology, EERL 76-01.

2. Abdel-Ghaffar, A. M., and Housner, G. W. 1979. *J. Eng. Mech. Div., Proc. ASCE*, 104:983–999.

3. Abdel-Ghaffar, A. M. 1979. Personal communication.

4. Abramowitz, M., and Stegun, I. A. 1965. *Handbook of Mathematical Functions*. New York: Dover.

5. Barakat, R. 1968. *J. Math. Phys.*, 47:327–349.

6. Baron, F., et al., 1976. Report. Earthqu. Eng. Res. Cent., University of California, Berkeley, EERC 76-31.

7. Baron, F. 1979. Ann. Conv. Am. Soc. Civ. Engrs., Boston, preprint no. 3590.

8. Berteaux, H. O. 1976. *Buoy Engineering*. New York: Wiley.

9. Bleich, F., et al. 1950. *The Mathematical Theory of Vibration in Suspension Bridges*. Washington, D.C.: Government Printing Office.

10. Blevins, R. D. 1977. *Flow-induced Vibration*. New York: Van Nostrand Reinhold.

11. Buckland, P. G., et al. 1979. *J. Struct. Div., Proc. ASCE*, 105:859–874.

12. Bundock, M., 1979. Master of Science thesis. University of Waikato, Hamilton, New Zealand.

13. Chugh, A. K., and Biggers, S. B. 1978. *Int. J. Comp. Struct.*, 8:125–133.

14. Crosthwaite, C. D. 1947. *Proc. Instn. Civ. Engrs.*, paper no. 5604.

15. Davenport, A. G. 1959. *Trans. Eng. Inst. Canada*, 3:119–141.

16. Davenport, A. G., and Steels, G. N. 1965. *J. Struct. Div., Proc. ASCE*, 91: 43–70.

17. Evans, J. H., and Adamchak, J. C. 1972. *Ocean Engineering Structures*. Cambridge, Mass.: The MIT Press.

18. Finn, L. D. 1976. Offshore Technology Conference, paper no. 2688.

19. Finn, L. D., and Young, K. E. 1978. Offshore Technology Conference, paper no. 3131.

20. Fung, Y. C. 1969. *Theory of Aeroelasticity*. New York: Dover.

21. Grant, E. D. 1918. Ph.D. dissertation. University of Chicago.

22. Housner, G. W. 1970. *Earthquake Engineering*. Edited by R.L. Wiegel. Englewood Cliffs, N.J.: Prentice-Hall, pp. 93–106.

23. Howson, W. P. 1975. Ph.D. dissertation. University of Birmingham, Birmingham, England.

24. Irvine, H. M. 1974. *J. Struct. Div., Proc. ASCE*, 100:789–812.

25. Irvine, H. M. 1974. *Int. J. Earthqu. Eng. Struct. Dyn.*, 3:203–214.

26. Irvine, H. M., and Jennings, P. C. 1975. *J. Eng. Mech. Div., Proc. ASCE*, 101:403–416.

27. Irvine, H. M. 1975. *J. Eng. Mech. Div., Proc. ASCE*, 101:429–446.

28. Irvine, H. M. 1978. *J. Eng. Mech. Div., Proc. ASCE*, 104:491–497.

29. Irvine, H. M. 1978. *J. Waterways Harb. Div., Proc. ASCE*, 104:464–465.

30. Irvine, H. M. 1979. *Proc. Roy. Soc.* (Lond.), A365:345–347.

31. Irvine, H. M., and O'Sullivan, M. J. 1979. *Int. J. Appl. Ocean Res.*, 1:203–207.

32. Irvine, H. M. 1980. *Int. J. Earthqu. Eng. Struct. Dyn.*, 8:267–273.

33. Jennings, A., and Mairs, J. E. 1972. *J. Struct. Div., Proc. ASCE*, 98:2433–2454.

34. Kolousek, V. 1947. *Int. Assoc. Bridge Struct. Engrs.*, 8.

35. Livesley, R. K. 1975. *Matrix Methods of Structural Analysis*. Oxford: Pergamon Press, pp. 203–207.

36. McCullough, D. 1972. *The Great Bridge*. New York: Simon and Schuster.

37. Milne-Thompson, L. M. 1968. *Theoretical Hydrodynamics*. 5th ed. London: Macmillan, ch. 7.

38. Møllmann, H. 1974. *Analysis of Hanging Roofs by the Displacement Method*. Lyngby, Denmark: Polyteknisk Forlag ch. 5 (see also *J. Struct. Div., Proc. ASCE*, 96:2059–2082).

39. Newman, J. N. 1977. *Marine Hydrodynamics*. Cambridge, Mass.: The MIT Press.

40. Nielsen, J. N. 1963. *J. Appl. Mech., Trans. ASME*, 30:435–442.

41. Parsons, M. F. 1979. Personal communication.

42. Pugsley, A. G. 1968. *The Theory of Suspension Bridges*. 2nd ed. London: Edward Arnold.

43. Richardson, A. S. 1979. Personal communication.

44. Scanlan, R. H. 1971. Vibrations Conference, *ASME*, Toronto, Canada, paper no. 71-Vibr-38.

45. Schleyer, F.-K 1969. *Tensile Structures*. Vol. 2. (Edited by F. Otto.) Cambridge, Mass.: The MIT Press, ch. 4.

46. Shears, M. 1968. Structural Engineering Laboratory, College of Engineering, University of California, Berkeley, report no. 68–6.

47. Shipp, J. G. 1977. *J. Waterways Harb. Div., Proc. ASCE*, 103:537–546.

48. Simitses, G. J. 1976. *Elastic Stability of Structures*. New York: Wiley, pp. 68–75.

49. Simiu, E., and Scanlan, R. H. 1978. *Wind Effects on Structures*. New York: Wiley.

50. Steinman, D. B. 1959. *Ann. N.Y. Acad. Sci.*, 79.

51. Subcommittee on Cable-Suspended Structures, 1971. *J. Struct. Div., Proc. ASCE*, 97:1715–1761.

52. Task Committee on Cable-Suspended Structures, Commentary on the Tentative Recommendations for Cable-Stayed Bridge Structures. 1977. *J. Struct. Div., Proc. ASCE*, 103:941–959.

53. Thwaites, B. 1961. *Proc. Roy. Soc.* (Lond.), A261:402–442.

54. Timoshenko, S. P., and Young, D. H. 1965. *Theory of Structures*. 2nd ed. New York: McGraw-Hill.

55. Voelz, K. 1950. *Zeitschr. f. angew. Math. u. Mechanik*, 30:22–33.

56. Von Kármán, T., and Biot, M. A. 1940. *Mathematical Methods in Engineering*. New York: McGraw-Hill, pp. 220–228.

57. Wiegel, R. L. 1964. *Oceanographical Engineering*. Englewood Cliffs, N.J.: Prentice-Hall.

58. Ye, Y. 1959. *Large Span Suspension Roof Structures.* People's Republic of China: Architectural Engineering Publishing Service (in Chinese).

59. Zetlin, L. 1964. *Eng. J. AISC*, 1:1–11.

5

Three-Dimensional Surfaces

5.1 Introduction

A single cable resists applied loads by changing both its geometry and tension—an inevitable consequence of a need to satisfy two equations of equilibrium. In a flexible, three-dimensional tension surface three in-plane forces may develop. Expressed in terms of forces per unit representative length, these stress resultants consist of two direct tensile actions and a shear stress resultant. Since there are three equations of equilibrium for these three unknown stress resultants, the following conclusions may be drawn.

First, a curved surface does not necessarily require a change of geometry to resist smoothly spread loads. Therefore constitutive relations do not need to be introduced before obtaining the distributions of internal actions within the membrane to at least a reasonable degree of accuracy. (This conclusion would not hold when the materials are rubberlike because large elastic deformations are possible [5]. The strains we deal with are infinitesimal.) The second (associated) conclusion is that, because the problem is statically determinate, the stress resultants may in theory be found by integrating differential equations and adjusting the constants of integration to satisfy the boundary conditions [17].

Having said this much, we must mention the inevitable qualifications. In the first place few surfaces permit simple analysis—the sphere is one of those few. Moreover, since tension structures cannot resist compression, the combination of tensile stresses and shear stresses must be such as to preclude this, no matter where an element of the cable is situated or in what direction it is viewed. When this condition is not met, wrinkles form, and the solution in that region could properly be judged suspect. If the surface is relatively flat to start with, as in a pretensioned cable-suspended roof, it must deform to resist applied load. Fabrics and cable networks often have rather low in-plane shear moduli, and shear stresses are not developed to any appreciable degree. Two of the equations of equilibrium then do not supply any directly useful information, and we arrive back at a three-dimensional analog of the single cable where in general changes in the tensions and the geometry are essential to the load-carrying function. The elements stretch, requiring the simultaneous consideration of stress-strain behavior in all but lightly loaded shallow membranes. This mode of behavior is quite different from, say, a shallow cylindrical reinforced concrete shell where shear stress resultants are usually very much in evidence and the associated changes in geometry are practically imperceptible.

In this final chapter we show what can be done by exploiting these trends in behavior. We start with the sphere and work an idealized problem of a lighter-than-air balloon. With slight extensions the equations of equilibrium of a tension shell of revolution may be established and then used to explore surfaces in which constraints are placed on the stress resultants. Examples are drawn from pneumatic domes, parachutes, and hot air balloons. An engineering theory of the static and dynamic response of suspended membranes is developed with applications to simple forms of cable-suspended roofs.

5.2 A Sphere under Hydrostatic Pressure

Figure 5.1 a shows an element of the surface of a sphere bounded by two parallel circles (or lines of constant latitude) and two meridians (or lines of constant longitude). The geometry of this element may be described by referring to figures 5.1b and c. The radius of curvature is a, and the angle ϕ is the colatitude. The longitude is θ, and the parallel circle is of radius a sin ϕ, so that the area of the element is $ad\phi(a \sin \phi)d\theta$.

The loading on this membrane is taken to be a hydrostatic pressure p, directed outward, which is constant around any parallel circle but may vary with latitude. The self-weight of the membrane is judged negligible in comparison to this fluid pressure.

The stress resultants are shown in figure 5.2a: T_ϕ is the meridional stress resultant, T_θ is the hoop stress resultant, and $T_{\theta\phi}(=T_{\phi\theta})$ is the shear stress resultant. Both T_ϕ and T_θ may vary with ϕ, but they are clearly independent of θ for p is axisymmetric. The shear stress resultant is identically zero everywhere: if a free body is isolated by cutting around any hoop, and $T_{\theta\phi}$ is summed for every element adjacent to that hoop, we obtain a resisting torque about the axis of revolution of $2\,a \sin \phi\; T_{\theta\phi}$; but there is no applied torque, and it follows that $T_{\theta\phi}$ is zero. This is characteristic of surfaces of revolution under axisymmetric loads, and it is important to realize that this absence of shear stress resultants does not in this case necessitate a change in profile geometry.

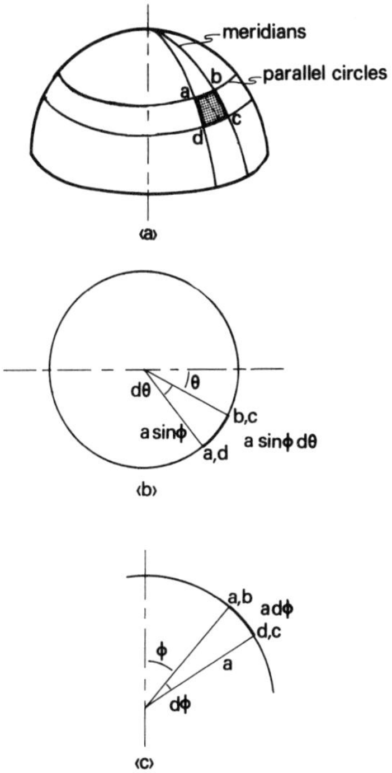

5.1 Definition diagrams for the geometry of an element of the surface of a sphere

Internal actions that allow the possibility of variations in the loading are shown in figure 5.2b. The term $d/d\phi\,(T_\phi a \sin\phi)d\phi d\theta$ permits variations in T_ϕ, while at the same time allowing for the variation with ϕ in the elemental length over which T_ϕ acts (namely, $(a \sin\phi)d\theta$); the terms $(T_\theta\, ad\phi)d\theta$ and $(T_\phi a \sin\phi d\theta)d\phi$ are the inward-directed components of the two hoop forces; and the sole applied load $pad\phi(a \sin\phi)d\theta$ is directed along the outward normal.

Resolving along the tangent gives

$$\frac{d}{d\phi}(T_\phi \sin\phi) - T_\theta \cos\phi = 0, \tag{5.1}$$

while along the normal we obtain

$$T_\phi + T_\theta = p(\phi)a. \tag{5.2}$$

Substituting (5.2) into (5.1) and rearranging yield

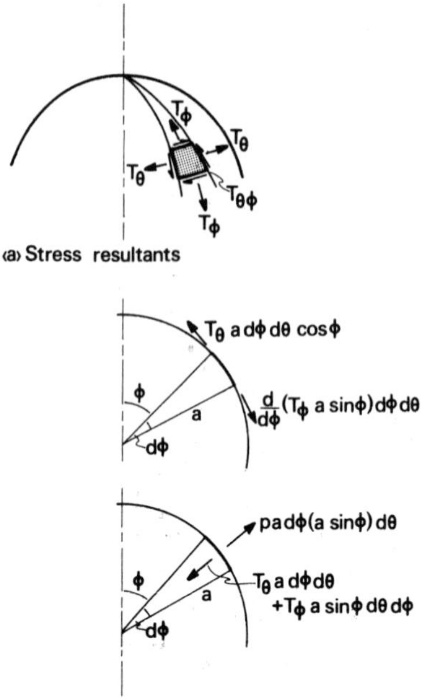

(a) Stress resultants

(b) Force equilibrium along ϕ-direction and along outward normal

5.2

$$\frac{d}{d\phi}(T_\phi \sin^2 \phi) = p(\phi)a \sin \phi \cos \phi, \tag{5.3}$$

from which

$$T_\phi = \frac{a}{\sin^2 \phi}\left\{\int p(\phi) \sin \phi \cos \phi d\phi + A\right\}, \tag{5.4}$$

where A is a constant of integration. This is but a special case of the result for a tension shell of revolution under axially symmetric loads [4].

At the crown of the sphere there can be no distinction between meridians and hoops. At this point from (5.2) we have

$$T_\phi = T_\theta = p(0)\frac{a}{2}. \tag{5.5}$$

This allows the constant of integration to be determined: in the present case

$$T_\phi = \frac{a}{\sin^2 \phi}\int_0^\phi p(\phi) \sin \phi \cos \phi \, d\phi, \tag{5.6}$$

and T_θ is found from (5.2). An alternative derivation, considering the overall equilibrium of a free body, is also straightforward.

When $p(\phi)$ is constant, we obtain the well-known result that the stress resultants are equal and constant everywhere. The sphere is the funicular surface for this loading.

Before working an example, it is instructive to consider the change in profile brought about by the elastic response of this spherical surface to the loading. Since the stress resultants are known, and Hooke's law is assumed, the only potential complication is in relating material strains to surface displacements. This presents no problem for a sphere, and the extension to a surface of revolution is not demanding [25].

The diagram in figure 5.3 shows a meridional element AB, originally of length $ad\phi$, which has moved to a new position and is now of length $A'B'$. If w is the radial displacement measured outward, and u is the tangential displacement in the direction of increasing ϕ, the change in length is $du + wd\phi$ and the meridional strain is

$$\varepsilon_\phi = \frac{1}{a}\left(\frac{du}{d\phi} + w\right). \tag{5.7}$$

The radius of a parallel circle increases by $u \cos \phi + w \sin \phi$, so the hoop strain is

$$\varepsilon_\theta = \frac{1}{a}(u \cot \phi + w), \tag{5.8}$$

because a hoop and its radius increase by the same fractional amount.

These strains are related to the stress resultants by

$$\begin{aligned} \varepsilon_\phi &= \frac{1}{Et}(T_\phi - \mu T_\theta), \\ \varepsilon_\theta &= \frac{1}{Et}(T_\theta - \mu T_\phi), \end{aligned} \tag{5.9}$$

where t is the membrane thickness and μ is Poisson's ratio. Substitution and rearrangement then yield the linear first-order differential equation

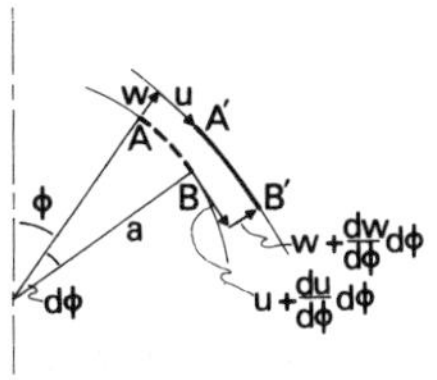

5.3 Definition diagram for the displacements of a meridional element

$$\frac{du}{d\phi} - \cot \phi u = \frac{a(1 + \mu)}{Et}(T_\phi - T_\theta),$$

or

$$\frac{d}{d\phi}(u \operatorname{cosec} \phi) = \frac{a(1 + \mu)}{Et}(T_\phi - T_\theta) \operatorname{cosec} \phi. \tag{5.10}$$

The solution is

$$u = \sin \phi \left\{ \int \frac{a(1 + \mu)}{Et}(T_\phi - T_\theta) \operatorname{cosec} \phi d\phi + B \right\}, \tag{5.11}$$

where B is a constant of integration. The radial displacement is found by combining (5.8) and the second of (5.9) with (5.11).

When, for example, the sphere is inflated by air pressure, the meridional displacements are everywhere zero. The radial displacement is $pa^2(1 - \mu)/2Et$.

Example 5.1 An Idealized Trans-Continental Balloon

A spherical tension shell of radius a is made of a light, flexible material, filled with helium, and inflated to a mean pressure above mean atmospheric of Δp, the difference in specific weights between the surrounding air and the confined fluid being $\Delta \gamma = \gamma_a - \gamma_h$. Along a parallel circle denoted by the colatitude $\phi_0(> \pi/2)$ a tangential line load is applied by means of numerous suspension cords that support the ballast (consisting of crew, equipment, and so on, see figure 5.4). The balloon is assumed to be in static equilibrium, and Δp and $\Delta \gamma$ are specified with reference to an operating altitude. (This is a good model of Columbine II, the first gas-filled balloon to cross the continental divide of the United States—the crossing was made in July 1978.) Eventualities such as that depicted in figure 5.5 are not foreseen!

To obtain the distributions of stress resultants and associated deflections, we proceed as follows. If the reference level is set at the equatorial diameter, the pressure loading is

$$p = \Delta p + \Delta \gamma a \cos \phi.$$

Substitution of this into (5.6) and integration give

$$T_\phi = \frac{\Delta pa}{2} + \frac{\Delta \gamma a^2}{3} \frac{1 - \cos^3 \phi}{\sin^2 \phi},$$

or

$$T_\phi = \frac{\Delta pa}{2} + \frac{\Delta \gamma a^2}{3} \frac{1 + \cos \phi + \cos^2 \phi}{1 + \cos \phi}.$$

This holds up to the immediate vicinity of the line load, and T_ϕ never changes sign. The hoop stress in this region is

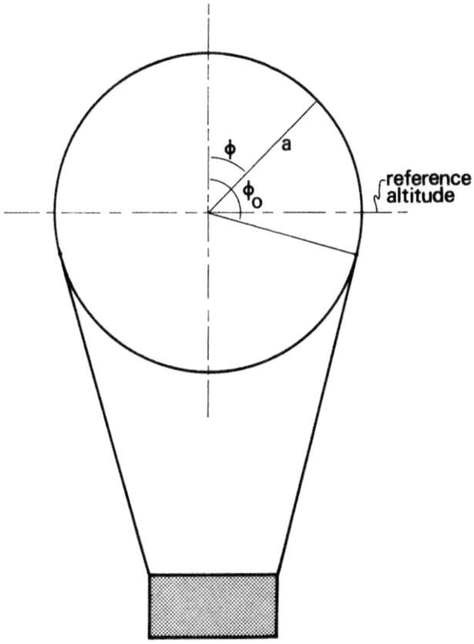

5.4 An idealized lighter-than-air balloon

$$T_\theta = \frac{\Delta pa}{2} + \frac{\Delta\gamma a^2}{3}\frac{-1 + 2\cos\phi + 2\cos^2\phi}{1 + \cos\phi},$$

and it will change sign if $\Delta\gamma a/\Delta p$ is sufficiently large. As might be expected, the stress resultants at the crown are both $\Delta pa/2 + \Delta\gamma a^2/2$.

The line load is of intensity $W/(2\pi a \sin^2 \phi_0)$, where W is the ballast (equal to $4\pi a^3 \Delta\gamma/3$). Thus for $\phi > \phi_0$ we must add the following increments to the stress resultants:

$$\Delta T_\phi = -\frac{W}{2\pi a \sin^2\phi} = -\frac{2\Delta\gamma a^2}{3\sin^2\phi},$$

and, since $\Delta T_\phi + \Delta T_\theta = 0$,

$$\Delta T_\theta = \frac{2\Delta\gamma a^2}{3\sin^2\phi}.$$

Therefore

$$T_\phi = \frac{\Delta pa}{2} - \frac{\Delta\gamma a^2}{3}\frac{1 - \cos\phi + \cos^2\phi}{1 - \cos\phi}, \quad \phi_0 < \phi \le \pi.$$

$$T_\theta = \frac{\Delta pa}{2} - \frac{\Delta\gamma a^2}{3}\frac{-1 - 2\cos\phi + 2\cos^2\phi}{1 - \cos\phi}, \quad \phi_0 < \phi \le \pi.$$

The correct values of $\Delta pa/2 - \Delta\gamma a^2/2$ are recovered at the bottom of the balloon.

A wrinkle-free surface certainly requires $\Delta p > \Delta\gamma a$. The avoidance of meridional wrinkles just above the line load requires

5.5 "Little boy! Your change!" (from Saturday Review, February 3, 1979)

$$\Delta p > \frac{2\Delta\gamma a}{3}\frac{1 - 2\cos\phi_0 - 2\cos^2\phi_0}{1 + \cos\phi_0},$$

and this is a more stringent criterion if ϕ_0 is such that

$$2\cos^2\phi_0 + \frac{7}{2}\cos\phi_0 + \frac{1}{2} < 0,$$

which it is if $\phi_0 > 99°$—when $\phi_0 = 120°$, the surface will be wrinkle-free only if $\Delta p > 2\Delta\gamma a$.

In calculating the displacement field above the line load, we first obtain an expression for the meridional component. Substitution of the present results for the stress resultants in (5.11) yields

$$u = \sin\phi\left[\frac{\Delta\gamma a^3(1 + \mu)}{3Et}\int\left\{\frac{\sin\phi}{(1 + \cos\phi)^2} + \frac{\sin\phi}{(1 + \cos\phi)}\right\}d\phi + B\right],$$

or

$$u = \sin\phi\left[\frac{\Delta\gamma a^3(1 + \mu)}{3Et}\left\{\frac{1}{1 + \cos\phi} - \ln(1 + \cos\phi)\right\} + B\right].$$

Use of this result together with (5.8) and (5.9) gives w. The constant B is found from the condition that the radial displacement at the crown must be $(\Delta p + \Delta\gamma a)a^2(1 - \mu)/2Et$. Therefore

$$u = \frac{\Delta\gamma a^3(1 + \mu)}{3Et}\sin\phi\left\{\frac{1}{1 + \cos\phi} - \frac{1}{2} - \ln\left(\frac{1 + \cos\phi}{2}\right)\right\},$$

and

$$w = -u \cot \phi + \frac{\Delta p a^2(1 - \mu)}{2Et}$$

$$+ \frac{\Delta \gamma a^3}{3Et} \frac{1}{1 + \cos \phi} \{(-1 + 2\cos \phi + 2\cos^2 \phi) - \mu(1 + \cos \phi + \cos^2 \phi)\},$$

both of which hold for $0 \leq \phi < \phi_0$.

By specifying again the radial displacement expected at the base, namely, $(\Delta p - \Delta \gamma a)a^2(1 - \mu)/2Et$, we obtain for the region below the line load

$$u = \frac{\Delta \gamma a^3(1 + \mu)}{3Et} \sin \phi \left\{ \frac{1}{1 - \cos \phi} - \frac{1}{2} + \ln\left(\frac{1 - \cos \phi}{2}\right) \right\}$$

and

$$w = -u \cot \phi + \frac{\Delta p a^2(1 - \mu)}{2Et}$$

$$- \frac{\Delta \gamma a^3}{3Et} \frac{1}{1 - \cos \phi} \{(-1 - 2\cos \phi + 2\cos^2 \phi) - \mu(1 - \cos \phi + \cos^2 \phi)\}.$$

These expressions do not hold right at the line load because the displacements are incompatible there due to the jump in the stress resultants. This is a well-known shortcoming of the membrane theory.

Figure 5.6 shows a grossly exaggerated view of the perturbed profile of a balloon in which $\phi_0 \simeq 120°$, $\Delta p = 2\Delta \gamma a$, and Poisson's ratio is ignored. (The meridional displacements are for the most part an order of magnitude smaller than the radial displacements. If u is ignored, w may be found directly from equation 5.8 and the second equation of 5.9: this approximation has at least the advantage of simplicity.) The profile in the immediate vicinity of ϕ_0 has simply been sketched in—the disturbances caused by the enforcement of compatibility there will in this instance be largely local in character and will not be felt farther away.

Under operating conditions the increase in the equatorial diameter is usually then less than 1 percent. This lends support to the view that the effects of stretch do not amount to much in membranes of appreciable curvature.

5.3 Surfaces of Revolution

A knowledge of two radii of curvature is necessary in a general surface of revolution. The generating meridian AB has a radius of curvature r_1, and the second radius r_2 arises because normals to the same point on adjacent meridians intersect on the axis of revolution (see figure 5.7). These radii are equal only for the sphere, or where the surface is locally spherical. The angle ϕ is akin to the colatitude, and r and z are cartesian coordinates often used to describe the profile.

Together with the longitude θ there are four quantities that describe the surface. But they are not independent. One condition of constraint exists

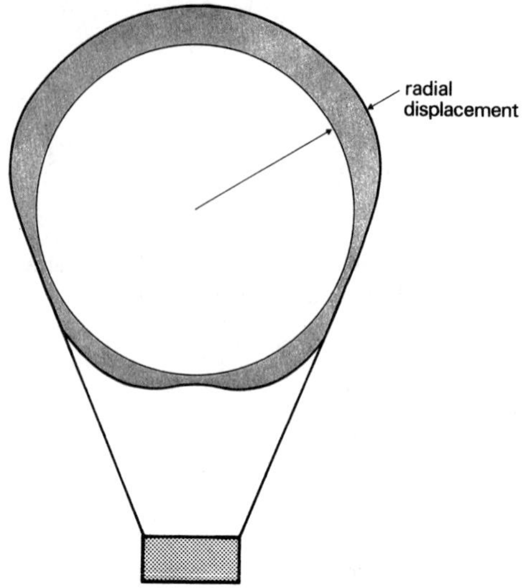

5.6 View on a meridional section of an idealized lighter-than-air balloon (scale of displacements is exaggerated)

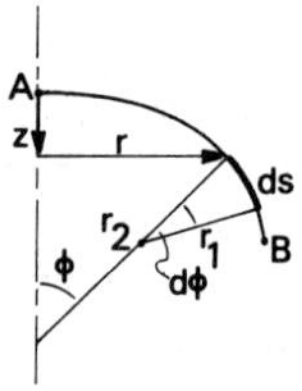

5.7 Definition diagram for a surface of revolution

which, since $r = r_2 \sin\phi$, $r_1 = ds/d\phi$, and $dr/ds = \cos\phi$, may be written in the form

$$\frac{d}{d\phi}(r_2 \sin\phi) = r_1 \cos\phi. \tag{5.12}$$

(A general surface has three geometrical constraints given by the two Codazzi equations and Gauss's equation. Equation 5.12 is a special case of one of the Codazzi equations [6].)

The internal actions on a representative element of the surface are shown in figure 5.8. The three equations of equilibrium are

$$\frac{\partial}{\partial \phi}(r_2 \sin \phi T_\phi) + r_1 \frac{\partial T_{\theta\phi}}{\partial \theta} - r_1 \cos \phi T_\theta = -p_\phi r_1 r_2 \sin \phi \tag{5.13}$$

in the meridional direction,

$$r_1 \frac{\partial T_\theta}{\partial \theta} + \frac{\partial}{\partial \phi}(r_2 \sin \phi T_{\theta\phi}) + r_1 \cos \phi T_{\theta\phi} = -p_\theta r_1 r_2 \sin \phi \tag{5.14}$$

in the hoop direction, and

$$\frac{T_\phi}{r_1} + \frac{T_\theta}{r_2} = p_n \tag{5.15}$$

along the normal.

The occurrence of the terms involving the direct stress resultants is adequately covered by the discussion in the previous section. By analogy one may see how the second terms on the left of (5.13) and (5.14)—those involving the shear stress resultants—arise. The third term on the left of (5.14) requires explanation. The shear resultants along the two meridional portions of the element are shown in plan view in figure 5.9. These faces are inclined at $d\theta$ to one another, so the shear stress resultants combine to give a component in the θ direction of $(T_{\theta\phi} r_1 d\phi) \cos \phi d\theta$.

There is an extensive literature on solutions to these membrane equations, so much in fact that even a treatise devoted solely to it could not reasonably cover all the applications. Nevertheless considerable guidance

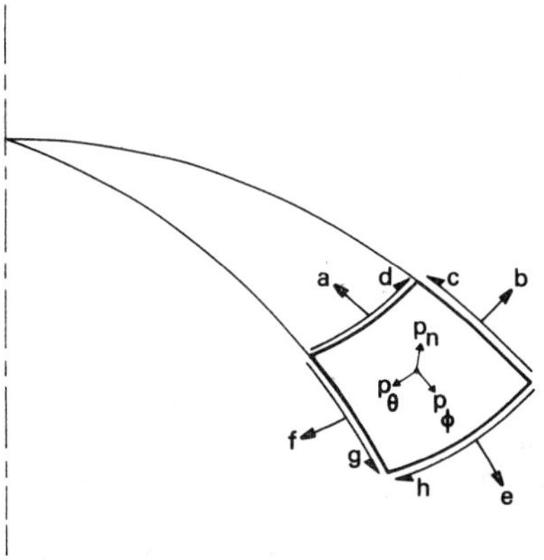

5.8 Internal actions on an element of a surface of revolution. Legend: $a \equiv T_\phi r_2 \sin \phi d\theta$, $b \equiv T_\theta r_1 d\phi$, $c \equiv T_{\theta\phi} r_1 d\phi$, $d \equiv T_{\theta\phi} r_2 \sin \phi d\theta$, $e \equiv a + \partial/\partial\phi (T_{\theta\phi} r_2 \sin \phi) d\phi d\theta$, $f \equiv b + (\partial T_\theta / \partial \theta) d\theta r_1 d\phi$, $g \equiv c + (\partial T_{\theta\phi} / \partial \theta) d\theta r_1 d\phi$, $h \equiv d + \partial/\partial\phi (T_{\theta\phi} r_2 \sin \phi) d\phi d\theta$.

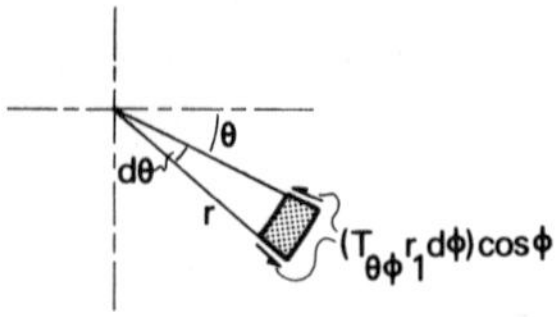

5.9 Plan view of disposition of shear stress resultants along meridional faces of the element

is available from such standard works as Timoshenko and Woinowsky-Krieger [25] and Flügge [4].

Our objectives will be served by a treatment much narrower in scope. We shall consider surfaces of revolution under loadings generated by fluid pressure, and the magnitudes of these loadings will be assumed to be substantially in excess of that due to self-weight.

Also, rather than specify a geometry and so determine the stress resultants, we shall place constraints on the stress resultants and endeavor to find the associated profile. This method was first introduced by G. I. Taylor in a study of parachute profiles undertaken during the First World War [24]. Around the same time Biezeno employed the same technique in the design of tanks of constant strength [3]—an application that is still of some practical interest. A more recent, and somewhat more exotic application, involves the search for suitable shapes for the reentry shields of rockets [1]. The three examples that follow illustrate the power of the technique.

Example 5.2 The Shallowest Pneumatic Dome

A hemisphere inflated by air pressure has tensile stress resultants that are everywhere equal and constant. Its rise to span is 1:2. On the other hand, a pneumatic oblate ellipsoid of revolution has tensile meridional stress resultants, but the hoop stress resultants may change sign near the major axis if the ellipticity is too pronounced. Meridional wrinkles will form there if the rise to span is less than $1:2\sqrt{2}$ (see exercise 5.1).

One might infer from this that any pneumatic dome with a meridional profile vertical around the perimeter (to make best use of the clear space) and a rise less than about 35 percent of the clear span will display meridional wrinkles near ground level. However, photographs of experiments conducted by Kawaguchi, and shown in figure 5.10, suggest otherwise. There a dome with the rise to span of about 0.30 is free of wrinkles, while wrinkles are plainly evident in the ellipsoid of the same dimensions [11].

Following Kawaguchi, it is noted that the stress resultants in a surface of revolution inflated by air pressure are

$$T_\phi = \frac{1}{2} p r_2$$

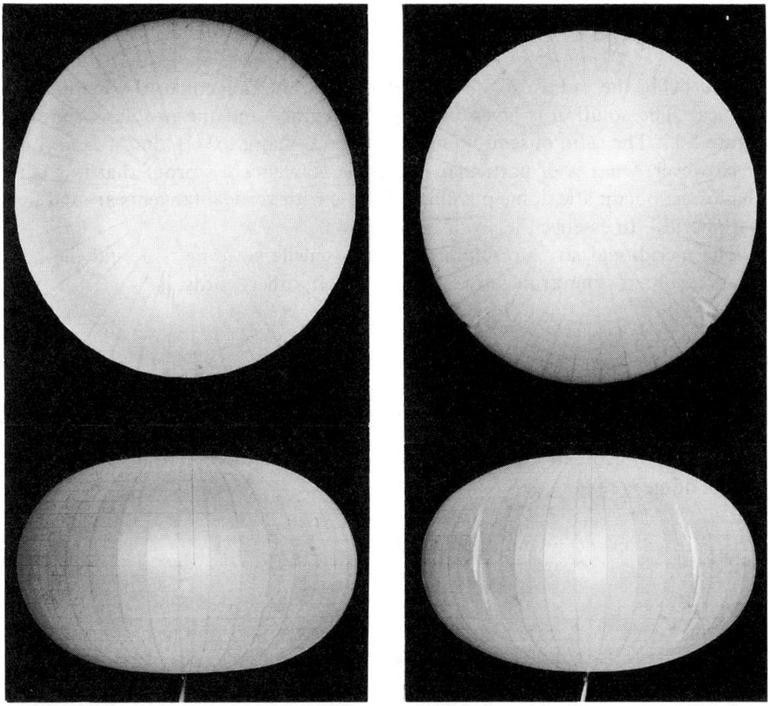

5.10 Kawaguchi's photographs of inflated models shown in plan and in elevation (courtesy M. Kawaguchi). The left-hand pair are of the wrinkle-free shallowest surface; the right-hand pair are of an ellipsoid of equal dimensions, in which wrinkles are plainly evident.

and

$$T_\theta = \frac{1}{2}pr_2\left(2 - \frac{r_2}{r_1}\right).$$

The first result is readily established by balancing forces across a horizontal plane cutting the dome. The second then follows from (5.15). The surface for which T_θ is everywhere zero is thus given by

$$r_1 = \frac{1}{2}r_2,$$

or, after some working, by

$$\frac{dz}{dr} = \frac{r^2}{(a^4 - r^4)^{1/2}},$$

with a being the radius of the perimeter where the tangent to the meridians is vertical. The solution is given by elliptic integrals, and the profile is shown in figure 5.11. The ratio of semi-minor axis to semi-major axis is almost exactly 0.6.

However, what is of particular interest is Kawaguchi's proof that this is the shallowest pneumatic dome possible (of those with vertical tangents around their peripheries). In essence the proof is as follows.

The meridional stress resultant is always tensile since $r_2 > 0$, and the hoop stress resultant is never negative if $r_1 \geq r_2/2$. In other words, if

$$r_1 = \frac{1}{2}r_2 + f(\phi),$$

where $f(\phi) \geq 0$. The span of the dome is $2a$, and its rise is $\int_0^{\pi/2} r_1 \sin\phi \, d\phi$, because from geometry $dz = r_1 \sin\phi \, d\phi$. Therefore the condition for the shallowest dome is that

$$\frac{1}{2a}\int_0^{\pi/2} \frac{1}{2}r_1 \sin\phi \, d\phi \rightarrow \text{minimum},$$

or

$$\int_0^{\pi/2} \frac{1}{2}r_2 \sin\phi \, d\phi + \int_0^{\pi/2} f(\phi) \sin\phi \, d\phi \rightarrow \text{minimum}.$$

But r_2, $\sin\phi$, and $f(\phi)$ are never negative in the domain of integration, so a minimum is achieved when $f(\phi) = 0$. As a result the profile in figure 5.11 is the shallowest possible.

The profile is flat at the crown, which means that there is theoretically a singularity in the meridional stress resultant there. This of course does not happen in practice, for the finite extensibility inherent in the fabric allows some curvature. This curvature is locally spherical at the crown because the two stress resultants must be equal there. Kawaguchi has also developed satisfactory ways of handling such analyses. Nevertheless stress resultants are still high in this vicinity.

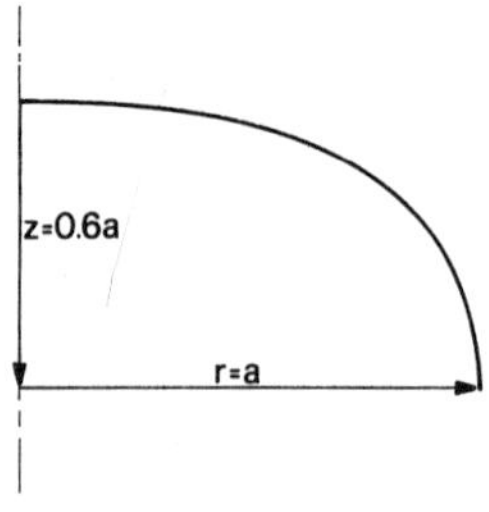

5.11 Taylor's profile (from Taylor [24], courtesy Cambridge University Press)

Inflatable canvas membranes with this profile have been used as formwork for the construction of fiber-reinforced plastic shells. The general use of pneumatic structures is becoming quite widespread to house sporting and recreational activities, although problems with inflation and wind loading occasionally crop up. An analysis of the free vibrations of a simple pneumatic structure is set as problem 5.6.

Example 5.3 On the Shape of a Parachute

Over the last 50 to 70 years parachutes have evolved along the following lines. First-generation parachutes consist of approximately hemispherical canopies, with diameters ranging between 8 and 11 m; they were originally constructed of silk, but nylon is now used. They are inherently unstable and prone to oscillation, although venting in the apex reduces the likelihood of instability. Steering is done with the aid of slots in the canopy walls. Typical rates of vertical descent are in the range 5 to 7 m/sec.

The major difference with second-generation parachutes is in the alteration of the hemispherical canopy to produce a lifting surface: the apex is pulled down by extra suspension lines. Large drive openings in the canopy allow horizontal speeds of 5 to 7 m/sec to be achieved, with corresponding vertical rates of descent in the vicinity of 3 to 5 m/sec.

Third-generation systems, actually nonrigid gliders with upper and lower wing surfaces, were developed in the early 1970s. The lifting surfaces are retained in aerofoil shape by vertical partitions, and a negative angle of attack permits horizontal speeds of 7 to 14 m/sec and vertical speeds of 3 to 9 m/sec.

Our concern is with the much simpler first-generation parachute which is nevertheless still used extensively. We shall investigate only the steady-state, when the form drag on the canopy balances the payload and the parachute descends with uniform speed.

The profile in figure 5.11 was that produced by G. I. Taylor on the understanding that the pleats, or gores, in the parachute were so closely spaced as to justify the assumption that the meridional stress resultants were the principal means of satisfying equilibrium [24]. Implicit in this analysis was the further assumption that the pressure distribution over the falling parachute is roughly constant: on reflection this is probably justifiable (see, for example, plots of the pressure distribution on a plate normal to a free stream).

There is, however, an alternative way of considering the parachute, and that is to assume that the ribs that separate the pleats take all the meridional stress, leaving the pleats to transfer the form drag forces to them by developing hoop stress resultants and shear stress resultants [9].

We concentrate on an individual pleat $ABDC$ which is now no longer a surface of revolution (see figure 5.12). Equations (5.13), (5.14), and (5.15) reduce to

$$\frac{\partial T_{\theta\phi}}{\partial \theta} - T_\theta \cos\phi = 0,$$

$$r_1 \frac{\partial T_\theta}{\partial \theta} + \frac{\partial}{\partial \theta}(r_2 \sin\phi T_{\theta\phi}) + r_1 \cos\phi T_{\theta\phi} = 0,$$

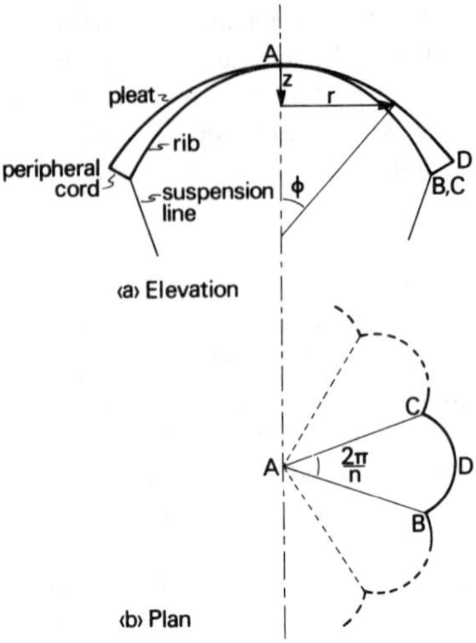

5.12 Definition diagram for a representative pleat of a parachute

and

$$\frac{T_\theta}{r_2} = p.$$

Consequently

$$T_\theta = pr \operatorname{cosec} \phi,$$

and so

$$T_{\theta\phi} = pr \cot \phi \; \theta, \quad -\frac{\pi}{n} \leq \theta \leq \frac{\pi}{n},$$

where the included angle of the pleat is $2\pi/n$, in which n is the number of pleats. In obtaining this latter result, use has been made of the symmetry condition that the shear stress resultant must be zero along the high point of the pleat AD.

It is reasonable to assume that the hoop stress resultant varies little with longitude—we may therefore ignore $\partial T_\theta/\partial\theta$. Since it is also evident that r varies little with θ—the scalloping effect will be small compared with the plan dimensions—we may consider r to measure distance to the high point of the pleat AD. Equation (5.14) is finally reduced to

$$\frac{d}{d\phi}(T_{\theta\phi} r) + T_{\theta\phi}\frac{dr}{d\phi} = 0,$$

or

$$\frac{d}{d\phi}(T_{\theta\phi} r^2) = 0.$$

Integrating this and substituting for $T_{\theta\phi}$ lead to

$$\tan \phi = Ar^3,$$

where A is a constant of integration. But $\tan \phi = dz/dr$, so in cartesian coordinates the pleat profile is found to be

$$z = \frac{Ar^4}{4},$$

where we have met the boundary condition at the crown of $z = 0$ at $r = 0$. The constant A may be determined by specifying conditions at the free edge of the pleat. The associated pleat stress resultants are

$$T_\theta = \frac{p}{Ar^2}(1 + A^2 r^6)^{1/2}$$

and

$$T_{\theta\phi} = \frac{p}{Ar^2}\theta, \quad -\frac{\pi}{n} \leq \theta \leq \frac{\pi}{n},$$

which are singular at the apex. These expressions provide an interesting complement to Taylor's problem.

If one considers an element of the pleat off the line AD, it is clear that the combination of hoop stress and shear stress gives rise to principal stresses tensile in one direction and compressive in the other. Since the fabric cannot sustain compression, wrinkles will form aligned with the local direction of the principal tension. This is observed in practice.

However, the main point is that Taylor's profile and this alternative one may be combined to give in all probability a fairly accurate picture of the whole parachute profile. We simply let the rib assume the form given by Taylor's profile and fit to it the pleat profile. Matching may be done by considering equilibrium at the free edge where a suspension line, the peripheral cord, and a rib meet (see figure 5.13). One possibility would be to assume that AD is the same length as AB and that BD is perpendicular to AD at D. This then locates D and permits determination of the pleat profile.

Example 5.4 Profiles for Hot Air Balloons

The purpose of this example is to describe some aspects of the development of new profiles for hot air balloons [10]. The technique of enforcing constraints on the

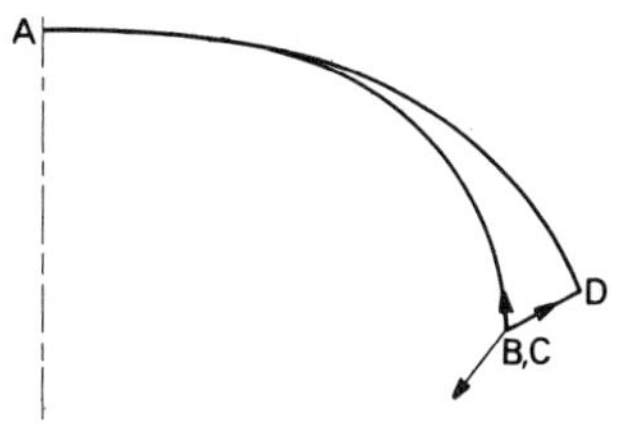

5.13 Suggested combination of Taylor's profile and pleat profile showing force balance at the junction of a suspension line, rib, and peripheral cord

stress resultants is particularly useful. Although the goal of a complete analytical solution cannot be achieved, the procedure is partly analytical in that the stress resultants may be obtained in closed form. However, the associated meridional profile is described by a differential equation which is hopelessly nonlinear and must be solved numerically.

The definition diagram in figure 5.14 shows the geometry of a typical meridian. The most convenient way of describing the profile is by way of the functional form $r = r(z)$, so that the following geometrical relations are useful

$$r = r_2 \sin\phi, \quad \frac{dr}{d\phi} = r_1 \cos\phi,$$

and the principal curvatures are

$$\frac{1}{r_1} = -\frac{d^2r/dz^2}{\{1 + (dr/dz)^2\}^{3/2}}$$

and

$$\frac{1}{r_2} = \frac{1/r}{\{1 + (dr/dz)^2\}^{1/2}}.$$

If we assume that the lift force balances the payload W, and use as a datum the crown of the balloon, we can readily show that the pressure loading along an outward normal is

$$\Delta p = \Delta\gamma(h - z), \quad 0 \leq z \leq h,$$

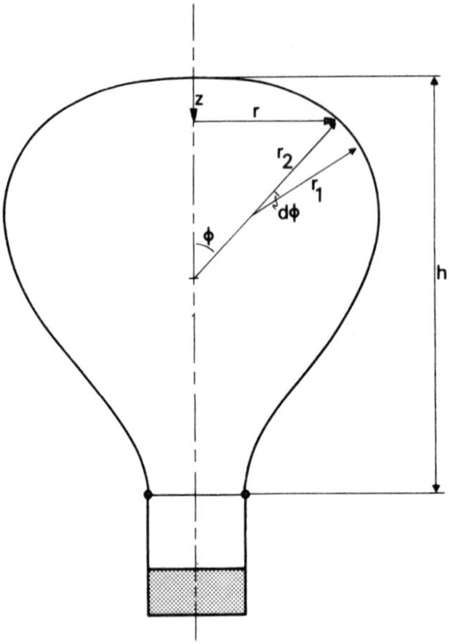

5.14 Geometry of a hot air balloon

where $\Delta\gamma$ is the difference in specific weights between the colder ambient air and the heated contained volume, h is the height of the balloon, and z is the vertical distance from the crown.

Implicit in this is that there is no temperature gradient in the balloon. The heating process is similar to a large, highly turbulent buoyant jet in which cooler air is both ingested into the flaming column and expelled down around the outer rim of the opening. The assumption of the absence of any meaningful temperature gradient at the end of a heating cycle is therefore probably as good as can be made in the circumstances. Our particular concern is with the basic statics of the problem and not with attendant problems such as heat loss or jet noise (which in fact makes operation of larger balloons rather uncomfortable).

The overall equilibrium of the balloon is given by

$$W = \int_0^L 2\pi r \Delta p \cos\phi \, ds,$$

where ds is an element of the meridian and L is thus the meridional length from crown to opening. Alternative expressions for the righthand side are $\int_0^h 2\pi r \Delta\gamma (h - z) \cot\phi \, dz$, or simply $\int_0^h \pi r^2 \Delta\gamma \, dz$.

In deciding on a suitable relation between the stress resultants, three things had to be accomplished: first, equality of the stress resultants at the crown; second, avoidance of a change of sign in the hoop stress resultant; and, third, a pinched meridional profile at the base of the balloon. This suggested a relation between the stress resultants of the form

$$T_\theta = T_\phi\{1 - f(r)\},$$

where $0 \leq f(r) < 1$.

The functional form $f(r)$ was chosen because this allows (5.13) to be integrated directly to give

$$T_\phi = \frac{\Delta\gamma h r_a}{2} \exp\left\{-\int_0^r \frac{f(r)}{r} dr\right\},$$

$$T_\theta = \frac{\Delta\gamma h r_a}{2}\left\{1 - f(r)\right\} \exp\left\{-\int_0^r \frac{f(r)}{r} dr\right\},$$

where r_a is the radius of curvature at the crown (which is locally spherical). Among an obvious multitude of possibilities the best form for $f(r)$ turned out to be the simple relation

$$f(r) = Ar^{1/2},$$

so after substitutions into (5.15) and some rearrangements the differential equation governing the profile may be established:

$$\frac{-d^2\mathbf{r}/d\mathbf{z}^2}{\{1 + (d\mathbf{r}/d\mathbf{z})^2\}^{3/2}} + \frac{(1 - \mathbf{A}\mathbf{r}^{1/2})}{\mathbf{r}\{1 + (d\mathbf{r}/d\mathbf{z})^2\}^{1/2}} = \frac{2}{\mathbf{r}_a}(1 - \mathbf{z})\exp(2\mathbf{A}\mathbf{r}^{1/2}),$$

where $\mathbf{r} = r/h$, $\mathbf{r}_a = r_a/h$, and $\mathbf{z} = z/h$. (Further discussion may be found in reference [10].) Standard subroutines are available to solve equations such as these. It is an initial value problem because we know in advance only conditions at the crown that $\mathbf{r}(0) = 0$ and $d\mathbf{r}(0)/d\mathbf{z} = \infty$; some care has to be exercised with the second of these. Conditions at the throat are dependent on the value of **A** chosen. Clearly many possibilities exist.

However, the largest value of $\mathbf{r}_a$ that still avoids a change of sign in the hoop stress at the balloon's widest point is found to be $\mathbf{r}_a = 3.20$, and the associated value of **A**, which also gives a vertical tangent to the meridian at the throat, is $\mathbf{A} = 1.44$. The profile of this balloon is shown in figure (5.15), and next to it is a lune which is a development of one-eighth of the surface of the balloon. Several other profiles are presented in the article from which this example has been drawn. There also is presented the results of a short experimental investigation using small-scale models filled with water and suspended from their throats. The models behaved much as expected and meridional wrinkles were absent [10].

The present balloon has the following characteristics:

Radius of curvature of crown, $r_a = 3.20\text{h}$,
Maximum diameter, $2r_{\max} = 0.96\text{h}$,
Throat diameter, $2r_b = 0.20h$,
Surface area, $S = 2.72h^2$,
Contained volume, $V = 0.40h^3$,
Payload, $W = \Delta\gamma V$.

The ratio of contained volume to surface area for a given payload is

$$\frac{V}{S} = 0.29\left(\frac{W}{\pi\Delta\gamma}\right)^{1/3},$$

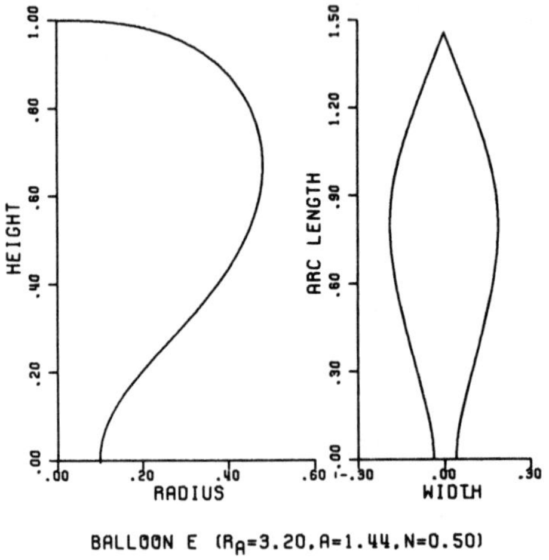

5.15 Typical meridional profile of a hot air balloon and associated lune (representing one-eighth of its surface area)

and this compares favorably with the best result possible, which is for a sphere when the coefficient is 0.30. The balloon is efficient with respect to both material used and heat loss characteristics.

Peak stress resultants occur at the crown, but these do not usually govern fabric choice. Ease of handling and the need to avoid spiking or ripping are more pressing considerations. Special high-strength polyesters and nylon are the materials currently favored. They are tear resistant as well as heat and flame resistant.

As an example, suppose a volume of 2,000 m^3 is specified. Under normal operating conditions such a volume would be required to lift a payload of 2 to 3 kN. From the previous formulas suitable balloon dimensions are

Height, 17.1 m,

Crown radius, 54.6 m,

Maximum diameter, 16.4 m,

Throat diameter, 3.4 m,

Surface area, 793 m^2.

5.4 Equations of Equilibrium for a General Surface

When there is no axis of symmetry, the equations of equilibrium are more complicated, but solutions can still be found. In fact the approximate analytical solutions developed for relatively flat tension surfaces of regular plan form are straightforward to establish and may be viewed as natural extensions of the engineering theory of the flat-sag single cable.

Following Heyman [6], we assume the surface to have a given geometry of the form

$$z = f(x, y), \tag{5.16}$$

where x, y, and z are Cartesian coordinates that locate an element in the surface of area dA with sides ds_1 and ds_2 (see figure 5.16). The coordinate z is vertical, and the element is inclined to the horizontal plane at angles α and β, respectively. From geometry

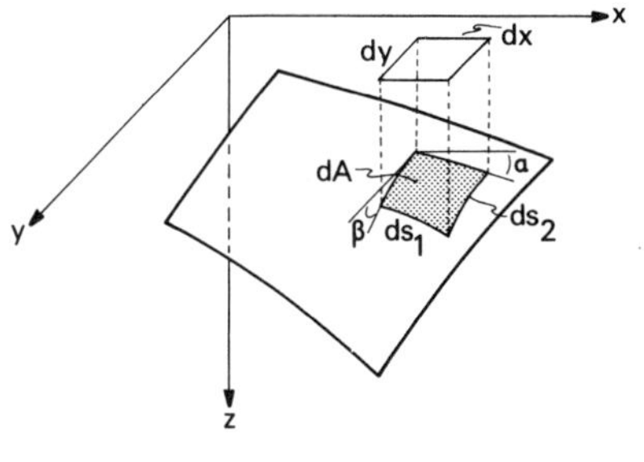

5.16 Geometry of an element in a general surface

$$\tan\alpha = \frac{\partial z}{\partial x}, \quad \tan\beta = \frac{\partial z}{\partial y},$$

and (5.17)

$$ds_1 \cos\alpha = dx, \quad ds_2 \cos\beta = dy.$$

The three stress resultants are T_x, T_y, and T_{xy} $(= T_{yx})$, and, when these are projected onto the horizontal plane through the origin, it may be shown that

$$\bar{T}_x = \frac{\cos\alpha}{\cos\beta} T_x,$$

$$\bar{T}_y = \frac{\cos\alpha}{\cos\beta} T_y, \tag{5.18}$$

$$\bar{T}_{xy} = T_{xy},$$

where $\bar{T}_x$, $\bar{T}_y$, and $\bar{T}_{xy}$ refer to the projected stress resultants. The applied loads per unit surface area are p_x, p_y, and p_z, and their projected equivalents are $\bar{p}_x$, $\bar{p}_y$, and $\bar{p}_z$, where $p_x dA = \bar{p}_x dxdy$, and so on (see figure 5.17).

The three equations of equilibrium are

$$\frac{\partial \bar{T}_x}{\partial x} + \frac{\partial \bar{T}_{xy}}{\partial y} + \bar{p}_x = 0 \tag{5.19}$$

in the x direction,

$$\frac{\partial \bar{T}_y}{\partial_y} + \frac{\partial \bar{T}_{xy}}{\partial x} + \bar{p}_y = 0 \tag{5.20}$$

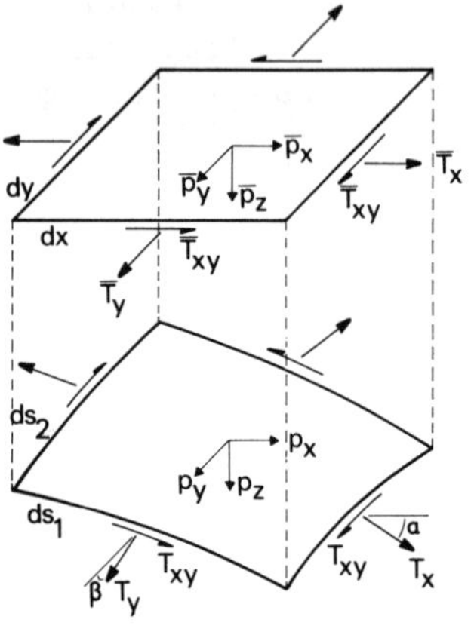

5.17 Applied loads and stress resultants and their projected equivalents in the y direction, and

$$\frac{\partial}{\partial x}\left(\bar{T}_x\frac{\partial z}{\partial x}\right) + \frac{\partial}{\partial x}\left(\bar{T}_{xy}\frac{\partial z}{\partial y}\right) + \frac{\partial}{\partial y}\left(\bar{T}_{xy}\frac{\partial z}{\partial x}\right) + \frac{\partial}{\partial y}\left(\bar{T}_y\frac{\partial z}{\partial y}\right) + \bar{p}_z = 0 \quad (5.21)$$

in the vertical direction. The first two may be used to bring the third into the form

$$\bar{T}_x\frac{\partial^2 z}{\partial x^2} + 2\bar{T}_{xy}\frac{\partial^2 z}{\partial x dy} + \bar{T}_y\frac{\partial^2 z}{\partial y^2} = -\bar{p}_z + \bar{p}_x\frac{\partial z}{\partial x} + \bar{p}_y\frac{\partial z}{\partial y}. \quad (5.22)$$

Before obtaining solutions for networks of closely spaced pretensioned cables, we look at the assumptions that need to be made and the approximations that seem justifiable to achieve that goal. First, an orthogonal network of closely spaced, uniform, flexible cables is replaced by an equivalent membrane. Owing to the discrete nature of the actual network, the in-plane shear resistance of the membrane is taken to be zero. Second, the only applied load is vertical. Therefore from (5.19) and (5.20) it is seen that $\bar{T}_x$ and $\bar{T}_y$ are at most functions of y or of x, respectively, and that (5.22) reduces to

$$\bar{T}_x\frac{\partial^2 z}{\partial x^2} + \bar{T}_y\frac{\partial^2 z}{\partial y^2} = -\bar{p}_z, \quad (5.23)$$

which is a general form of Poisson's equation. However, since the tensions and the profile will change under applied load, we are far from reaching a solution to this equation. A useful approach is to assume an initial profile and consider subsequent perturbations of it.

If in the initial profile the self-weight is negligible (which is certainly plausible in many cable-suspended roofs prior to the placing of the cladding), the equation reduces to

$$H_1\frac{\partial^2 z_0}{\partial x^2} + H_2\frac{\partial^2 z_0}{\partial y^2} = 0, \quad (5.24)$$

where H_1 and H_2 are the equivalent pretensions per unit length and $z_0 = f(x, y)$ describes this initial surface. One obvious solution is $z_0 = 0$, in other words, the membrane is initially taut and flat, but this is only one possibility among many. There are numerous surfaces of negative Gaussian curvature (or, loosely, saddle-backed) that satisfy (5.24). In these surfaces $\partial^2 z_0/\partial x^2$ and $\partial^2 z_0/\partial y^2$ are of opposite sign, and the geometry of the surface can be found once the profile of the perimeter is known. The individual cables need not be straight and in fact are usually not straight, for considerable stiffness may be developed by straining cables against their neighbors that span in the other direction. Because the slopes are usually small, $\partial^2 z_0/\partial x^2$ and $\partial^2 z_0/\partial y^2$ are accurate expressions for the initial curvatures in the x and y directions, respectively. It is also not necessary that the pretensions be the same in each direction since, by changing the scale of the plan dimensions in the ratio $(H_2/H_1)^{1/2}$, solutions that hold for cases of equal pretension (which we now treat) may with a little extra effort be adapted for the more common case where H_1 and H_2 differ (see problem 5.3 and example 5.5).

5.5 Statics of Suspended Membranes

Numerous studies of the behavior of cable networks have been published, the vast majority of the work being computer oriented. A review paper by the Subcommittee on Cable-Suspended Structures of the American Society of Civil Engineers cites many references to numerical work published up to 1971 [23]. The books by Otto [18], Møllmann [16], and the recent work by Krishna [13] on the design of cable-suspended roofs contain further references and accounts of the methods of solution employed (see figure 5.18). The development of numerical techniques for the solution of network problems is still an active area of research. (Møllmann's book includes an extensive bibliography (up to 1973) of

5.18 View of 1964 Tokyo Olympic Swimming Pool (courtesy Y. Tsuboi and M. Kawaguchi)

methods, both analytical and numerical, for the static and dynamic response of cable roofs. There are also numerous published reports describing the design and construction of cable roofs in various parts of the world. A particularly good example is the report by Tsuboi and Kawaguchi on the analysis, model testing, design, and construction of the cable roof covering the Tokyo Olympic Swimming Pool [27]. This structure ranks with the Munich Olympic Stadium net as one of the most exciting cable roofs so far built.)

There have been very few analytical studies published. Because of the mathematical difficulties that can arise, numerical methods are by far the most popular. Nevertheless some analytical work has been attempted. Trostel [26] and Schleyer [21] have given comprehensive mathematical formulations for membrane problems. Shore and Bathish have presented a method by which the solution of certain membranes can be reduced to the solution of a system of nonlinear algebraic equations [22]. Mϕllmann has produced linearized solutions for hyperbolic paraboloids which are attractive on account of their simplicity and yet sufficiently accurate for engineering purposes.

Linearized Response of a Hyperbolic Paraboloid of Rectangular Plan

Suppose that the initial profile of the surface is given by (see figure 5.19)

$$z_0 = -\left\{4d_1 \frac{x}{2a}\left(1 - \frac{x}{2a}\right) - 4d_2 \frac{y}{2b}\left(1 - \frac{y}{2b}\right)\right\}. \tag{5.25}$$

If the pretensions are the same in each direction, (5.25) satisfies (5.24) if

$$\frac{d_1}{a^2} = \frac{d_2}{b^2}, \tag{5.26}$$

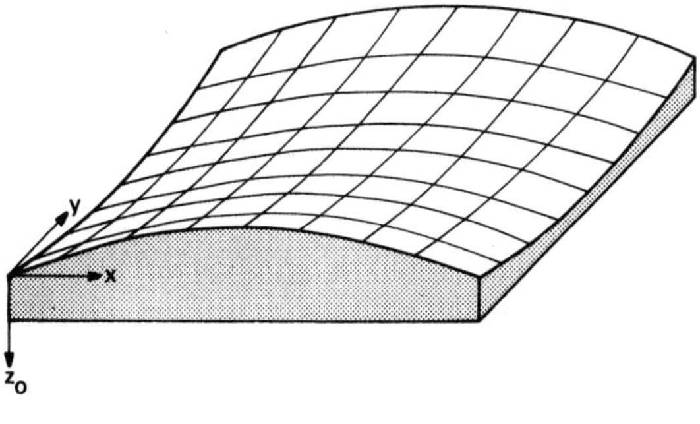

5.19 A hyperbolic paraboloidal net over a rectangular plan

where $2a$ and $2b$ are the lengths of the sides and d_1 and d_2 are the rise and fall of the sides, respectively.

Under a uniformly distributed applied load of p per unit area, vertical equilibrium of an element is specified by

$$\{H + h(y)\}\frac{\partial^2}{\partial x^2}(z_0 + w) + \{H + h(x)\}\frac{\partial^2}{\partial y^2}(z_0 + w) = -p, \tag{5.27}$$

or, to first order,

$$H\left(\frac{\partial^2 w}{\partial x^2} + \frac{\partial^2 w}{\partial y^2}\right) = -p - \frac{8d_1}{(2a)^2}h(y) + \frac{8d_2}{(2b)^2}h(x), \tag{5.28}$$

where $h(y)$ and $h(x)$ are the increments in additional tension for cables spanning the x and y directions, respectively. This equation is a three-dimensional analog of the cable truss equations of the previous chapter because $h(y)$ is negative (the bracing cables generally lose tension under applied load) while $h(x)$ is positive (the tension in the hanging cables usually increases).

The linearized equations of compatibility are

$$\begin{aligned} \frac{h(y)2a}{EA} &= \int_0^{2a} \frac{\partial z_0}{\partial x}\frac{\partial w}{\partial x}dx, \\ \frac{h(x)2b}{EA} &= \int_0^{2b} \frac{\partial z_0}{\partial y}\frac{\partial w}{\partial y}dy, \end{aligned} \tag{5.29}$$

where we have assumed that the cable areas per unit length are the same in each direction. This assumption, like that of the constancy of the pretensions, may be relaxed with little added complication (see problem 5.3).

Although Møllmann has solved these equations, they are somewhat unmanageable, and we present here a simplified solution method which nonetheless captures all the features of interest. The averaging technique employed is widely applicable to problems of this sort and is particularly straightforward to apply.

Define

$$\begin{aligned} h_1 &= \frac{1}{2b}\int_0^{2b} h(y)dy, \\ h_2 &= \frac{1}{2a}\int_0^{2a} h(x)dx, \end{aligned} \tag{5.30}$$

and replace $h(y)$ and $h(x)$ in (5.28) by h_1 and h_2. The governing equilibrium equation becomes

$$\frac{\partial^2 \mathbf{w}}{\partial \mathbf{x}^2} + \alpha^2 \frac{\partial^2 \mathbf{w}}{\partial^2 \mathbf{y}^2} = -1 + \frac{1}{\mathbf{p}}(\mathbf{h}_2 - \mathbf{h}_1), \tag{5.31}$$

with $\mathbf{w}(0, \mathbf{y}) = \mathbf{w}(1, \mathbf{y}) = \mathbf{w}(\mathbf{x}, 0) = \mathbf{w}(\mathbf{x}, 1) = 0$. In this nondimensionalization $\mathbf{x} = x/2a$, $\mathbf{y} = y/2b$, $\alpha = a/b$, $\mathbf{w} = w/(p(2a)^2/8d_1 H)$, $\mathbf{p} = (p(2a)^2/8d_1 H)$, $\mathbf{h}_1 = h_1/H$, and $\mathbf{h}_2 = h_2/H$.

The solution of (5.31) is (see reference [15] for the technique)

$$\mathbf{w}(\mathbf{x}, \mathbf{y}) = \left\{1 - \frac{1}{\mathbf{p}}(\mathbf{h}_2 - \mathbf{h}_1)\right\}\left\{\frac{1}{2}\mathbf{x}(1 - \mathbf{x}) - \sum_{n,\ \text{odd}} \frac{4}{n^3\pi^3} \sin n\pi\mathbf{x} \frac{\cosh\{(1 - 2\mathbf{y})n\pi/2\alpha\}}{\cosh(n\pi/2\alpha)}\right\}. \tag{5.32}$$

Alternatively a solution could be obtained as a double Fourier series, but the single series (5.32) is more rapidly convergent. No more than two terms need be retained. A Fourier series solution is useful if the load is not uniformly distributed.

The cable equations for these averaged additional tensions are

$$\mathbf{h}_1 = -\lambda^2\, \mathbf{p} \int_0^1 \int_0^1 \mathbf{w}\, \mathrm{d}\mathbf{x}\mathrm{d}\mathbf{y},$$
$$\mathbf{h}_2 = \lambda^2\, \mathbf{p} \int_0^1 \int_0^1 \mathbf{w}\, \mathrm{d}\mathbf{x}\mathrm{d}\mathbf{y}, \tag{5.33}$$

where $\lambda^2 = (8d_1/2a)^2\, EA/H$. Thus

$$\mathbf{h}_1 = -\mathbf{h}_2. \tag{5.34}$$

Substitution, integration, and rearrangement lead to

$$\frac{\mathbf{h}_{1,2}}{\mathbf{p}} = \mp \frac{\{1 - f(\alpha)\}}{[2\{1 - f(\alpha)\} + 12/\lambda^2]}, \tag{5.35}$$

where $f(\alpha) = \sum_{n,\ \text{odd}} [\tanh(n\pi/2\alpha)/(n\pi/2\alpha)]96/(n\pi)^4$.

This result is similar in form to the result for a single suspended cable. Notice that, when λ^2 is large, $\mathbf{h}_{1,2} \to \mp\mathbf{p}/2$ and $\mathbf{w}(\mathbf{x}, \mathbf{y}) \to 0$. However, these results will hold only if the lengths of the sides are not substantially dissimilar—a point to be addressed at further length shortly—and we should not expect the results to reduce precisely to those of the single cable when the limit $\alpha \gg 1$ is taken, for example.

Corrections to the averaged additional tensions may be found by substituting (5.32) into (5.29). The corrected additional tension for the drape cables is

$$\mathbf{h}(\mathbf{y}) = -\lambda^2\, \mathbf{p} \int_0^1 \mathbf{w}\, d\mathbf{x},$$

or

$$\mathbf{h}(\mathbf{y}) = -\lambda^2\, \mathbf{p}\left(1 + \frac{2\mathbf{h}_1}{\mathbf{p}}\right)\left[\frac{1}{12} - \sum_{n,\,\mathrm{odd}} \frac{8}{n^4\pi^4} \frac{\cosh\{(1 - 2\mathbf{y})\, n\pi/2\alpha\}}{\cosh(n\pi/2\alpha)}\right]. \quad (5.36)$$

Therefore, in comparison to the averaged value for these cables,

$$\frac{\mathbf{h}(\mathbf{y})}{\mathbf{h}_1} = \frac{\{1 - f(\mathbf{y})\}}{\{1 - f(\alpha)\}}, \quad (5.37)$$

where $f(\mathbf{y}) = \sum_{n,\,\mathrm{odd}} [\cosh\{(1 - 2\mathbf{y})\, n\pi/2\alpha\}/\cosh(n\pi/2\alpha)]\, 96/(n\pi)^4$. For the hanging cables the corrected tensions are given by

$$\frac{\mathbf{h}(\mathbf{x})}{\mathbf{h}_2} = \frac{6\mathbf{x}(1 - \mathbf{x}) - \sum_{n,\,\mathrm{odd}} \frac{48}{n^3\pi^3} \sin n\pi\mathbf{x} \frac{\tanh(n\pi/2\alpha)}{(n\pi/2\alpha)}}{\{1 - f(\alpha)\}}. \quad (5.38)$$

In both cases the expressions yield the correct results of zero additional tension for the cables lying along the boundaries.

A practical range for α is $1 \le \alpha \le 2$, and in this range (5.37) is insensitive to it. In table 5.1 values of $\mathbf{h}(\mathbf{y})/\mathbf{h}_1$ and $\mathbf{h}(\mathbf{x})/\mathbf{h}_2$ are listed for $\alpha = 1$ and $\alpha = 2$. These trends are confirmed by the analyses of roof structures presented by Møllmann. Another hyperbolic surface is discussed in problem 5.5.

Example 5.5 Calculations for a Cable Roof

A cable roof of rectangular plan has an initial profile (under prestress alone) given by (5.25). Properties of the roof are

$2a = 90$ m, $\quad 2b = 60$ m,

$d_1 = 8$ m, $\quad d_2 = 3.55$ m,

$H_1 = 120$ kN/m, $\quad H_2 = 120$ kN/m,

Table 5.1 Values of corrected additional tension for a hyperbolic paraboloid

$\mathbf{h}(\mathbf{y})/\mathbf{h}_1$	$\mathbf{y} = 0$	0.1	0.2	0.3	0.4	0.5
$\alpha = 1$	0	0.59	0.99	1.25	1.40	1.44
$\alpha = 2$	0	0.58	0.97	1.24	1.39	1.44
$\mathbf{h}(\mathbf{x})/\mathbf{h}_2$	$\mathbf{x} = 0$	0.1	0.2	0.3	0.4	0.5
$\alpha = 1$	0	0.62	1.02	1.26	1.38	1.42
$\alpha = 2$	0	0.79	1.13	1.21	1.19	1.17

so that $H_1 d_1/a^2 = H_2 d_2/b^2$. The hanging cables span the shorter side and $EA_2 = 10 \times 10^4$ kN/m—the bracing cables have $EA_1 = 5 \times 10^4$ kN/m. The loading due to the cladding is $p = 0.3$ kN/m^2. The problem is to find the new equilibrium configuration under dead load. We adopt the solutions given by (5.32) and (5.35) as modified by the results suggested by problem 5.3. The pertinent dimensionless quantities are $\lambda_1^2 = (8d_1/2a)^2 EA_1/H_1 = 210$, $\lambda_2^2 = (8d_2/2b)^2 EA_2/H_2 = 180$, $\alpha_*^2 = H_2 a^2/H_1 b^2 = 2.25$, and $\mathbf{p} = 4a^2 p/8d_1 H_1 = 0.32$. The additional vertical deflection is

$$\mathbf{w}(\mathbf{x}, \mathbf{y}) = \left\{1 - \frac{\mathbf{h}_2}{\mathbf{p}}\left(1 + \frac{\lambda_1^2}{\alpha_*^2 \lambda_2^2}\right)\right\}\left\{\frac{1}{2}\mathbf{x}(1 - \mathbf{x}) - \sum_{n,\,\mathrm{odd}} \frac{4}{n^3\pi^3} \sin n\pi\mathbf{x} \frac{\cosh\{(1 - 2\mathbf{y})n\pi/2\alpha_*\}}{\cosh(n\pi/2\alpha_*)}\right\},$$

where

$$\frac{\mathbf{h}_2}{\mathbf{p}} = \frac{1 - f(\alpha_*)}{\{1 + \lambda_1^2/(\alpha_*^2\lambda_2^2)\}\{1 - f(\alpha_*)\} + 12/(\alpha_*^2\lambda_2^2)}$$

and

$$\mathbf{h}_1 = -\frac{\lambda_1^2}{\alpha_*^2\lambda_2^2}\mathbf{h}_2.$$

Here $f(\alpha_*) = \sum_{n,\,\mathrm{odd}} [\tanh(n\pi/2\alpha_*)/(n\pi/2\alpha_*)]96/(n\pi)^4 = 0.74$, so $\mathbf{h}_2/\mathbf{p} = 0.62$, and $\mathbf{h}_1/\mathbf{p} = -0.33$. The additional deflection at the midpoint is $\mathbf{w}(1/2, 1/2) = 0.0022$.

Converting these to dimensional quantities yields

$$H_1 + h_1 = 105 \text{ kN/m},$$

and

$$H_2 + h_2 = 145 \text{ kN/m}.$$

In the middle of the roof the corrected additional tensions are about 40 percent different from the averaged values, so that

$$H_1 + h(y)_{\max} = 100 \text{ kN/m}$$

and

$$H_2 + h(x)_{\max} = 155 \text{ kN/m}.$$

The addition of snow loading, for example, will bring the ratio of these peak tensions down to a value approaching 1:2, in keeping with the cable areas assigned to each direction, although cable stresses are low under all combinations of dead and live load. The maximum additional deflection under dead load is 0.05 m, which is low because λ_1^2 and λ_2^2 are both high.

The use of this linearized solution is both direct and accurate. The reason for its accuracy may be traced to the appreciable curvatures exhibited by the initial profile. The natural frequencies of the roof under dead load are calculated at the end of example 5.7.

Example 5.6 Wind Uplift on a Rectangular Roof

Figure 5.20 shows a typical distribution of pressure coefficients on a model of a low-rise rectangular roof as determined from boundary layer wind tunnel experiments. Calculation of the uplift pressure is by means of the formula

$$p = C_p \rho \frac{v^2}{2},$$

where C_p is the local pressure coefficient and $\rho v^2/2$ is the stagnation pressure. Generally speaking, peak pressure coefficients are recorded when the wind is along the diagonal. The roof tends to behave like a swept-back wing at angle of attack—lift is high along the leading edges where tip vortices form. But over much of the roof the uplift is much lower, being typically an order of magnitude less than values recorded at the leading edges. The uplift also drops off very rapidly away from these leading edges.

While figure 5.20 does give an indication of the probable behavior, adjacent buildings and other topographical features will modify the patterns of lift experienced by a roof. In such instances wind tunnel tests on models are the only reliable way of estimating loads.

Cable-suspended roofs are light and relatively flexible structures. In the absence of suitable model tests building code provisions for wind loading usually suggest loads substantially in excess of self-weight. If these uplifts did actually occur over a significant portion of the surface, the use of cable roofs would be difficult to justify. However, cable roofs are built, and there have been few recorded instances of undesirable behavior in the wind.

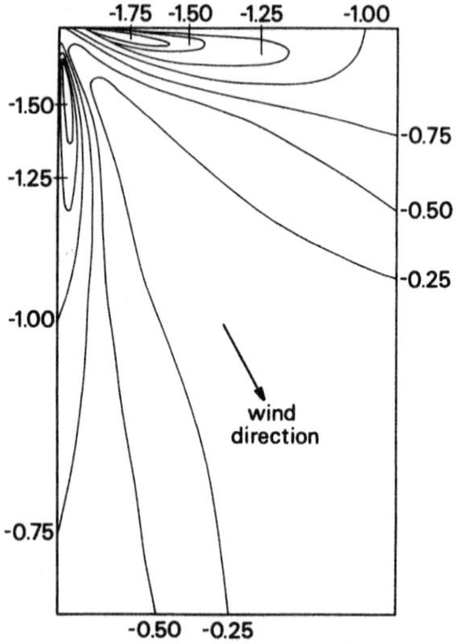

5.20 Pressure coefficients C_p on an unparapeted model roof in rough flow (courtesy Frank H. Durgin, Wright Brothers' Wind Tunnel, MIT)

The reasons for this appear to be twofold. First, as figure 5.20 shows, a large central protion of the roof experiences low uplift, and so deflection response is accordingly low. The second aspect involves the beneficial influence of the air contained within an airtight building. In the few seconds during which a gust passes over the roof the air within the building has insufficient time to adjust to the change in contained volume that a change in roof profile may bring about. The air rarefies slightly, and its temperature drops slightly, the process being adiabatic since the associated heat loss will be recovered much more slowly than the passage time of the gust. The contained air thus provides an elastic restraint to roof movement in the form of a uniformly distributed load of magnitude

$$\frac{E_v}{V}\int\int w\, dxdy,$$

where E_v = the bulk modulus of elasticity of the air (about 0.145 MPa for an adiabatic process), V is the contained volume, and the double integral represents the volume change.

For purposes of analysis let us suppose that the region of uplift on the roof may be idealized by the shaded region in figure 5.21, and suppose that there exists in this region of the roof a uniform uplift of magnitude p per unit area. The equation of equilibrium is

$$H\frac{\partial^2 w}{\partial x^2} + H\frac{\partial^2 w}{\partial y^2} = p + \frac{E_v}{V}\int_0^{2a}\int_0^{2b} w\, dxdy - \frac{8d_1}{(2a)^2}2h_1,$$

and the membrane equation is

$$\frac{h_1 2a2b}{EA} = -\frac{8d_1}{(2a)^2}\int_0^{2a}\int_0^{2b} w\, dxdy.$$

A solution to these equations in the form of a double Fourier series can be found.

To get a feel for the likely effect of air elasticity, we may treat a simple example to some degree representative of an actual cable-suspended roof. Consider the linearized response of a taut flat cable under a uniform uplift that extends from the lefthand support a distance x_3 along the span (see figure 5.22). The equations of equilibrium are

$$H\frac{d^2 w}{dx^2} = p + k, \quad 0 \le x \le x_1,$$

and

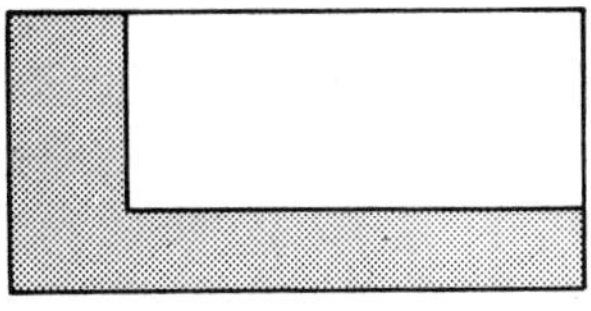

5.21 Idealized region of uplift on a rectangular roof

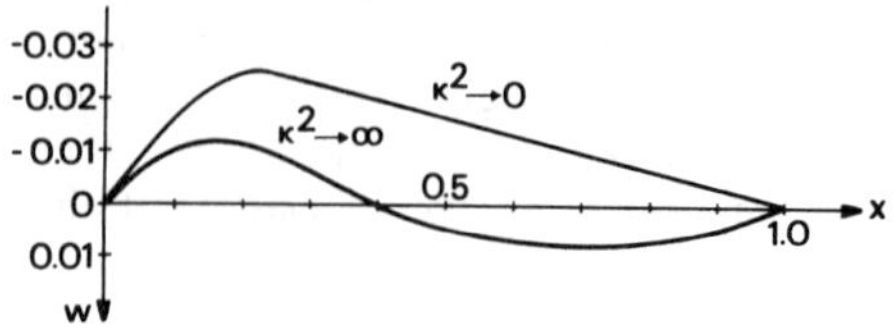

5.22 Effect of wind uplift on an idealized cable-suspended roof. In (b) the curve for $\kappa^2 \rightarrow \infty$ is more likely to represent actual behavior in an airtight building

$$H\frac{d^2w}{dx^2} = k, \quad x_1 \leq x \leq l,$$

where $k = E_v \int_0^l (w\,dx)/V$, and the cable is considered representative of a unit width of the roof. The equations have been delibarately written in this form to emphasize the analogy that exists with the linearized response of a flat-sag cable to static load. The additional tension (absent in the present case) is analogous to the restraint supplied by the air. As a result we may extract the solution directly from (2.38), (2.39), and (2.41). The solution we require is

$$\mathbf{w} = -\frac{\mathbf{x}}{2}\{\mathbf{x}_3(2 - \mathbf{x}_3) - \mathbf{x}\} + \frac{\mathbf{k}}{2}\mathbf{x}(1 - \mathbf{x}), \quad 0 \leq \mathbf{x} \leq \mathbf{x}_3,$$

and

$$\mathbf{w} = -\frac{1}{2}\mathbf{x}_3^2(1 - \mathbf{x}) + \frac{\mathbf{k}}{2}\mathbf{x}(1 - \mathbf{x}), \quad \mathbf{x}_3 \leq \mathbf{x} \leq 1,$$

where

$$\mathbf{k} = \frac{\mathbf{x}_3^2(3 - 2\mathbf{x}_3)}{(1 + 12/\kappa^2)},$$

in which $\mathbf{x} = x/l$, $\mathbf{x}_3 = x_3/l$, $\mathbf{k} = k/p$, $\mathbf{w} = w/(pl^2/H)$, and $\kappa^2 = (E_v l^3/VH)$.

The last parameter κ^2 determines the relative stiffness of the enclosed air and the cable and is analogous to the cable parameter λ^2. Calculations show that κ^2 typically lies in the range 10 to 100, or more, in which case the results are essentially independent of it because the expression for $\mathbf{k}$ is then

$$\mathbf{k} = \mathbf{x}_3^2(3 - 2\mathbf{x}_3),$$

and there is no volume change under the roof, $\int_0^l w\,dx = 0$.

Figure 5.22 shows the (dimensionless) deflection response when the uplift extends to the first quarter-span point and κ^2 is both small (the air is compressible) and large (the air is incompressible). Because κ^2 will tend to be large in practice, the restraint offered the roof by the contained air is highly beneficial to roof performance if λ^2 is low—in the previous example, where λ^2 was large, there is little additional benefit to be gained.

Nonlinear Response of Taut Flat Membranes

Because the profile is initially flat ($z_0 = 0$), the equilibrium of an element of the membrane is specified by

$$\{H + h(y)\} \frac{\partial^2 w}{\partial x^2} + \{H + h(x)\} \frac{\partial^2 w}{\partial y^2} = -p, \tag{5.39}$$

this being the appropriate form for (5.23). Recall that the additional tensions $h(x)$ and $h(y)$ are constant along lines parallel to the y axis and x axis, respectively. They are given by

$$\begin{aligned} \frac{h(y)(x_1 - x_2)}{EA} &= \frac{1}{2}\int_{x_2}^{x_1} \left(\frac{\partial w}{\partial x}\right)^2 dx, \\ \frac{h(x)(y_1 - y_2)}{EA} &= \frac{1}{2}\int_{y_2}^{y_1} \left(\frac{\partial w}{\partial y}\right)^2 dy, \end{aligned} \tag{5.40}$$

where x_1 and x_2 are values of x on the boundaries for given values of y, and so on. There are no first-order changes in additional tension.

The boundaries are assumed to be rigid, so there are no contributions from the in-plane components of displacement (u, v), although support compliance could be accounted for if necessary. Together with the condition that w is zero on the boundaries, (5.39) and (5.40) govern the nonlinear response of the membrane.

Unfortunately analytical solutions of these equations are far from straightforward, and the eventual use of the computer appears to be inevitable [22], in which case one might as well proceed with a numerical solution from the outset. The approach adopted here is an attempt to circumvent these analytical difficulties by rearranging the governing equations. In this way, and by expressing the solutions in suitable dimensionless forms, the mechanics of the response are readily portrayed [7]. The method is similar to that given earlier for hyperbolic surfaces.

For the purposes of a first approximation it is assumed that $h(x) = h(y) = \bar{h} =$ constant, which is reasonable only if the plan dimensions are not substantially dissimilar. Equation (5.39) becomes

$$(H + \bar{h})\,\nabla^2 w = -p. \tag{5.41}$$

The equation of elastic and geometric compatibility is written in terms of the area of an element and takes the form

$$\frac{\bar{h}dxdy}{EA} + \frac{\bar{h}dydx}{EA} = \left(\frac{\partial u}{\partial x}\right)dxdy + \left(\frac{\partial v}{\partial y}\right)dydx$$

$$+ \frac{1}{2}\left(\frac{\partial w}{\partial x}\right)^2 dxdy + \frac{1}{2}\left(\frac{\partial w}{\partial y}\right)^2 dydx. \tag{5.42}$$

In view of the assumed constancy of $\bar{h}$, and because u and v are zero on the boundaries, this relation may be expressed in integral form as

$$\frac{2\bar{h}S}{EA} = \frac{1}{2}\iint_S \left\{\left(\frac{\partial w}{\partial x}\right)^2 + \left(\frac{\partial w}{\partial y}\right)^2\right\} dxdy, \tag{5.43}$$

where S is the plan area of the membrane. However, w is also zero on the boundaries, and the righthand side may be integrated by parts to give

$$\frac{2\bar{h}S}{EA} = -\frac{1}{2}\iint_S \nabla^2 w\, w\, dxdy, \tag{5.44}$$

which, if p is constant and covers the whole area, may be further simplified to

$$\frac{2\bar{h}S}{EA} = \frac{p}{2(H + \bar{h})}\iint_S w\, dxdy. \tag{5.45}$$

Equations (5.41) and (5.45) have their counterparts in the membrane analogy for torsion in prisms. The membrane deflection is analogous to the shearing stress function, and (5.45) is analogous to the expression for the torsional rigidity of the prism [15].

For certain plan forms solutions of (5.41) may be obtained by standard methods. The solution is then substituted into (5.45), and, after the integration has been performed, a cubic in $\bar{h}$ is obtained. This cubic has only one positive real root and is of such a form that a simple solution exists. The method is thus an extension of the work on the taut flat cable presented in chapter 2.

However, $\bar{h}$ is an averaged value that will tend to underestimate the true additional tensions in those cables that intersect near the middle of the network but will grossly overestimate the additional tensions in those cables that lie near the boundaries. Consequently improved estimates of $h(x)$ and $h(y)$ are required. These may be obtained from (5.40), using the solutions of (5.41) and (5.45).

It is apparent that this approach is largely one of successive approximation which, as subsequent comparison with experiment will show, yields solutions of acceptable accuracy. The method is now applied to two particular problems.

The sides of the *rectangular membrane* shown in figure 5.23 are $2a$ and $2b$. The solution of (5.41) for a load uniformly distributed over the whole surface and subject to the boundary conditions $\mathbf{w}(0, \mathbf{y}) = \mathbf{w}(1, \mathbf{y}) =$

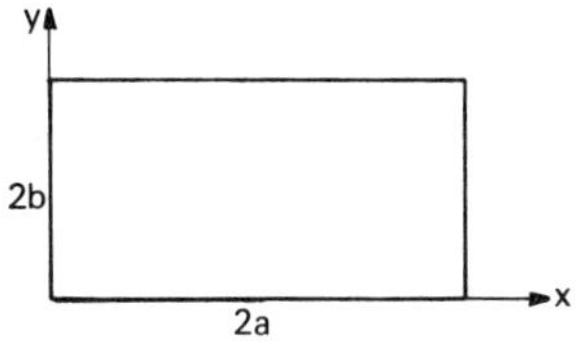

5.23 Definition diagram for a rectangular membrane

$\mathbf{w}(\mathbf{x}, 0) = \mathbf{w}(\mathbf{x}, 1) = 0$ may be written as (see equation 5.32)

$$\mathbf{w}(\mathbf{x}, \mathbf{y}) = \frac{1}{(1 + \bar{\mathbf{h}})}\left\{\frac{1}{2}\mathbf{x}(1 - \mathbf{x}) - \sum_{n,\,\mathrm{odd}} \frac{4}{n^3\pi^3}\sin n\pi\mathbf{x}\frac{\cosh\{(1 - 2\mathbf{y})n\pi/2\alpha\}}{\cosh(n\pi/2\alpha)}\right\}, \tag{5.46}$$

where in this dimensionless formulation $\mathbf{x} = x/2a$, $\mathbf{y} = y/2b$, $\alpha = a/b$, $\bar{\mathbf{h}} = \bar{h}/H$, and $\mathbf{w} = w/(p(2a)^2/H)$.

The maximum deflection occurs at midspan and is

$$\mathbf{w}_{\max} = \frac{1}{8(1 + \bar{\mathbf{h}})}\left\{1 - \sum_{n,\,\mathrm{odd}} \frac{32(-1)^{(n-1)/2}}{n^3\pi^3}\operatorname{sech}\left(\frac{n\pi}{2\alpha}\right)\right\}. \tag{5.47}$$

When $\alpha \ll 1$, $\mathbf{w}_{\max} \to 1/(8(1 + \bar{\mathbf{h}}))$, and, when $\alpha \gg 1$, $\mathbf{w}_{\max} \to 1/(8(1 + \mathbf{h})\alpha^2)$, which is the same if the nondimensionalization is changed to $\mathbf{w} = w/(p(2b)^2/H)$. These results indicate that for limiting plan geometries the network behaves like a single cable spanning the shorter side.

The cubic from which the additional tension $\bar{\mathbf{h}}$ may be found is obtained by substituting (5.46) into (5.45), performing the integration, and carrying out some rearrangement. The result is

$$\bar{\mathbf{h}}(1 + \bar{\mathbf{h}})^2 = \frac{\lambda^2}{48}\{1 - f(\alpha)\}, \tag{5.48}$$

where $\lambda^2 = (2pa/H)^2(EA/H)$ and $f(\alpha) = \sum_{n,\,\mathrm{odd}} [\tanh(n\pi/2\alpha)/(n\pi/2\alpha)]96/(n\pi)^4$. Its solution is

$$\bar{\mathbf{h}} = \frac{1}{\sqrt[3]{B}}\left(\sqrt[3]{B} - \frac{1}{3}\right)^2, \tag{5.49}$$

where $B = -q/2 + (q^2/4 + r^3/27)^{1/2}$, in which $q = -2/27 - \{1 - f(\alpha)\}\lambda^2/48$ and $r = -1/3$.

This solution depends on two parameters λ^2 and $f(\alpha)$. When $\alpha \ll 1$, $f(\alpha) \to 0$, and, when $\alpha \gg 1$, $f(\alpha) \to 1 - 1/\alpha^2$. But again, by changing the nondimensionalization in the second case to $\lambda^2 = (2pb/H)^2(EA/H)$, the results are seen to be identical. However, for these limiting values of $f(\alpha)$ (5.48) does not reduce to the result for a single taut cable obtained earlier (see equation 2.42). The right-hand side of (5.48) is too small by a factor of two. The reason for this discrepancy may be traced to the left-hand side of (5.45). For these limits on the plan geometry it is clearly inappropriate to assume an overall average value for the additional tension. The corrected additional tensions (namely, $\mathbf{h}(\mathbf{x})$ and $\mathbf{h}(\mathbf{y})$) reduce this discrepancy.

Substituting (5.46) into the first of (5.40) and performing the integration yield

$$\mathbf{h}(\mathbf{y}) = \frac{\lambda^2}{24(1+\bar{\mathbf{h}})^2}\left[1 - \sum_{n,\text{odd}} \frac{96}{n^4\pi^4}\{f(\mathbf{y})(2 - f(\mathbf{y}))\}\right], \tag{5.50}$$

where $\mathbf{h}(\mathbf{y}) = h(y)/H$ and $f(\mathbf{y}) = \cosh\{(1 - 2\mathbf{y})n\pi/2\alpha\}/\cosh(n\pi/2\alpha)$.

A further rearrangement is possible after (5.50) is divided by (5.48), namely,

$$\frac{\mathbf{h}(\mathbf{y})}{\bar{\mathbf{h}}} = \frac{2\left[1 - \sum_{n,\text{odd}} \frac{96}{n^4\pi^4}\{f(\mathbf{y})(2 - f(\mathbf{y}))\}\right]}{1 - f(\alpha)}, \tag{5.51}$$

and this is a function only of geometry. As is required by the nature of the deflected surface, $\mathbf{h}(0) = \mathbf{h}(1) = 0$, and, when $\mathbf{y} = 1/2$, $\mathbf{h}(\mathbf{y})$ has a maximum that ranges in value between

$$0 < \frac{\mathbf{h}_{\text{max}}}{\bar{\mathbf{h}}} < 2,$$

where the upper limit corresponds to $\alpha \ll 1$ and the lower limit to $\alpha \gg 1$. A similar expression may be derived for $\mathbf{h}(\mathbf{x})$, but this is unnecessary since it may be assumed that $\alpha \leq 1$ without loss of generality, in which case the largest increase in $\bar{\mathbf{h}}$ will be for the cable parallel to the x axis and located at $\mathbf{y} = 1/2$.

A further example is afforded by an *elliptical membrane*. The boundary of an ellipse with semi-axes of a and b, respectively (see figure 5.24) is given by

$$\mathbf{x}^2 + \mathbf{y}^2 = 1, \tag{5.52}$$

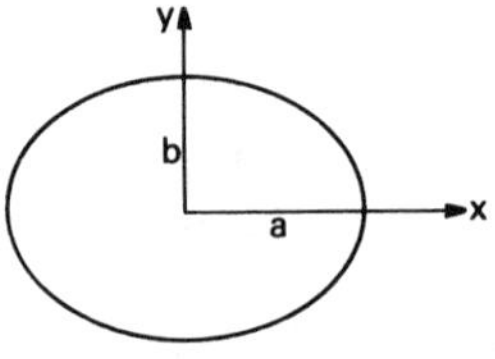

5.24 Definition diagram for an elliptical membrane

where $\mathbf{x} = x/a$ and $\mathbf{y} = y/b$. The solution of (5.41) subject to the condition of zero deflection on the boundary is (see, for example, reference [15])

$$\mathbf{w}(\mathbf{x}, \mathbf{y}) = \frac{1}{8(1 + \alpha^2)(1 + \bar{\mathbf{h}})}(1 - \mathbf{x}^2 - \mathbf{y}^2), \tag{5.53}$$

where $\mathbf{w} = w/(p(2a)^2/H)$, $\bar{\mathbf{h}} = \bar{h}/H$, and $\alpha = a/b$. Curves of constant deflection are concentric ellipses. The maximum deflection occurs at the origin and is

$$\mathbf{w}_{\max} = \frac{1}{8(1 + \alpha^2)(1 + \bar{\mathbf{h}})}. \tag{5.54}$$

The equation from which $\bar{\mathbf{h}}$ is found may be obtained by substituting (5.53) into (5.45) and carrying out the integration. This yields

$$\bar{\mathbf{h}}(1 + \bar{\mathbf{h}})^2 = \frac{\lambda^2}{64(1 + \alpha^2)}. \tag{5.55}$$

where $\lambda^2 = (2pa/H)^2\, EA/H$.

Finally, if it is assumed that $\alpha \leq 1$, only $\mathbf{h}(\mathbf{y})$ needs to be considered in the calculation of the corrected additional tensions. Substitution of (5.53) into the first equation of (5.40), integration, and rearrangement lead to

$$\frac{\mathbf{h}(\mathbf{y})}{\bar{\mathbf{h}}} = \frac{8}{3(1 + \alpha^2)}(1 - \mathbf{y}^2), \tag{5.56}$$

and $\mathbf{h}(1) = \mathbf{h}(-1) = 0$, as expected. The maximum value lies between

$$0 < \frac{\mathbf{h}_{\max}}{\bar{\mathbf{h}}} < \frac{8}{3}.$$

It may be noted that, when $\alpha = 1$, (5.53), (5.55), and (5.56) provide results for the circular membrane. Details of another simple plan form—the *equilateral triangular membrane*—are left as an exercise (see problem 5.4).

Comparison of Theory and Experiment

A comparison was made between the theory presented here and some experimental and theoretical results of Shore and Bathish [22] for a taut flat square network. The results of the comparison are listed in table 5.2.

Their experiment was conducted on an orthogonal network consisting of five cables each way, spanning a distance of 60 in (see figure 5.25). The cable used was steel aircraft cord of diameter 3/32 in. The pretension in each cable was 120 lb. From each of the twenty-five intersection points a concentrated load of 5 lb was hung. Deflections were measured at six

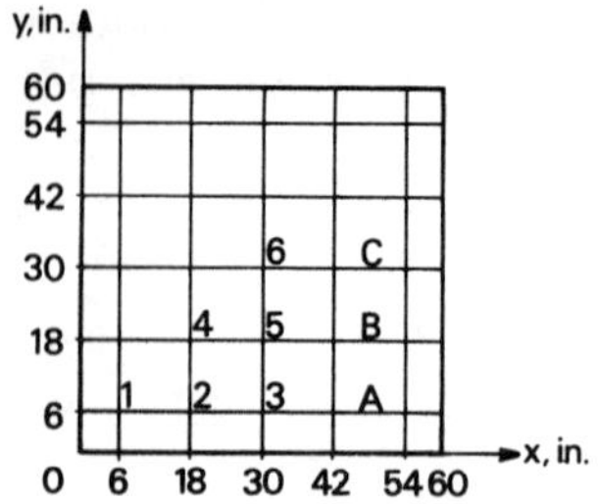

5.25 Shore and Bathish's experimental network showing six typical intersection points and three typical cables (from Shore and Bathish [22], courtesy Blackwell Scientific Publications)

typical intersection points, and additional tensions were measured in three typical cables.

In terms of the equivalent membrane the following quantities may be specified: $2a = 2b = 60$ in, $H = 10$ lb/in, $p = 3.5 \times 10^{-2}$ lb/in^2, $E = 15 \times 10^6$ lb/in^2, $A = 5.75 \times 10^{-4}$ in^2/in. Therefore $\lambda^2 = 38$, and $f(\alpha) = 0.577$. The solution of (5.48) gives $\bar{\mathbf{h}} = 0.224$, which compares favorably with Shore and Bathish's experimental and theoretical results of 0.218 and 0.205, respectively.

Results for the defection and corrected additional tensions were calculated using (5.46) and (5.51) and are presented in table 5.2 along with the corresponding experimental and theoretical data of Shore and Bathish. The corrected additional tensions were calculated by multiplying the relevant value of $\mathbf{h}(\mathbf{y})$ by H and the spacing between cables.

Table 5.2 Comparison of theories and experiment for a square network

Intersection point	1	2	3	4	5	6
Theoretical deflection (inches)						
Present theory	0.153	0.226	0.300	0.565	0.665	0.758
Shore and Bathish [22]	0.138	0.256	0.285	0.539	0.618	0.716
Experimental deflection (inches)						
Shore and Bathish [22]	0.169	0.290	0.325	0.589	0.667	0.773

Cable	A	B	C
Theoretical additional tension (pounds)			
Present theory	8.52	35.0	46.4
Shore and Bathish [22]	7.92	32.5	42.4
Experimental additional tension (pounds)			
Shore and Bathish [22]	9.96	33.0	45.0

Clearly good agreement exists between the theories and the experimental results. However, the present theory is a good deal simpler to apply than that of Shore and Bathish. The extent of agreement is surprising because the mesh is rather coarse—the cable spacing being one-fifth of the span. Nevertheless it is apparent that the use of an equivalent membrane as a model of this square network is more than satisfactory.

5.6 Dynamics of Suspended Membranes

This section is a companion to the previous one where we considered the static response of certain uniform pretensioned networks using the concept of an equivalent membrane. Here we follow a similar procedure and examine the free vibrations of the suspended network by replacing the network with an equivalent membrane. These small vibrations are assumed to occur about an equilibrium position in which the network supports, in addition to its self-weight, load uniformly distributed over its plan area. Consequently, provided that the profile of the network under this load is relatively shallow and the plan dimensions are not substantially dissimilar, the static equilibrium will be accurately expressed by (5.41). The term suspended membrane will therefore refer to a network that meets these requirements, and the self-weight will be taken to mean the total uniformly distributed load acting on the network. For example, it covers the situation where the pretensioned network is first hung in place and the cladding is then attached (giving rise to changes in the initial profile and the pretensions that may be accounted for with the theory of the previous section).

A history of the development of the linear theory of free vibrations of classical membranes—those initially taut and flat—has been presented in the first chapter. However, if the membrane's supports are horizontal, some sag will always exist owing to the self-weight of the membrane. Thus in this chapter we look at the effect that this sag has on the vibration theory. Although the theory applies only to suspended membranes with relatively shallow profiles, several phenomena arise that cannot be predicted by the classical theory [8].

Derivation of Governing Equations

The partial differential equation governing the static equilibrium of a uniform suspended membrane is

$$H\frac{\partial^2 w}{\partial x^2} + H\frac{\partial^2 w}{\partial y^2} = -p. \tag{5.57}$$

A comparison with (5.41) shows that the horizontal components of membrane tension are identical to the quantities $(H + \bar{h})$ used there. (Example 5.7 will illustrate how the theory may be applied to other situations.)

If a small disturbance is applied to the membrane, the dynamic equilibrium is expressed by

$$\left\{H + h(t)\right\}\frac{\partial^2}{\partial x^2}(w + \delta) + \left\{H + h(t)\right\}\frac{\partial^2}{\partial y^2}(w + \delta) = -p + \frac{p}{g}\frac{\partial^2 \delta}{\partial t^2}, \tag{5.58}$$

where δ is the additional dynamic vertical deflection, g the acceleration due to gravity, and $h(t)$ the additional horizontal component of membrane tension, assumed to be uniform throughout the membrane. In terms of the overall response this assumption is reasonable, provided that the transverse dimensions of the membrane are not substantially dissimilar. Expanding (5.58), substituting for (5.57), and dropping terms of the second order yield

$$H\frac{\partial^2 \delta}{\partial x^2} + H\frac{\partial^2 \delta}{\partial y^2} = \frac{p}{g}\frac{\partial^2 \delta}{\partial t^2} + \frac{p}{H}h(t). \tag{5.59}$$

The associated longitudinal dynamic equations of equilibrium have been omitted because they are of second order for the relatively flat profiles under consideration here.

To find first order, the additional tension is given by the expression

$$\frac{2h(t)S}{EA} = \iint_S \left(\frac{\partial w}{\partial x}\frac{\partial \delta}{\partial x} + \frac{\partial w}{\partial y}\frac{\partial \delta}{\partial y}\right) dxdy. \tag{5.60}$$

Again the assumed rigidity of the boundaries means that there are no contributions from the longitudinal movements of the membrane. The righthand side of (5.60) may be further simplified by integration by parts which, noting that δ vanishes on the boundaries, yields

$$\frac{2h(t)S}{EA} = -\iint_S \nabla^2 w\, \delta\, dxdy = \frac{p}{H}\iint_S \delta\, dxdy. \tag{5.61}$$

Equations (5.59) and (5.61) constitute a linear homogeneous system in δ. With the governing equations in this form the fundamental features of the linear theory of free vibrations may be explored. Here, however, only the special cases of the rectangular and circular membranes are considered. Nevertheless the equations hold for any plan form, and a general proof of the orthogonality of the modes of vibration may be constructed in the same way as indicated in problem 3.3.

Rectangular Membranes

The equations of motion of the suspended rectangular membrane shown previously in figure 5.23 may be written in the following dimensionless form

$$\frac{\partial^2 \tilde{\boldsymbol{\delta}}}{\partial \mathbf{x}^2} + \alpha^2 \frac{\partial^2 \tilde{\boldsymbol{\delta}}}{\partial \mathbf{y}^2} + \boldsymbol{\omega}^2\, \tilde{\boldsymbol{\delta}} = \tilde{\mathbf{h}},$$

$$\tilde{\mathbf{h}} = \frac{\lambda^2}{2} \int_0^1 \int_0^1 \tilde{\boldsymbol{\delta}}\, d\mathbf{x}\, d\mathbf{y}, \tag{5.62}$$

where $\mathbf{x} = x/2a$, $\mathbf{y} = y/2b$, $\alpha = a/b$, $\tilde{\boldsymbol{\delta}} = \tilde{\delta}/(p(2a)^2/H)$, $\tilde{\mathbf{h}} = \tilde{h}/H$, $\boldsymbol{\omega}^2 = p(2a)^2\omega^2/gH$, and $\lambda^2 = (2pa/H)^2(EA/H)$. Time has been removed from these equations by prior substitutions of $\delta(x, y, t) = \tilde{\delta}(x, y)e^{i\omega t}$ and $h(t) = \tilde{h}e^{i\omega t}$.

For the second equation of (5.62) it is seen that $\tilde{\mathbf{h}} = 0$ when

$$\int_0^1 \int_0^1 \tilde{\boldsymbol{\delta}}\, d\mathbf{x} d\mathbf{y} = 0. \tag{5.63}$$

Modes that meet this condition are antisymmetric; all other modes induce additional tension and are symmetric. (Locally, additional tensions are induced in some of the antisymmetric modes, and they along with the corrected additional tensions in the symmetric modes may be found from the respective component cable equations. Their presence does little to change the overall dynamic characteristics typified by the natural frequencies and the modes of vibration.)

To discover which modes induce additional tension, it proves convenient to solve the classical rectangular membrane problem ($\lambda^2 \to 0$). Equations (5.62) reduce to

$$\frac{\partial^2 \tilde{\boldsymbol{\delta}}}{\partial \mathbf{x}^2} + \alpha^2 \frac{\partial^2 \tilde{\boldsymbol{\delta}}}{\partial \mathbf{y}^2} + \boldsymbol{\omega}^2\, \tilde{\boldsymbol{\delta}} = 0. \tag{5.64}$$

A solution for the mode $\tilde{\boldsymbol{\delta}}_{mn}$ that satisfies the condition of zero deflection on the boundaries is

$$\tilde{\boldsymbol{\delta}}_{mn}(\mathbf{x}, \mathbf{y}) = \sin m\pi\mathbf{x} \sin n\pi\mathbf{y}, \tag{5.65}$$

where $m = 1, 2, 3, \ldots$, $n = 1, 2, 3, \ldots$, and, as an aid to the later discussion, the modal constant has been taken as unity. The associated natural frequency is

$$\boldsymbol{\omega}_{mn} = \pi(m^2 + \alpha^2 n^2)^{1/2}. \tag{5.66}$$

The question of multiple natural frequencies is omitted for the present.

Equation (5.65) shows that only in those cases where m and n are both odd is (5.63) not satisfied. This leads to the following.

The *antisymmetric modes* of a suspended rectangular membrane are those that do not induce additional tension. In its simplest form the antisymmetric mode $\tilde{\boldsymbol{\delta}}_{mn}$ is

$$\tilde{\boldsymbol{\delta}}_{mn} = \sin m\pi\mathbf{x} \sin n\pi\mathbf{y}, \tag{5.67}$$

and its associated natural frequency is

$$\omega_{mn} = \pi(m^2 + \alpha^2 n^2)^{1/2}, \tag{5.68}$$

where m and n are those positive integers, so that at least one of them is even.

It appears that multiple natural frequencies do not occur if α^2 is irrational, but they can occur if α^2 is rational [14]. Where multiple natural frequencies do not occur, the antisymmetric modes are given by (5.67) and, including the boundaries, the nodal lines are $(m + 1)$ equally spaced lines and $(n + 1)$ equally spaced lines parallel to the edges of the rectangle, respectively.

In the case of multiple natural frequencies a wide diversity of associated antisymmetric modes are possible in which the nodal lines may be quite complicated curves. For example, consider the variety of possible first antisymmetric modes of the square membrane ($\alpha = 1$) indicated by $k_1 \tilde{\boldsymbol{\delta}}_{12} + k_2 \tilde{\boldsymbol{\delta}}_{21}$, where k_1 and k_2 are arbitrary constants. Some of these are presented by Courant and Hilbert [2], and a few are shown later in figure 5.27. They are reminiscent of the strangely beautiful curves of Chladni's figures for a plate.

Turning now to the *symmetric modes*, we note that, since they are the only ones with additional tension, the governing partial differential and integral equations are given by (5.62). The Fourier series solution of the first equation of (5.62), which satisfies the condition of zero deflection on the boundary, is

$$\tilde{\boldsymbol{\delta}}(\mathbf{x}, \mathbf{y}) = 16\tilde{\mathbf{h}} \sum_{i,\, j\,\text{odd}}\sum \frac{\sin i\pi\mathbf{x} \sin j\pi\mathbf{y}}{(i\pi)(j\pi)\{\boldsymbol{\omega}^2 - (i\pi)^2 - \alpha^2(j\pi)^2\}}, \tag{5.69}$$

in which the value of $\boldsymbol{\omega}$ specifies the particular symmetric mode.

Use is now made of the second equation of (5.62) to eliminate $\tilde{\mathbf{h}}$ and thereby obtain the following transcendental equation from which the natural frequencies may be found:

$$\sum_{i,\, j\,\text{odd}}\sum \frac{1}{(i\pi)^2(j\pi)^2\{\omega^2 - (i\pi)^2 - \alpha^2(j\pi)^2\}} = \frac{1}{32\lambda^2}. \tag{5.70}$$

This equation is similar to that given in problem 3.2, where the Fourier series method was applied to the symmetric in-plane modes of the single cable and, likewise, its roots are critically dependent on the value of λ^2. Unfortunately in contrast to the single cable it does not appear possible to put (5.70) in a simple closed form.

To shorten the discussion, only the case of the square membrane ($\alpha = 1$) will be considered. Reference should be made to figure 5.26, where a graphical solution is presented for the first two roots of (5.70).

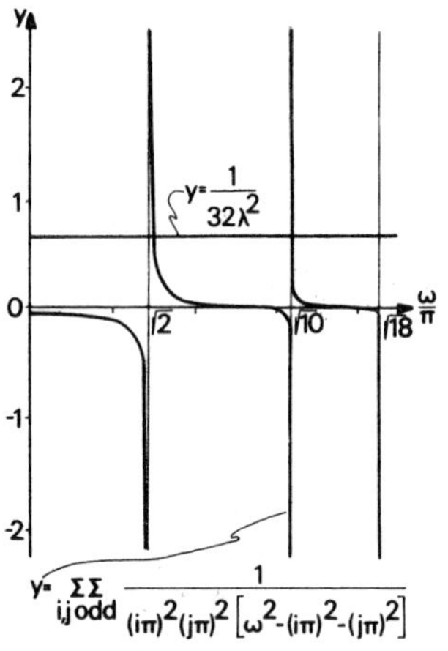

5.26 Graphical solution of first two roots of (5.70), with $\alpha = 1$

When λ^2 is very large (when the membrane is essentially inextensible) (5.70) becomes

$$\sum_{i,j\,\text{odd}}\sum \frac{1}{(i\pi)^2(j\pi)^2\{\omega^2 - (i\pi)^2 - (j\pi)^2\}} = 0, \tag{5.71}$$

the first root of which is

$$\omega_1 \simeq \left(\frac{94}{11}\right)^{1/2}\pi = 2.92\pi, \tag{5.72}$$

the second root

$$\omega_2 \simeq \left(\frac{334}{19}\right)^{1/2}\pi = 4.19\pi, \tag{5.73}$$

and so on.

At the other extreme, when λ^2 is very small, the natural frequencies of the classical square membrane are recovered. The first root is then

$$\omega_1 = \omega_{11} = \sqrt{2}\pi, \tag{5.74}$$

the second root

$$\omega_2 = \omega_{13}(= \omega_{31}) = \sqrt{10}\pi, \tag{5.75}$$

and so on.

Therefore the natural frequency of the first symmetric mode of the general suspended square membrane lies between

$$1.41\pi < \omega_1 < 2.92\pi,$$

the natural frequency of the second symmetric mode lies between

$3.16\pi < \omega_2 < 4.19\pi,$

and so on. Limits that incorporate weaker upper bounds are

$\omega_{11} < \omega_1 < \omega_{12}, \quad \omega_{13} < \omega_2 < \omega_{33}, \ldots,$

as is apparent from figure 5.26.

Several important cases may now be considered. First, define

$$F(\omega) = \frac{1/32}{\sum_{i,\,j \text{ odd}} \sum 1/[(i\pi)^2(j\pi)^2\{\omega^2 - (i\pi)^2 - j\pi)^2\}]}. \tag{5.76}$$

Therefore, if $\lambda^2 < F(\sqrt{5}\pi)$, the natural frequency of the first symmetric mode is less than that of the first antisymmetric mode $k_1\,\tilde{\boldsymbol{\delta}}_{12} + k_2\,\tilde{\boldsymbol{\delta}}_{21}$. When $\lambda^2 = F(\sqrt{5}\pi) \simeq 1.07 \times 10^2$ (the corresponding value for the circular membrane is 1.17×10^2), the natural frequencies of the first symmetric and first antisymmetric modes are identical. However, owing to the two-dimensional nature of the problem, this value of λ^2 does not give a crossover point because the first symmetric mode cannot be tangential to the static profile at the boundaries. If $\lambda^2 > F(\sqrt{5}\pi)$, the natural frequency of the first symmetric mode is greater than that of the first antisymmetric mode.

When $\lambda^2 = F(2\sqrt{2}\pi) \simeq 6.36 \times 10^2$, the frequencies of the first symmetric mode and second antisymmetric mode $\tilde{\boldsymbol{\delta}}_{22}$ are now equal. When $\lambda^2 = F(\sqrt{13}\pi) \simeq 1.94 \times 10^2$, the frequency of the second symmetric mode is the same as that of the third antisymmetric mode $k_1\,\tilde{\boldsymbol{\delta}}_{23} + k_2\,\tilde{\boldsymbol{\delta}}_{32}$. When $\lambda^2 = F(\sqrt{17}\pi) \simeq 3.99 \times 10^2$, the frequency of the second symmetric mode is identical to that of the fourth antisymmetric mode $k_1\,\tilde{\boldsymbol{\delta}}_{14} + k_2\,\tilde{\boldsymbol{\delta}}_{41}$, and so on.

Four situations in which a symmetric mode has its natural frequency identical to that of an antisymmetric mode are shown in figure 5.27. Where applicable, examples of the multiplicity of form of the antisymmetric modes are given.

Reference [8] contains further discussion of the classical square membrane. (The first treatment of the discretized membrane was given by Routh [20]. The complexity of the solution is great, so it seems unlikely that the inclusion of gravity will help the situation.) Perhaps the most interesting point is that, while multiple natural frequencies certainly exist for the symmetric modes, the modes themselves have just one fixed form each. This cannot be established from the classical theory: to show it, one must allow for the effects of gravity and then take the limit as λ^2 becomes small.

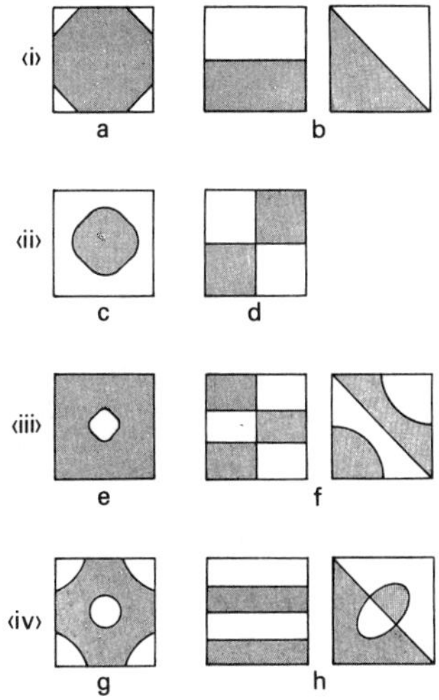

5.27 Four situations in which the natural frequency of a symmetric mode of a suspended square membrane is identical to that of an antisymmetric mode. Legend: (i) $\lambda^2 = F(\sqrt{5}\pi) \simeq 1.07 \times 10^2$, (a) first symmetric mode, (b) first antisymmetric mode $k_1\tilde{\boldsymbol{\delta}}_{12} + k_2\tilde{\boldsymbol{\delta}}_{21}$ ($k_1 = 1, k_2 = 0$ and $k_1 = 1, k_2 = 1$, respectively); (ii) $\lambda^2 = F(2\sqrt{2}\pi) \simeq 6.36 \times 10^2$, (c) first symmetric mode, (d) second antisymmetric mode $\tilde{\boldsymbol{\delta}}_{22}$; (iii) $\lambda^2 = F(\sqrt{13}\pi) \simeq 1.94 \times 10^2$, (e) second symmetric mode, (f) third antisymmetric mode $k_1\tilde{\boldsymbol{\delta}}_{23} + k_2\tilde{\boldsymbol{\delta}}_{32}$ ($k_1 = 1, k_2 = 0$, and $k_1 = 1, k_2 = 1$, respectively); (iv) $\lambda^2 = F(\sqrt{17}\pi) \simeq 3.99 \times 10^2$, (g) second symmetric mode, (h) fourth antisymmetric mode $k_1\tilde{\boldsymbol{\delta}}_{14} + k_2\tilde{\boldsymbol{\delta}}_{41}$ ($k_1 = 1, k_2 = 0$ and $k_1 = 1, k_2 = 1$, respectively).

Example 5.7 The Linear Theory of Free Vibrations of a Hyperbolic Paraboloid

In this example the theory presented for the rectangular suspended membrane is adapted to apply to a general hyperbolic paraboloidal membrane of rectangular plan. The method relies on work presented in section 5.5 and problem 5.3.

The static profile is assumed to be that corresponding to prestress plus dead load. The dead load tensions are $(H_1 + h_1)$ and $(H_2 + h_2)$, and the profile is given by $(z_0 + w)$. As we have seen in example 5.5, $w/z_0 \ll h_1/H_1$ or h_2/H_2, so we are justified in writing the linearized equation of dynamic equilibrium as

$$(H_1 + h_1)\frac{\partial^2 \delta}{\partial x^2} + (H_2 + h_2)\frac{\partial^2 \delta}{\partial y^2} = \frac{p}{g}\frac{\partial^2 \delta}{\partial t^2} - h_1^*\frac{\partial^2 z_0}{\partial x^2} - h_2^*\frac{\partial^2 z_0}{\partial y^2}.$$

The only new terms are h_1^* and h_2^* which are the averaged additional tensions due to the dynamic disturbance. Separating time from this equation gives

$$(H_1 + h_1)\frac{\partial^2 \tilde{\delta}}{\partial x^2} + (H_2 + h_2)\frac{\partial^2 \tilde{\delta}}{\partial y^2} + \frac{p}{g}\omega^2\tilde{\delta} = -\tilde{h}_1^*\frac{\partial^2 z_0}{\partial x^2} - \tilde{h}_2^*\frac{\partial^2 z_0}{\partial y^2},$$

where

$$\tilde{h}_1^* = -\frac{EA_1}{4ab}\int_0^{2a}\int_0^{2b}\frac{\partial z_0}{\partial x}\frac{\partial\tilde{\delta}}{\partial x}dxdy,$$

$$\tilde{h}_2^* = \frac{EA_2}{4ab}\int_0^{2a}\int_0^{2b}\frac{\partial z_0}{\partial y}\frac{\partial\tilde{\delta}}{\partial y}dxdy.$$

We convert these equations to dimensionless form as follows: let $\mathbf{x} = x/2a$, $\mathbf{y} = y/2b$, $\tilde{\boldsymbol{\delta}} = \tilde{\delta}/\{p(2a)^2/(H_1 + h_1)\}$, $\boldsymbol{\omega}^2 = \omega^2 p(2a)^2/g(H_1 + h_1)$, $\mathbf{h}_1 = h_1/H_1$, $\mathbf{h}_2 = h_2/H_2$, $\tilde{\mathbf{h}}_1^* = \tilde{h}_1^*/(H_1 + h_1)$, $\tilde{\mathbf{h}}_2^* = \tilde{h}_2^*/(H_2 + h_2)$, $\alpha_*^2 = (a/b)^2 H_2/H_1$, $\lambda_1^2 = (8d_1/2a)^2 EA_1/H_1$, $\lambda_2^2 = (8d_2/2b)^2 EA_2/H_2$, and $\mathbf{p} = p(2a)^2/8d_1H_1 = p(2b)^2/8d_2H_2$. The equations become

$$\frac{\partial^2\tilde{\boldsymbol{\delta}}}{\partial\mathbf{x}^2} + \frac{(1 + \mathbf{h}_2)}{(1 + \mathbf{h}_1)}\alpha_*^2\frac{\partial^2\tilde{\boldsymbol{\delta}}}{\partial\mathbf{y}^2} + \boldsymbol{\omega}^2\tilde{\boldsymbol{\delta}} = \frac{1}{\mathbf{p}}\{(1 + \mathbf{h}_2)\tilde{\mathbf{h}}_2^* - (1 + \mathbf{h}_1)\tilde{\mathbf{h}}_1^*\},$$

where

$$\tilde{\mathbf{h}}_1^* = -\frac{\lambda_1^2\mathbf{p}}{(1 + \mathbf{h}_1)^2}\int_0^1\int_0^1\tilde{\boldsymbol{\delta}}\,d\mathbf{x}d\mathbf{y},$$

$$\tilde{\mathbf{h}}_2^* = \frac{\alpha_*^2\lambda_2^2\mathbf{p}}{(1 + \mathbf{h}_1)(1 + \mathbf{h}_2)}\int_0^1\int_0^1\tilde{\boldsymbol{\delta}}\,d\mathbf{x}d\mathbf{y},$$

from which

$$\frac{\tilde{\mathbf{h}}_1^*}{\tilde{\mathbf{h}}_2^*} = -\frac{\lambda_1^2}{\alpha_*^2\lambda_2^2}\,\frac{1 + \mathbf{h}_2}{1 + \mathbf{h}_1}$$

This permits the final step to be taken, which is to write the governing equations in a form reminiscent of (5.62), namely,

$$\frac{\partial^2\tilde{\boldsymbol{\delta}}}{\partial\mathbf{x}^2} + \alpha_e^2\frac{\partial^2\tilde{\boldsymbol{\delta}}}{\partial\mathbf{y}^2} + \boldsymbol{\omega}^2\tilde{\boldsymbol{\delta}} = \tilde{\mathbf{h}}_e,$$

where

$$\tilde{\mathbf{h}}_e = \frac{\lambda_e^2}{2}\int_0^1\int_0^1\tilde{\boldsymbol{\delta}}\,d\mathbf{x}d\mathbf{y},$$

in which

$$\alpha_e^2 = \frac{(1 + \mathbf{h}_2)}{(1 + \mathbf{h}_1)}\alpha_*^2,$$

$$\tilde{\mathbf{h}}_e = \tilde{\mathbf{h}}_2^*\left[\frac{(1 + \mathbf{h}_2)}{\mathbf{p}}\left\{1 + \frac{\lambda_1^2}{\alpha_*^2\lambda_2^2}\right\}\right],$$

and

$$\frac{\lambda_e^2}{2} = \left\{\frac{\lambda_1^2 + \alpha_*^2\lambda_2^2}{1 + \mathbf{h}_1}\right\}.$$

Knowing the particular values of α_e^2 and λ_e^2, we may now use directly the theory developed in the present section. This method of attack is different from that suggested by Møllmann [16].

Turning now to the 90 m × 60 m roof of example 5.5, we note that λ_e^2 is very large, since $\lambda_1^2 = 210$ and $\lambda_2^2 = 180$. We may regard the roof as inextensible, so that the frequency of the first symmetric mode is adequately determined by the first two terms of (5.70), namely,

$$\frac{1}{\pi^4\{\omega^2 - (1 + \alpha_e^2)\pi^2\}} + \frac{1}{9\pi^4\{\omega^2 - (9 + \alpha_e^2)\pi^2\}} = 0.$$

In the present case $\alpha_e^2 = 3.0$, and the solution is

$$\omega_1 \simeq \left(\frac{118}{10}\right)^{1/2} \pi = 3.46\pi.$$

The first antisymmetric mode is $\tilde{\boldsymbol{\delta}}_{21}$ (recall equation 5.67), and its frequency is

$$\omega_{21} = \sqrt{7}\pi = 2.64\pi,$$

indicating that $\tilde{\boldsymbol{\delta}}_{21}$ is the fundamental mode of the roof. Thus the period of the first antisymmetric mode is 1.2 sec, while that of the first symmetric mode is 0.9 sec. Even though the roof covers a large area, it has a surprising degree of stiffness.

Circular Membrane

If (5.59) and (5.61) are transformed to polar coordinates, and the substitutions $\delta(r, \theta, t) = \tilde{\delta}(r, \theta)e^{i\omega t}$, $h(t) = \tilde{h}e^{i\omega t}$, are made, the partial differential and integral equations governing the free vibrations of the suspended circular membrane of diameter $2R$ (see figure 5.28) may be written in the following dimensionless form:

$$\frac{1}{\mathbf{r}^2}\frac{\partial^2\tilde{\boldsymbol{\delta}}}{\partial\theta^2} + \frac{1}{\mathbf{r}}\frac{\partial}{\partial\mathbf{r}}\left(\mathbf{r}\frac{\partial\tilde{\boldsymbol{\delta}}}{\partial\mathbf{r}}\right) + \boldsymbol{\omega}^2\tilde{\boldsymbol{\delta}} = \frac{\tilde{\mathbf{h}}}{4},$$

$$\tilde{\mathbf{h}} = \frac{\lambda^2}{2\pi}\int_0^1\int_0^{2\pi}\tilde{\boldsymbol{\delta}}\,\mathbf{r}d\mathbf{r}d\theta, \qquad (5.77)$$

where $\mathbf{r} = r/R$, $\tilde{\boldsymbol{\delta}} = \tilde{\delta}/\{p(2R)^2/H\}$, $\tilde{\mathbf{h}} = \tilde{h}/H$, $\boldsymbol{\omega}^2 = pR^2\omega^2/gH$, and $\lambda^2 = (2pR/H)^2(EA/H)$.

Again, $\tilde{\mathbf{h}} = 0$ when

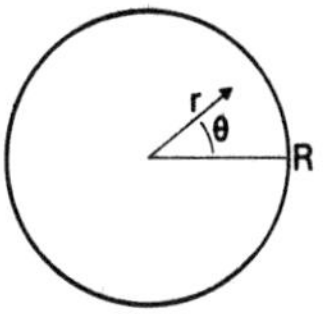

5.28 Definition diagram for a circular membrane

$$\int_0^1 \int_0^{2\pi} \tilde{\boldsymbol{\delta}}\, \mathbf{r} d\mathbf{r} d\theta = 0, \tag{5.78}$$

and modes of vibration that meet this condition may be called antisymmetric. All other modes are radially symmetric, and additional tension is induced.

We may solve the classical drumskin equation to find out which modes induce additional tension. Equations (5.77) reduce to

$$\frac{1}{\mathbf{r}^2}\frac{\partial^2 \tilde{\boldsymbol{\delta}}}{\partial \theta^2} + \frac{1}{\mathbf{r}}\frac{\partial}{\partial \mathbf{r}}\left(\mathbf{r}\frac{\partial \tilde{\boldsymbol{\delta}}}{\partial \mathbf{r}}\right) + \omega^2 \tilde{\boldsymbol{\delta}} = 0. \tag{5.79}$$

The general solution continuous at the origin is (see, for example, Courant and Hilbert [2])

$$\tilde{\boldsymbol{\delta}}(\mathbf{r}, \theta) = (A \cos n\theta + B \sin n\theta) J_n(\omega \mathbf{r}), \tag{5.80}$$

where A and B are constants and J_n is the nth-order Bessel function of the first kind. In order that $\tilde{\boldsymbol{\delta}}(\mathbf{r}, \theta)$ be single valued, n must be a nonnegative integer. The boundary condition $\tilde{\boldsymbol{\delta}}(1, \theta) = 0$ requires that

$$J_n(\omega) = 0, \quad n = 0, 1, 2, \ldots. \tag{5.81}$$

The natural frequencies are the zeros of $J_n(\omega)$ and may be denoted by $\omega_{n,m}$, $m = 1, 2, 3, \ldots$. The associated modes of vibration may then be written as

$$\tilde{\boldsymbol{\delta}}_{n,m}(\mathbf{r}, \theta) = (A_{n,m} \cos n\theta + B_{n,m} \sin n\theta) J_n(\omega_{n,m} \mathbf{r}), \tag{5.82}$$

where $A_{n,m}$ and $B_{n,m}$ are modal constants and $m = 1, 2, 3, \ldots, n = 0, 1, 2, \ldots$. Therefore, only in the cases $n = 0$ is (5.78) not satisfied.

The antisymmetric modes of the suspended circular membrane are those that do not induce any additional tension. The natural frequencies are found from the zeros of $J_n(\omega)$, namely, $\omega_{n,m}$. The associated modes are

$$\tilde{\boldsymbol{\delta}}_{n,m}(\mathbf{r}, \theta) = (A_{n,m} \cos n\theta + B_{n,m} \sin n\theta) J_n(\omega_{n,m} \mathbf{r}), \tag{5.83}$$

where $A_{n,m}$, $B_{n,m}$ are model constants and $m = 1, 2, 3, \ldots, n = 1, 2, 3, \ldots$. The nodal lines for these modes are combinations of m concentric circles (including the boundary) and n equidistant diameters. A few antisymmetric modes are shown later in figure 5.30.

Because of their radial symmetry the symmetric modes of the suspended circular membrane are governed by

$$\frac{1}{\mathbf{r}}\frac{d}{d\mathbf{r}}\left(\mathbf{r}\frac{d\tilde{\boldsymbol{\delta}}}{d\mathbf{r}}\right) + \omega^2 \tilde{\boldsymbol{\delta}} = \frac{\tilde{\mathbf{h}}}{4},$$

$$\tilde{\mathbf{h}} = \lambda^2 \int_0^1 \tilde{\boldsymbol{\delta}}\, \mathbf{r} d\mathbf{r}. \tag{5.84}$$

The solution of the first equation that satisfies the conditions of $d\tilde{\boldsymbol{\delta}}(0)/d\mathbf{r} = \tilde{\boldsymbol{\delta}}(1) = 0$ is

$$\tilde{\boldsymbol{\delta}}(\mathbf{r}) = \frac{\tilde{\mathbf{h}}}{4\omega^2}\left\{1 - \frac{J_0(\omega\mathbf{r})}{J_0(\omega)}\right\}, \tag{5.85}$$

where the value of ω specifies the particular symmetric mode.

The transcendental equation from which the natural frequencies may be found is

$$\frac{J_1(\omega)}{J_0(\omega)} = \frac{1}{2}\,\omega - \frac{4}{\lambda^2}\,\omega^3. \tag{5.86}$$

This equation is of a similar form to that obtained for the single suspended cable. Figure 5.29 presents a graphical solution of (5.86) for the first nonzero root.

When λ^2 is very large (5.86) reduces to

$$\frac{J_1(\omega)}{J_0(\omega)} = \frac{1}{2}\,\omega. \tag{5.87}$$

This equation appears in other branches of mechanics: for example, it arises in the propagation of torsional waves along a cylinder [12], and we may therefore infer that the corresponding results for the rectangular membrane apply to the propagation of torsional waves along a rectangular bar. The membrane analogy for torsion of a prism is well-known and this is another manifestation of it.

The roots of (5.87) are tabulated in handbooks. However, a recurrence relation exists of the form

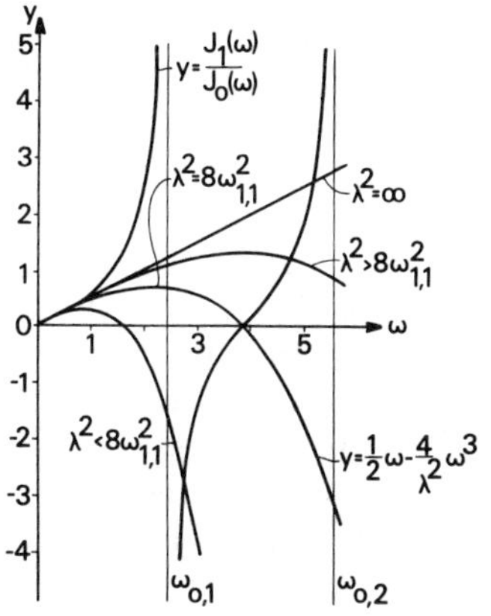

5.29 Graphical solution for the first nonzero root of (5.86)

$$J_0(\omega) + J_2(\omega) = \frac{2}{\omega} J_1(\omega), \tag{5.88}$$

so that (5.87) takes the simpler form

$$J_2(\omega) = 0. \tag{5.89}$$

Thus the symmetric modes of the inextensible suspended circular membrane have natural frequencies given by the zeros of $J_2(\omega)$, namely, $\omega_{2,m}$, $m = 1, 2, 3, \ldots$. They are therefore identical to the frequencies of the antisymmetric mode $\tilde{\delta}_{2,m}$.

When λ^2 is very small, the symmetric modes of the classical drumskin are recovered. The frequency equation is

$$J_0(\omega) = 0, \tag{5.90}$$

and the natural frequencies are $\omega_{0,m}$, $m = 1, 2, 3, \ldots$.

For the general suspended circular membrane the natural frequencies of the symmetric modes lie between

$$\omega_{0,m} < \omega_m < \omega_{2,m}, \quad m = 1, 2, 3, \ldots. \tag{5.91}$$

In other words,

$$2.40 < \omega_1 < 5.14,$$

$$5.52 < \omega_2 < 8.42,$$

and so on. A weaker upper bound for (5.91) is $\omega_{0,m+1}$, as is apparent from figure 5.29.

Therefore if $\lambda^2 < 8\omega_{1,1}^2$, the first symmetric mode has no internal nodes, and its natural frequency is less than that of the first antisymmetric mode $\tilde{\delta}_{1,1}$. If $\lambda^2 = 8\omega_{1,1}^2 \simeq 1.17 \times 10^2$, the first symmetric mode has no internal nodes, but around the perimeter it is tangential to the static profile. Its frequency is identical to that of the first antisymmetric mode. This value of λ^2 gives the first crossover point. If $\lambda^2 > 8\omega_{1,1}^2$, the first symmetric mode has an internal nodal circle, and its frequency is greater than that of the first antisymmetric mode. In fact, if $8\omega_{1,1}^2 < \lambda^2 < 8\omega_{1,2}^2$, both the first and second symmetric modes have one internal nodal circle.

When $\lambda^2 = -8\omega_{3,1}^2 J_0(\omega_{3,1})/J_2(\omega_{3,1}) \simeq 2.63 \times 10^2$, the frequencies of the second symmetric mode and the third antisymmetric mode $\tilde{\delta}_{3,1}$ are equal. When $\lambda^2 = 8\omega_{1,2}^2 \simeq 3.94 \times 10^2$, the frequency of the second symmetric mode is the same as that of the fourth antisymmetric mode $\tilde{\delta}_{1,2}$. This represents the second crossover point, because for larger values of λ^2 the second symmetric mode then has two internal nodal circles, and so on.

Six such situations in which equality of the natural frequencies is predicted are shown in figure 5.30.

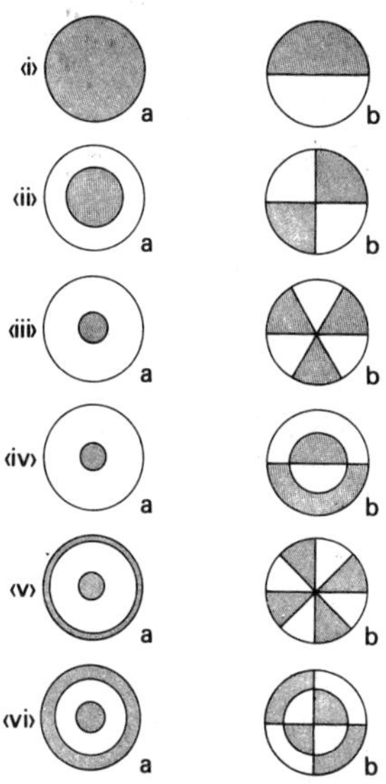

5.30 Six situations in which a symmetric mode of a suspended circular membrane has its natural frequency identical to that of an antisymmetric mode. Legend: (i) $\lambda^2 = 8\,\omega_{1,1}^2 \simeq 1.17 \times 10^2$, (a) first symmetric mode, (b) first antisymmetric mode $\tilde{\delta}_{1,1}$; (ii) $\lambda^2 \to \infty$, (a) first symmetric mode, (b) second antisymmetric mode $\tilde{\delta}_{2,1}$; (iii) $\lambda^2 = -8\,\omega_{3,1}^2\; J_0(\omega_{3,1})/J_2(\omega_{3,1}) \simeq 2.63 \times 10^2$, (a) second symmetric mode, (b) third antisymmetric mode $\tilde{\delta}_{3,1}$; (iv) $\lambda^2 = 8\,\omega_{1,2}^2 \simeq 3.94 \times 10^2$, (a) second symmetric mode, (b) fourth antisymmetric mode $\tilde{\delta}_{1,2}$; (v) $\lambda^2 = -8\,\omega_{4,1}^2\, J_0(\omega_{4,1})/J_2(\omega_{4,1}) \simeq 5.50 \times 10^2$, (a) second symmetric mode, (b) fifth antisymmetric mode $\tilde{\delta}_{4,1}$; (vi) $\lambda^2 \to \infty$, (a) second symmetric mode, (b) sixth antisymmetric mode $\tilde{\delta}_{2,2}$.

Application of this theory to the case of a suspended elliptical membrane is more involved, since Mathieu functions will arise. However, for the approximate analysis of ellipses that differ only slightly from a circle, we may use a technique first introduced by Lord Rayleigh [19]. Rayleigh proves that the frequencies of a classical circular membrane, in which the area is the same as that of the proposed planform, accurately represent the frequencies of that plan form. We should expect therefore that this result will be little altered by the introduction of sag, and so in the case of the elliptical membrane the frequency equations will be (5.81) and (5.86) if we replace in them R by $\sqrt{ab}$, where a and b are the semi-axes of the ellipse. In the modes nodes that are equi-spaced diameters remain so, while nodes that are concentric circles are stretched to form concentric ellipses.

Dynamic Response

Calculations for the linearized dynamic response may be performed using the normal mode technique. The general approach is the same as for the single suspended cable, although the manipulations required are somewhat more involved. In the interests of brevity, dynamic response calculations are here omitted although one or two comments of a general nature are in order.

Cable-suspended roofs have low values of λ^2, unless specific measures are taken to increase the curvatures in the initial roof profile (as in example 5.5). From this one might infer that displacement response under dynamic disturbances might pose a problem. However, if the building does not have large openings, the restraints supplied by the elasticity of the contained air can be decidedly beneficial to roof performance (recall example 5.6).

A situation where this is conceivably of importance is, for example, in a roof circular in plan under wind loading. The roof is akin to a circular wing at angle of attack, and regions of high lift will occur along the two edge portions parallel to the wind. A linearized dynamic analysis could easily be performed. Provided that the roof does not have unfavorable aeroelastic characteristics, response would be at most twice as high as if the gust loading were applied as a static load. There is in fact some evidence that aerodynamic damping and added mass effects are factors that may need to be contended with, although such information can only be supplied by wind tunnel tests on a full aeroelastic model of the structure [13].

Earthquake loading is unlikely to pose a problem for the cables of light cable-suspended roofs. However, careful consideration may need to be given to the supporting structural system to ensure an adequate level of performance in seismically active regions.

Nonlinear dynamic analyses seem for the most part to be analytically intractable. An exception is the case where the initial profile is taut and flat. There the techniques outlined in chapter 3 for the taut flat cable may be applied. The equations are of the form

$$(H + \bar{h})\nabla^2\delta - m\frac{\partial^2\delta}{\partial t^2} = f(x, y, t),$$

$$\frac{2\bar{h}S}{EA} = -\frac{1}{2}\int\int_S \nabla^2\delta\,\delta dxdy, \tag{5.92}$$

with $\delta = 0$ on the boundaries. In the case of the free vibrational response the modes of the linear problem uncouple these equations. Solutions may be found for the free vibrations and for response to simple forcing functions using the two-variable expansion procedure.

While the geometrically nonlinear static response of cable structures does not pose major analytical difficulties, insofar as the accuracy is consistent with engineering approximation, the same cannot be said of geometrically nonlinear dynamic response. It is probably the only area in which hand methods of solution do not presently offer a viable alternative to purely numerical techniques for the preliminary analysis of cable structures. Even so, this shortcoming is more apparent than real because nonlinear dynamic response analyses are a refinement of, rather than essential to, preliminary calculations.

Exercises

5.1 Show that a pneumatic tension membrane that is an ellipsoid of revolution has principal radii of curvature given by

$$r_1 = \frac{a^2 b^2}{(a^2 \sin^2 \phi + b^2 \cos^2 \phi)^{3/2}}$$

and

$$r_2 = \frac{a^2}{(a^2 \sin^2 \phi + b^2 \cos^2 \phi)^{1/2}}.$$

Hence show that meridional wrinkles will be on the verge of forming around the perimeter when

$$b = \frac{a}{\sqrt{2}}.$$

5.2 Because the shape with the smallest ratio of surface area to volume is a sphere, it is logical to consider as a possible hot air balloon profile a shape which is in part spherical. The balloon shown in figure 5.31 is spherical for $\phi < \phi_0$, and in this region $r_1 = r_2$ and $\Delta p = \Delta\gamma a(\beta + \cos\phi)$. Obtain the following expressions for the stress resultants, namely,

$$T_\phi = \Delta\gamma a^2 \left\{\frac{2\cos^2\phi + (2 + 3\beta)(1 + \cos\phi)}{6(1 + \cos\phi)}\right\}$$

and

$$T_\theta = \Delta\gamma a^2 \left\{\frac{4\cos^2\phi + (4 + 3\beta)\cos\phi + (3\beta - 2)}{6(1 + \cos\phi)}\right\}.$$

As might be expected the meridional stress resultants are always tensile, but the hoop stress resultants are tensile only if

$$\phi < \cos^{-1}\left\{\frac{-(4 + 3\beta) + \sqrt{9\beta^2 - 24\beta + 48}}{8}\right\},$$

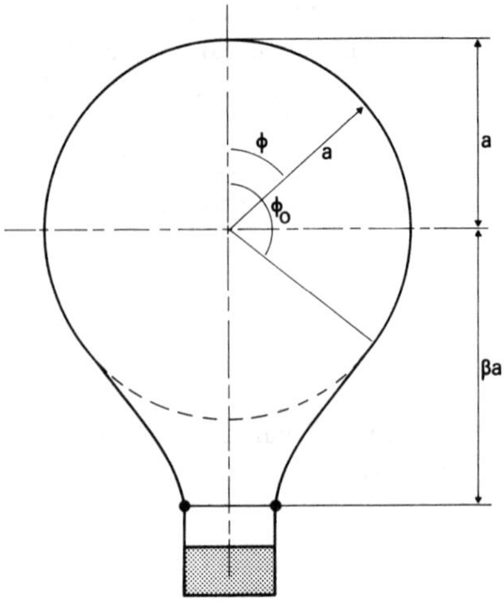

5.31 A partially spherical hot air balloon

so the maximum permissible value for ϕ_0 is given by this upper limit. When $\beta = 1$, $\phi_{0,\max} = 99°$, and, when $\beta = 2$, $\phi_{0,\max} = 120°$. In this range adequate shapes for partially spherical balloons may be found. But there is always the problem of matching a different curve for the lower part.

5.3 Consider the linearized static response of a hyperbolic paraboloid of rectangular plan to uniformly applied load. The initial profile is given by (5.25). However, in contrast to section 5.5 we no longer assume that the pretensions and the cable areas are equal but are H_1 and H_2 and A_1 and A_2, respectively.

Show that the equations governing the problem are in this case

$$\frac{\partial^2 \mathbf{w}}{\partial \mathbf{x}^2} + \alpha_*^2 \frac{\partial^2 \mathbf{w}}{\partial \mathbf{y}^2} = -1 + \frac{1}{\mathbf{p}}(\mathbf{h}_2 - \mathbf{h}_1),$$

and

$$\mathbf{h}_1 = -\lambda_1^2 \mathbf{p} \int_0^1 \int_0^1 \mathbf{w}\, d\mathbf{x} d\mathbf{y},$$

$$\mathbf{h}_2 = \alpha_*^2 \lambda_2^2 \mathbf{p} \int_0^1 \int_0^1 \mathbf{w}\, d\mathbf{x} d\mathbf{y},$$

where $\mathbf{x} = x/2a$, $\mathbf{y} = y/2b$, $\alpha_*^2 = H_2 a^2/H_1 b^2$, $\mathbf{w} = w/(p(2a)^2/H_1)$, $\mathbf{p} = 4a^2 p/8d_1 H_1 = 4b^2 p/8d_2 H_2$, $\mathbf{h}_1 = h_1/H_1$, $\mathbf{h}_2 = h_2/H_2$, $\lambda_1^2 = (8d_1/2a)^2 EA_1/H_1$, and $\lambda_2^2 = (8d_2/2b)^2 EA_2/H_2$. Therefore

$$\frac{\mathbf{h}_1}{\mathbf{h}_2} = -\frac{\lambda_1^2}{\alpha_*^2 \lambda_2^2},$$

and the solutions given in the text ($\lambda_1^2 \equiv \alpha_*^2 \lambda_2^2$) may with simple modification be used for this general situation (see example 5.5). Expressions for the corrected additional tensions $\mathbf{h}(\mathbf{y})$ and $\mathbf{h}(\mathbf{x})$ are readily found.

5.4 Relative to the coordinate system shown in figure 5.32, the boundaries of an equilateral triangular membrane (of side $2\sqrt{3}a$) are given by

$$(\mathbf{x} - 1)(\mathbf{x} - \sqrt{3}\mathbf{y} + 2)(\mathbf{x} + \sqrt{3}\mathbf{y} + 2) = 0,$$

where $\mathbf{x} = x/a$ and $\mathbf{y} = y/b$. Show that a solution of (5.39) that satisfies the condition of zero deflection on these boundaries is

$$\mathbf{w}(\mathbf{x}, \mathbf{y}) = \frac{1}{48(1 + \bar{\mathbf{h}})}(1 - \mathbf{x})(\mathbf{x} - \sqrt{3}\mathbf{y} + 2)(\mathbf{x} + \sqrt{3}\mathbf{y} + 2),$$

where $\mathbf{w} = w/(p(2a)^2/H)$. The maximum deflection occurs at the origin. In addition show that

$$\bar{\mathbf{h}}(1 + \bar{\mathbf{h}})^2 = \frac{3}{320}\lambda^2$$

and that the corrected additional tension for the most heavily loaded cables is

$$\frac{\mathbf{h}(\mathbf{x})}{\bar{\mathbf{h}}} = \frac{10}{27}(1 - \mathbf{x})^2(2 + \mathbf{x})^2.$$

A point of interest is that $\mathbf{h}_{max}$ ($=15\bar{\mathbf{h}}/8$) occurs for the cable lying along $\mathbf{x} = -1/2$ and not at the origin.

5.5 Show that, in the case of a roof circular in plan, an initial profile given by

$$z_0 = d\frac{r^2}{R^2}\sin 2\theta,$$

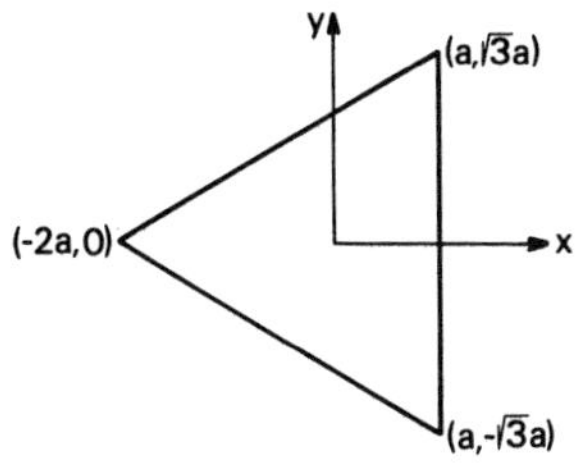

5.32 Definition diagram for an equilateral triangular membrane satisfies (5.24). The polar coordinates (r, θ) are related to the cartesian coordinates (x, y) by

$$x = r \cos \theta, \quad y = r \sin \theta,$$

where the axes are located at the midpoint.

In this hyperbolic surface the pretensions would obviously be the same in each direction, and the cables would be laid out parallel to the x and y axes. The cables are straight in the initial profile, although along diameters other than through $\theta = 0, \pi/2$, the curves are parabolas. The theories developed in sections 5.5 and 5.6 are directly applicable to this surface. Again, note that numerous other types of hyperbolic surfaces can be analyzed by resorting to approximate methods similar to those discussed in this book.

5.6 A typical cross section of a long pneumatic structure is a segment of a circle. The ends of the structure exert little influence over most of its length, so that a unit slice is considered representative. Static equilibrium is given by

$$T = pa,$$

where p is the inflating pressure, a is the radius of curvature, and T is the hoop stress resultant. Figure 5.33 shows the displaced configuration for free radial vibrations ($2\beta\pi$ is the included angle of the profile).

The curvature of a profile that differs little from a circle may be shown to be

$$\frac{1}{r} - \frac{1}{r^2}\frac{\partial^2 r}{\partial \phi^2},$$

where in the present case $r = a + w(\phi, t)$. Show that the equation of radial motion is

$$\frac{T}{a^2}\frac{d^2\tilde{w}}{d\phi^2} + \left(\frac{T}{a^2} + m\omega^2\right)\tilde{w} = \frac{\Delta\tilde{T}}{a},$$

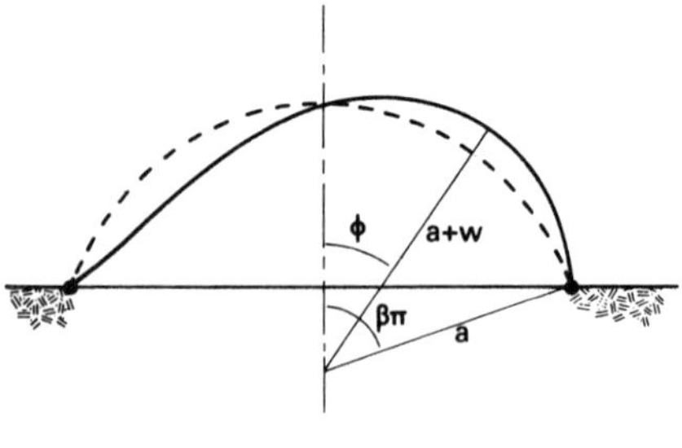

5.33 Definition diagram for radial vibrations of a pneumatic structure where m is the mass of the membrane per unit length and $\Delta\tilde{T}$ is the additional hoop tension and is essentially constant around the profile. The inertia of the contained air has been ignored, although its elastic effect is to force symmetric modes in which no change in volume occurs. The equation of tangential motion is of no significance regardless of β.

Show that the frequencies of the antisymmetric modes are given by

$$\omega_n = \left(\frac{n^2}{\beta^2} - 1\right)^{1/2}\left(\frac{p}{ma}\right)^{1/2}, \quad n = 1, 2, 3, \ldots.$$

These are likely to be of most importance in practice. It may be noted that, when $\beta \ll 1$, the results for the flat-sag cable are recovered if we define $p \equiv mg$. In the special case, when $\beta = 1$ (the profile is a full free circle), the first mode is missing because the membrane cannot then provide a dynamic reaction to the change in profile. These full circle modes are in fact symmetric about a diameter at $\phi = \pi/2n$, $n = 2, 3, 4, \ldots$.

The frequencies of symmetric modes are found from the familiar transcendental equation, $\tan(x/2) = x/2$, but we must exclude the limiting case $\beta = 1$.

References

1. Anderson, M. S., et al. 1965. NASA Tech. Note D-2675.

2. Courant, R., and Hilbert, D. 1953. *Methods of Mathematical Physics.* Vol. 1. New York: Interscience, ch. 5.

3. Den Hartog, J. P. 1952. *Advanced Strength of Materials.* New York: McGraw-Hill, ch. 3.

4. Flügge, W. 1967. *Stresses in Shells.* New York: Springer-Verlag, ch. 2.

5. Green, A. E., and Adkins, J. E. 1970. *Large Elastic Deformations.* 2nd ed. Oxford: Oxford University Press.

6. Heyman, J. 1977. *Equilibrium of Shell Structures.* Oxford: Oxford University Press, chs. 2 and 3.

7. Irvine, H. M. 1976. *J. Eng. Mech. Div., Proc. ASCE,* 102:43–57.

8. Irvine, H. M. 1976. *Proc. Roy. Soc.* (Lond.), A350:317–334.

9. Irvine, H. M. 1980. *Proc. Seventh Australasian Conf. Mech. Struct. Mats.,* Perth., 225–228.

10. Irvine, H. M., and Montauban, P. H. 1980. *Int. J. Mech. Sci.,* 22:637–649.

11. Kawaguchi, M. 1977. *Bull Int. Assoc. Shell Struct.,* 18(63):3–11.

12. Kolsky, H. 1963. *Stress Waves in Solids.* New York: Dover, ch. 3.

13. Krishna, P. 1977. *Cable-Suspended Roofs,* New York: McGraw-Hill.

14. Lamb, H. 1925. *The Dynamical Theory of Sound.* 2nd ed. London: Edward Arnold, ch. 5.

15. Love, A. E. H. 1944. *The mathematical theory of elasticity*. 4th ed. New York: Dover, ch. 14.

16. Møllmann, H. 1974. *Analysis of Hanging Roofs by Means of the Displacement Method.* Lyngby: Polyteknisk Forlag.

17. Norris, C. H., and Wilbur, J. B. 1976. *Elementary Structural Analysis.* 3rd ed. New York: McGraw-Hill, pp. 19–23.

18. Otto, F. 1967. *Tensile Structures.* Vols. 1 and 2. Cambridge, Mass.: The MIT Press.

19. Rayleigh, Lord 1945. *The Theory of Sound.* 2nd ed. Vol. 1. New York: Dover, ch. 9.

20. Routh, E. J. 1955. *Advanced Dynamics of Rigid Bodies.* 6th ed. New York: Dover, ch. 9.

21. Schleyer, F.-K 1967. See vol. 2, pt. 2 of reference [18].

22. Shore, S., and Bathish, G. N. 1967. *Space Structures.* Edited by R. M. Davies. London: Blackwell, ch. 76.

23. Subcommittee on Cable-Suspended Structures. 1971. *J. Struct. Div., Proc. ASCE*, 97:1715–1761.

24. Taylor, G. I. 1963. *The Scientific Papers of G. I. Taylor.* Vol. 3. Edited by G. K. Batchelor. Cambridge: Cambridge University Press, p. 26.

25. Timoshenko, S. P., and Woinowsky-Krieger, S. 1959. *Theory of Plates and Shells.* 2nd ed. New York: McGraw-Hill, ch. 14.

26. Trostel, R. 1967. See vol. 1, pt. 2 of reference [18].

27. Tsuboi, Y. and Kawaguchi, M. 1965. *Rep. Inst. Ind. Sci. Univ. Tokyo.* vol. 15, no. 2 (serial no. 101).

Index

A CATALOG OF SELECTED

DOVER BOOKS

IN SCIENCE AND MATHEMATICS

A CATALOG OF SELECTED

DOVER BOOKS

IN SCIENCE AND MATHEMATICS

QUALITATIVE THEORY OF DIFFERENTIAL EQUATIONS, V.V. Nemytskii and V.V. Stepanov. Classic graduate-level text by two prominent Soviet mathematicians covers classical differential equations as well as topological dynamics and ergodic theory. Bibliographies. 523pp. 5⅜ × 8½. 65954-2 Pa. $10.95

MATRICES AND LINEAR ALGEBRA, Hans Schneider and George Phillip Barker. Basic textbook covers theory of matrices and its applications to systems of linear equations and related topics such as determinants, eigenvalues and differential equations. Numerous exercises. 432pp. 5⅜ × 8½. 66014-1 Pa. $9.95

QUANTUM THEORY, David Bohm. This advanced undergraduate-level text presents the quantum theory in terms of qualitative and imaginative concepts, followed by specific applications worked out in mathematical detail. Preface. Index. 655pp. 5⅜ × 8½. 65969-0 Pa. $13.95

ATOMIC PHYSICS (8th edition), Max Born. Nobel laureate's lucid treatment of kinetic theory of gases, elementary particles, nuclear atom, wave-corpuscles, atomic structure and spectral lines, much more. Over 40 appendices, bibliography. 495pp. 5⅜ × 8½. 65984-4 Pa. $11.95

ELECTRONIC STRUCTURE AND THE PROPERTIES OF SOLIDS: The Physics of the Chemical Bond, Walter A. Harrison. Innovative text offers basic understanding of the electronic structure of covalent and ionic solids, simple metals, transition metals and their compounds. Problems. 1980 edition. 582pp. 6⅛ × 9¼. 66021-4 Pa. $14.95

BOUNDARY VALUE PROBLEMS OF HEAT CONDUCTION, M. Necati Özisik. Systematic, comprehensive treatment of modern mathematical methods of solving problems in heat conduction and diffusion. Numerous examples and problems. Selected references. Appendices. 505pp. 5⅜ × 8½. 65990-9 Pa. $11.95

A SHORT HISTORY OF CHEMISTRY (3rd edition), J.R. Partington. Classic exposition explores origins of chemistry, alchemy, early medical chemistry, nature of atmosphere, theory of valency, laws and structure of atomic theory, much more. 428pp. 5⅜ × 8½. (Available in U.S. only) 65977-1 Pa. $10.95

A HISTORY OF ASTRONOMY, A. Pannekoek. Well-balanced, carefully reasoned study covers such topics as Ptolemaic theory, work of Copernicus, Kepler, Newton, Eddington's work on stars, much more. Illustrated. References. 521pp. 5⅜ × 8½. 65994-1 Pa. $11.95

PRINCIPLES OF METEOROLOGICAL ANALYSIS, Walter J. Saucier. Highly respected, abundantly illustrated classic reviews atmospheric variables, hydrostatics, static stability, various analyses (scalar, cross-section, isobaric, isentropic, more). For intermediate meteorology students. 454pp. 6⅛ × 9¼. 65979-8 Pa. $12.95

CATALOG OF DOVER BOOKS